Digital System Design With LSI Bit-Slice Logic

Other books by
GLENFORD J. MYERS

Reliable Software Through Composite Design, 1975
Software Reliability: Principles and Practices, 1977
Composite/Structured Design, 1978
Advances in Computer Architecture, 1978
The Art of Software Testing, 1979

DIGITAL SYSTEM DESIGN WITH LSI BIT-SLICE LOGIC

Glenford J. Myers
Senior Staff Member
IBM Systems Research Institute

A Wiley-Interscience Publication

JOHN WILEY & SONS
New York • Chichester • Brisbane • Toronto • Singapore

Library of Congress Cataloging in Publication Data:

Myers, Glenford J 1946–
Digital system design with LSI bit-slice logic.

"A Wiley-Interscience publication."
Includes bibliographical references and index.
1. Electronic digital computers—Design and construction. 2. Digital electronics. 3. Logic circuits. 4. Logic design. 5. Integrated circuits—Large scale integration. 6. Microprogramming. I. Title.

TK888.3.M93 621.3819'58'3 79-26258
ISBN 0-471-05376-7

Printed in the United States of America

10 9 8 7 6 5 4 3 2

To Jennifer, Andy, and Jeff

Preface

If one takes the nature of the fundamental unit of design or atomic building block as a measure of "generations" of digital engineering, the field can be said to be in its fourth generation. In the first generation, the building blocks were discrete components, such as vacuum tubes and, later, transistors. The second generation can be categorized as the use of integrated-circuit gates as the basic units of design. In the third generation, the engineer dealt with more powerful building blocks, such as registers, multiplexers, and arithmetic and logic units. In the fourth generation, the units of design grew into such sophisticated single components as microprocessors, CRT controllers, and analog-to-digital converters. A more recent component in the fourth generation is bit-slice logic, a powerful and flexible digital building block. This is the subject of this book.

The motivation for writing this book was that, in comparison to the large amount of literature on other fourth-generation building blocks such as microprocessors, there is relatively little literature (except for semiconductor manufacturers' specifications) on bit-slice logic. A further motivation was that, beginning in 1978, a variety of manufacturers announced that new large-scale computers were being produced using bit-slice logic.

The main thrust of the book is as a tutorial and reference for the system architect and digital design engineer. The book also has application in electrical-engineering and computer-science curricula as a supplemental text in such courses as computer organization, digital system design, digital logic design, microprocessors, and microprogramming.

Chapter 1 introduces the concept of bit-slice logic. It also introduces the 2901, the most widely used bit-slice component. Since bit-slice logic is usually used with the control concept of microprogramming, this concept is introduced in Chapter 2. Chapters 3 and 4 describe many of the bit-slice components available from semiconductor manufacturers. In addition to covering these components in detail, Chapters 3 and 4 put the slices into perspective and discuss problems in the designs of some of the slices.

Chapter 5 discusses more advanced microprogramming topics, such as pipelining, encoding, and optimization. Chapter 6 illustrates some auxiliary LSI logic components, some of which are designed as bit slices. Complementing the devices discussed in Chapter 7 are general devices that can be customized or "programmed" as specialized logic components; these devices, such as programmable logic arrays and gate arrays, are discussed in Chapter 7.

Given that the microprogramming concept is closely allied with the use of bit-slice logic, the last two chapters discuss tools, facilities, and principles for the development of microprograms.

I gratefully acknowledge the help of two colleagues—Dan O'Donnell of the IBM Systems Research Institute and Poughkeepsie Laboratory, and Dave Hocker of the IBM Poughkeepsie Laboratory—in reviewing the manuscript of this book and providing many helpful suggestions. I also acknowledge the cooperation of the following companies for providing material and permission to use selected parts of it:

Advanced Micro Devices
Fairchild Camera and Instrument Corp.
Intel
Monolithic Memories
Motorola
Texas Instruments

The views and opinions herein are solely those of the author, who also takes responsibility for any errors.

GLENFORD J. MYERS

New York, New York
January 1980

Contents

Digital System Design With LSI Bit-Slice Logic

1

An Introduction to Bit-Slice Logic

Bit-slice logic is the most recent generation of fundamental building blocks available to the digital design engineer. Not only do such devices give the engineer a set of powerful, fast, and flexible building blocks, but also they attempt to solve the major problem of LSI technology, namely, the potential proliferation of unique part types.

When one looks at the history of the engineer's basic building blocks, one can identify roughly four generations of components, although the process has really been one of evolution, rather than a clearly separated sequence of distinct steps. Not only did each generation introduce significant improvements in component speed, cost, and reliability, but they also significantly improved the productivity of the system design and manufacturing processes. In reviewing the latter, it is helpful to see the effects of the generations on the traditional types of design that must be performed, namely:

Circuit design —the interconnection of components such as transistors, resistors, and capacitors, to form logic devices such as ANDs and ORs.

Logic design —the interconnection of logic devices to form combinatorial and sequential devices such as registers, counters, and adders.

System design —the interconnection of such devices as adders, registers, and memory arrays to form digital systems such as processors and I/O device controllers.

Physical design—the physical layout of the components, for instance, on printed-circuit or wire-wrap boards.

The building blocks of the first technology generation, beginning in the 1940s and lasting into the early 1960s, were discrete components such as transistors (earlier, vacuum tubes), diodes, resistors, and capacitors. Here the engineer was faced with the full tasks of circuit, logic, system, and physical design.

The second generation, occurring in the early- and mid-1960s, saw the building blocks grow into integrated circuits, each of which contained perhaps 10 to 50 elementary components and performed such logic functions as AND and NOR. The jobs of logic and system design remained the same (actually, they grew because of the increased sophistication of systems), but the task of circuit design was greatly reduced (except, of course, for the few people designing the integrated circuits themselves), and the task of physical design was reduced somewhat, since the integrated circuit reduced the total part count of a system.

The building blocks of the third generation, occurring in the late 1960s and early 1970s, were integrated circuits representing logic devices (MSI) containing perhaps 50 to 200 elementary components and forming such devices as registers, counters, multiplexers, and arithmetic-and-logic units. This generation continued the reduction of the task of physical design (again because of a reduction in the number of physical parts needed to build a system) and lessened the task of logic design.

The fourth generation, which began in the early 1970s, brought about an immense leap in the size and scope of the building blocks. One notable change occurred in storage devices; single chips containing 4096 bits of storage, and later, 16,384 and 65,536 bits, became widely available. Another notable change was the microprocessor; on one chip, embodied in 20,000 or more elementary components, is a full-fledged central processing unit and perhaps a small amount of memory.

Given that a microprocessor and its support chips (e.g., disk controllers, keyboard controllers, communication interfaces, memory-refresh logic, bus controllers) can be used, the tasks of digital system design and physical layout are substantially reduced. However, the microprocessor is not a viable solution to all design problems (i.e., the problem of designing a high-speed, single-processor computer); microprocessors are relatively slow and have static and primitive instruction sets. This precipitated a requirement for a more flexible set of fourth-generation building blocks. The bit-slice device is an answer to this requirement.

THE EVOLUTION OF BIT-SLICE DEVICES

Fourth-generation building blocks are LSI devices. The motivations of LSI are lower costs (large amounts of circuitry are mass produced on a single silicon chip, reducing the costs of the primitive components and eliminating most of the human-assembly costs of prior generations), higher speeds (by reducing transistor sizes, path lengths between components, and circuit capacitance), higher reliability (by reducing the number of mechanical interconnections among components), and shorter design times. However, the designer of LSI devices now faces two new problems: the "pin-out" problem and the "part-proliferation" problem.

The pin-out, or pin-count, problem is the easiest to understand. A typical LSI silicon chip may have a size of 0.15 x 0.2 inches. Obviously only a limited number of external connections can be made between this chip and the outside world. Currently the feasible upper limit is in the neighborhood of 100 connections (pins on the package holding the chip), and it is unlikely that significant increases in this limit will be made. This restricts the type of circuitry that can be placed on a chip. For instance, it is not feasible to construct a 32-bit ALU chip (in spite of the fact that the amount of circuitry is well within the state of the art), since such a device would require over 100 pins (32 outputs, 64 inputs, several control inputs, and several status outputs). Hence the pin-count limitation is one of the great barriers to the use of LSI.

The second problem is the potential proliferation of unique LSI devices. In producing an LSI device, the design costs are extremely high, but the production cost per unit is extremely low. Hence the economics of LSI are attractive only when a large number (e.g., tens of thousands or more) of units of each type can be used.

Here one is faced with a dilemma. The large number of circuits on an LSI device could easily imply that such devices (with the exception of memory arrays and microprocessors) are specialized toward a particular system and unusable in other designs. For instance, suppose that a computer company were developing a family of processors, each with distinct cost and performance objectives. If one could design several LSI chips for use in processor A, it is unlikely that they could be used in processor B, because the processors are likely to have dissimilar internal designs (e.g., different data-path widths, different ALU functions, different degrees of internal parallelism). That is, the mere nature of an LSI chip (containing thousands of gates) means that it is likely to absorb much of the nature of the system in which it is a part, rendering it unusable in other designs, and hence restricting the volume in which it can be manufactured. Thus one is faced with another dilemma that could prevent the designer from taking advantage of LSI. Note that the pin-count restriction compounds the part-proliferation problem, since it might require one to develop a larger number of LSI chip types, increasing the proliferation of specialized chips.

The road to a solution is the realization that, in LSI-based design, one should not be preoccupied with minimizing the number of elementary components or gates in a system but with minimizing the number of chip types. What is needed is a small set of universal chip types. Rather than containing static functions, the functions of these devices should be capable of being controlled externally (i.e., by logic external to the device). To serve as building blocks in a large number of systems, these devices should be capable of performing a large number of functions, many of which might not be used in a particular design. At the same time, these universal devices should place few, if any, restrictions on the designs in which they might be used (e.g., restrictions on data-path sizes). Last, such devices must conform to the pin-count limitations. The bit-slice device is an answer to these requirements.

THE NATURE OF A SLICE

The question then is determining how to partition a system into a set of LSI building blocks such that they can be used in a variety of other designs. Considering central processing units for the moment, the answer is the realization that the heart of most processors, independent of their instruction-set architecture, is similar to that illustrated in Figure 1.1. That is, the heart of a processor normally consists of an array of registers, a multifunction ALU, and shift logic. One thought might be to incorporate all this on a chip, but this is not a solution for several reasons. First, one is likely to encounter the pin-count problem. Second, the chip will not be universal, because it will contain too many dependencies (such as data-path width) on the original design. For instance, if Figure 1.1 is the heart of a 16-bit processor, such a chip would prove to be of no use in 32-bit, 36-bit, and 8-bit designs.

The solution is to view Figure 1.1 as the three-dimensional equivalent in Figure 1.2, and then make vertical slices through the design. That is, one creates devices having the appearance of Figure 1.3. Figure 1.3 illustrates a device that might be termed a 4-bit ALU/register slice. By designing such a chip and bringing out appropriate signals (e.g., ALU carry-in and carry-out, both sides of the shift register), one can devise a universal building block by allowing a set of these devices to be interconnected to yield an ALU/register section of arbitrary width

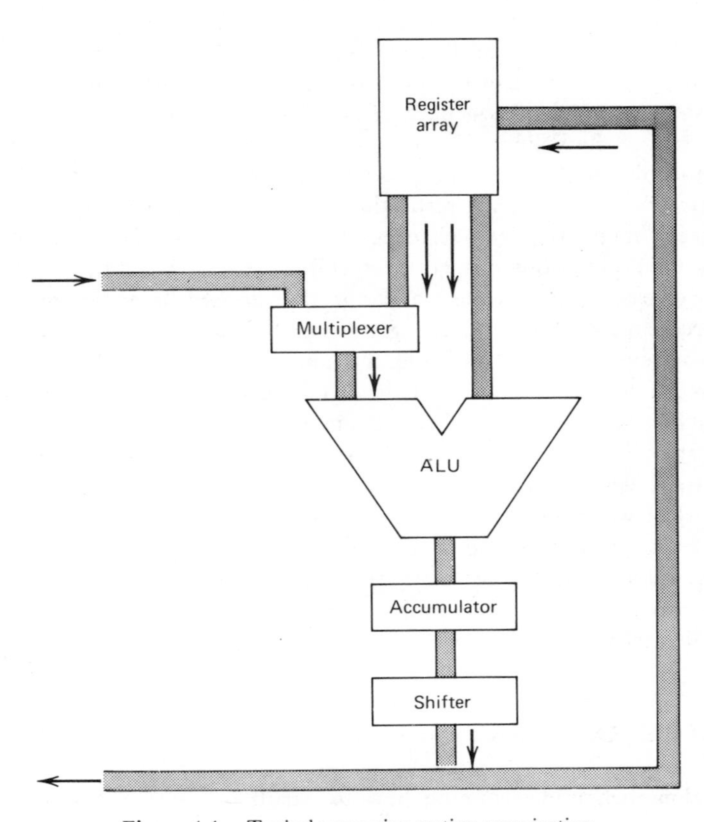

Figure 1.1. Typical processing-section organization.

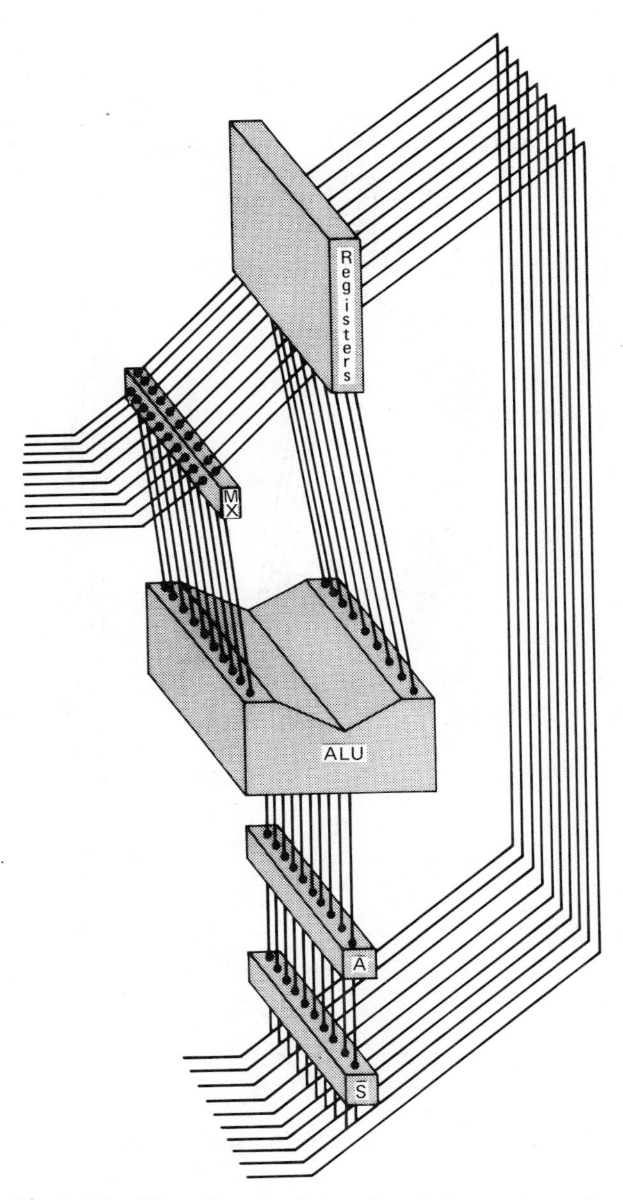

Figure 1.2. Three-dimensional view of a typical processing section.

(e.g., 8, 12, 16, . . . bits). Such a device solves the problems mentioned earlier by having the following attributes:

1. The pin-count problem is not present because of the "narrowness" of the device. A device of the form in Figure 1.3 might have 24 to 40 pins.

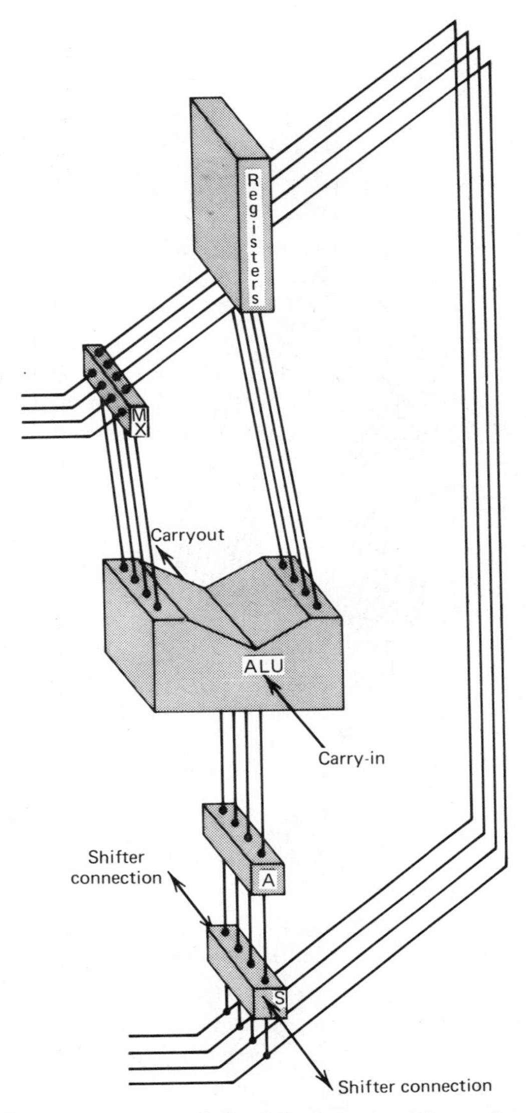

Figure 1.3. A vertical slice through the processing section.

2. It is an LSI device because, although it is a narrow slice through a set of registers, an ALU, shifters, and so on, it contains a large number of circuits.
3. It serves as a universal building block, because the device is designed to perform a large number of functions, some of which are likely not to be used in a particular application (the cost of the unused functions is virtually zero, assuming the device is manufactured in large quantities), and the devices can be cascaded together to form a processing section of any width.

Note that only one type of slice—the ALU/register slice—has been introduced at this point. Other types of slices have been devised. These will be introduced in later chapters.

THE 2901 ALU/REGISTER SLICE

The best way to gain an understanding of bit-slice logic is to examine an actual device, and the best device to start with is the 2901 ALU/register slice. The 2901, first produced by Advanced Micro Devices and now second-sourced by many other firms, is the most widely used bit-slice device. The 2901 is to bit-slice logic what the 8080 has been to microprocessors. The 2901 is used as the heart of such CPU products as DEC's DECsystem-2020, Data General's Nova 4, National Semiconductor's System/400, Functional Automation's F6400, and the Ampex Model 12.

The 2901 is a 40-pin LSI chip; most versions of the 2901 employ low-power Schottky TTL technology. The chip contains about 500 gates. Only an introduction to the 2901 is presented here; it is discussed in more detail in Chapter 3.

Figure 1.4 illustrates the organization of the 2901. Its data paths are 4-bits wide. The basic sections are a 16-word by 4-bit, 2-port RAM, a working register (Q), an ALU, and shifting, decoding, and multiplexing logic.

Any of the 16 registers can be read onto the A bus. Likewise, any of the 16 registers can be read onto the B bus. Each of these busses contains a latch, used to prevent race conditions when the output of the ALU is being written back into the register array.

Both inputs of the ALU are fed from multiplexers. The R input of the ALU can be selected to be the register value on the A bus or a value on the D bus, an external input bus. The S input of the ALU can be selected from the A bus, the B bus, or the Q register. Both multiplexers have an inhibit capability, meaning that the value zero can also be fed into the R and/or S ALU inputs.

The Q register is a working register available for general use, although the motivation for its existence is for use in implementing multiplication and division algorithms. Looking at the connections to the Q register, one can feed it from the ALU output bus or from itself. If Q is fed from itself, the shifter allows one to feed Q with its current value shifted right one bit, shifted left one bit, or not shifted (no change).

Figure 1.4 shows that the output of the ALU can be gated to three places—to the Y bus (an external output bus), to the Q register, or to the register array. When it is gated to the register array, the data moves into the register that was designated as the register to be gated to the B bus. Between the ALU and the register array is a shifter, allowing the ALU output to be shifted right or left one bit, or not shifted, before the value is placed in the register array. Note that the external output bus Y is controlled by a multiplexer, meaning that the data placed on the Y bus can be the ALU output or the A bus. Y is a three-state output, meaning that if nothing is to be moved onto the Y bus during the current operation, it can be held in the high-impedence state.

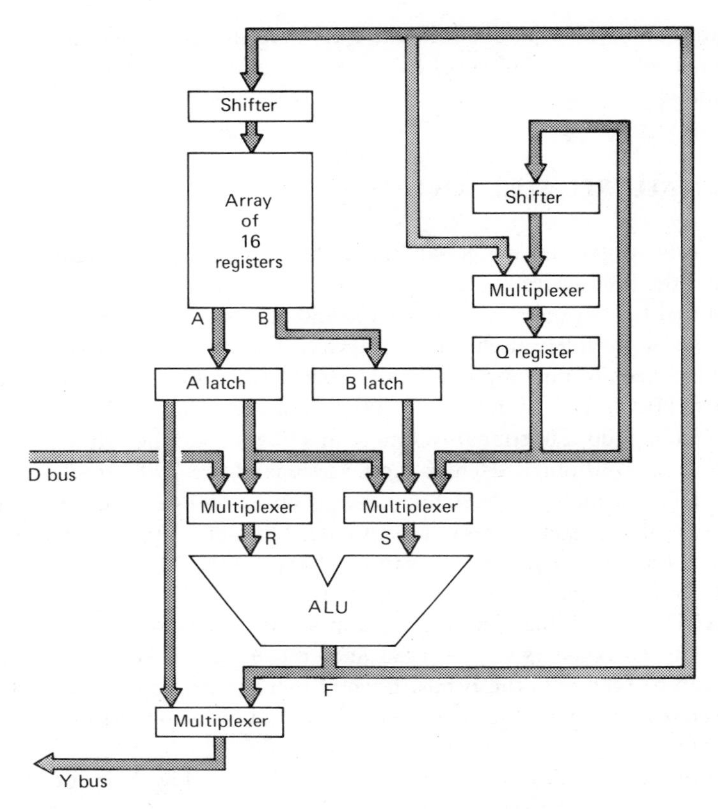

Figure 1.4. Organization of the 2901.

The 2901's external connections are shown in Figure 1.5. The inputs are the 4-bit D bus and 20 control lines. The outputs are the 4-bit Y bus and six status or condition signals. The four lines at the bottom are sometimes inputs, sometimes outputs, and sometimes neither (i.e., in a high-impedence state), depending on the functions specified by the I controls. For instance, if the I signals specify that the Q shifter should shift left by one bit, Q_0 is an input (allowing the low-order bit to be generated externally) and Q_3 is an output (allowing the displaced bit to be tested or fed into another 2901).

As a convention here, bits in registers or busses are numbered such that the least significant bit is numbered 0. Also, the overscore (e.g., $\overline{X}$) is used to indicate either (a) the one's complement of a value, (b) that an input control is active when in the low state, or (c) that an output condition is present when in the low state. The meaning that pertains in a particular situation should be obvious from the context.

Most of the control inputs should be obvious by inspecting Figure 1.5, with the exception of the I controls. The I lines are subdivided into three groups of three lines each. One group controls the multiplexers feeding the ALU. The second group controls the function of the ALU, and the third controls the two shifters, the

Q-register multiplexer, the Y-bus multiplexer, and the gating of the F bus into the register array.

Group one specifies the inputs to the ALU. The items that can be fed into the R side of the ALU are the A register (i.e., the word in the array selected by the A inputs), the D bus, and zero. The items that can be fed into the S side of the ALU are the A, B, and Q registers, and zero. However, only eight combinations are permitted (the combinations A-A, 0-0, A-0, and D-B are not provided)..

Group two (three of the I lines) specifies the function to be performed by the ALU. The ALU can perform three arithmetic and five logic functions, but since the C_n (carry-in) input is used during the arithmetic functions, the ALU can perform six different arithmetic functions if C_n is used as a control input. The functions are

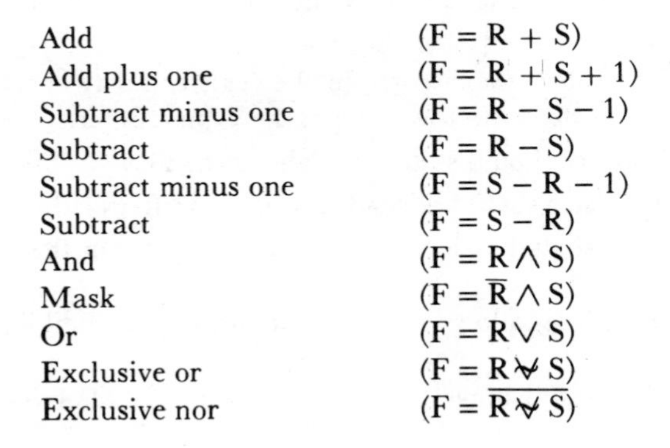

Add	$(F = R + S)$
Add plus one	$(F = R + S + 1)$
Subtract minus one	$(F = R - S - 1)$
Subtract	$(F = R - S)$
Subtract minus one	$(F = S - R - 1)$
Subtract	$(F = S - R)$
And	$(F = R \wedge S)$
Mask	$(F = \overline{R} \wedge S)$
Or	$(F = R \vee S)$
Exclusive or	$(F = R \veebar S)$
Exclusive nor	$(F = \overline{R \veebar S})$

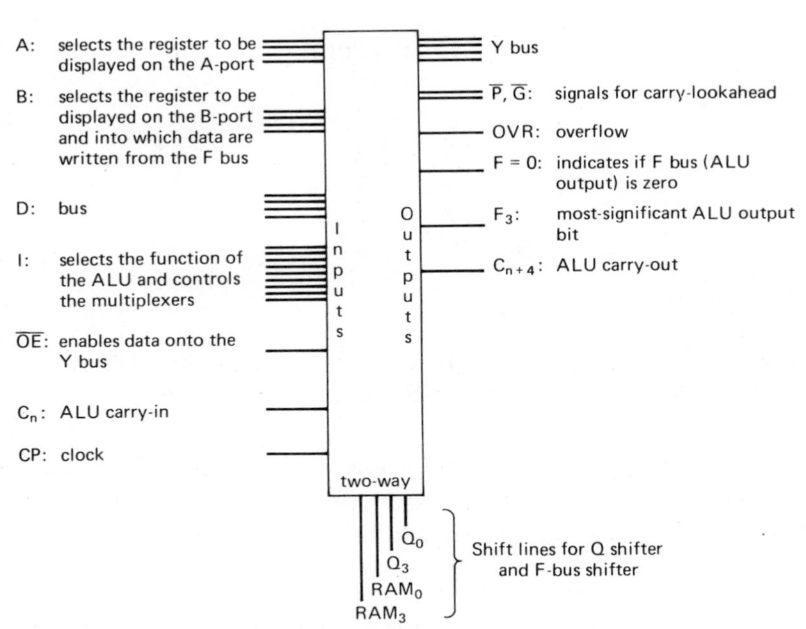

Figure 1.5. 2901 external connections.

By specifying combinations of groups 1 and 2, one can perform such functions as

Increment ($F=R+S+1$, R or $S = 0$)
Decrement ($F=S-R-1$, $R=0$ or $F=R-S-1$, $S=0$)
Invert, or take the one's complement
($F=S-R-1$, $S=0$, or $F=R-S-1$, $R=0$)
Negate, or take the two's complement
($F=S-R$, $S=0$, or $F-R-S$, $R=0$)

The third group of I lines controls the following operations:

1. Whether the F-bus shifter should shift the ALU output left one bit, right one bit, or not shift.
2. Whether the B register (the register in the register array specified by the B inputs) should be loaded with the value in the F-bus shifter.
3. Whether the Q shifter should shift left, right, or neither.
4. Whether the Q register should be loaded from its shifter or the F bus.
5. Whether the Y bus should be loaded from the A register or the F bus.

Group three with three lines allows one to specify only eight combinations of these operations.

As an example, by establishing the following input signals (where 0 is a low state and 1 is a high state)

$A = 0010$
$B = 0011$
$I = 110000001$
$\overline{OE} = 0$
$C_n = 0$
$Q_0 = 1$
$RAM_0 = 0$

the following operations occur

1. The values in registers 2 and 3 are added, the sum is shifted left by one bit, the injected low-order bit is a 0, and the value from the shifter is stored in register 3.
2. The value in the Q register is shifted left by one bit, the injected low-order bit is a 1, and this value is stored in the Q register.
3. The sum of the values in registers 2 and 3 is brought off the chip via the Y bus.

Figure 1.5 also includes a small number of output signals; these signals are used to feed other cascaded 2901 slices and for the testing of conditions. Two of the outputs are used for carry-lookahead logic, where a number of 2901s are cascaded and one wishes to use this technique to increase the speed of arithmetic operations. OVR indicates an overflow during an arithmetic operation (i.e., the carry-in and carry-out of the most significant bit of the ALU are unequal values). The F=0 output signifies whether the result on the F bus is zero. F_3 is the high-order bit of the result on the F bus and can be used to test the sign of the result.

Since the primary advantage of bit slices is the ability to interconnect them, it will be helpful to explore a simple example showing an interconnection. Figure 1.6 is such an example; it illustrates the interconnection of two 2901s to form an 8-bit processing section.

By connecting the two 2901s, we have a processing section with a data-path width of eight bits. The 16 registers and the Q register are 8-bits wide and reside

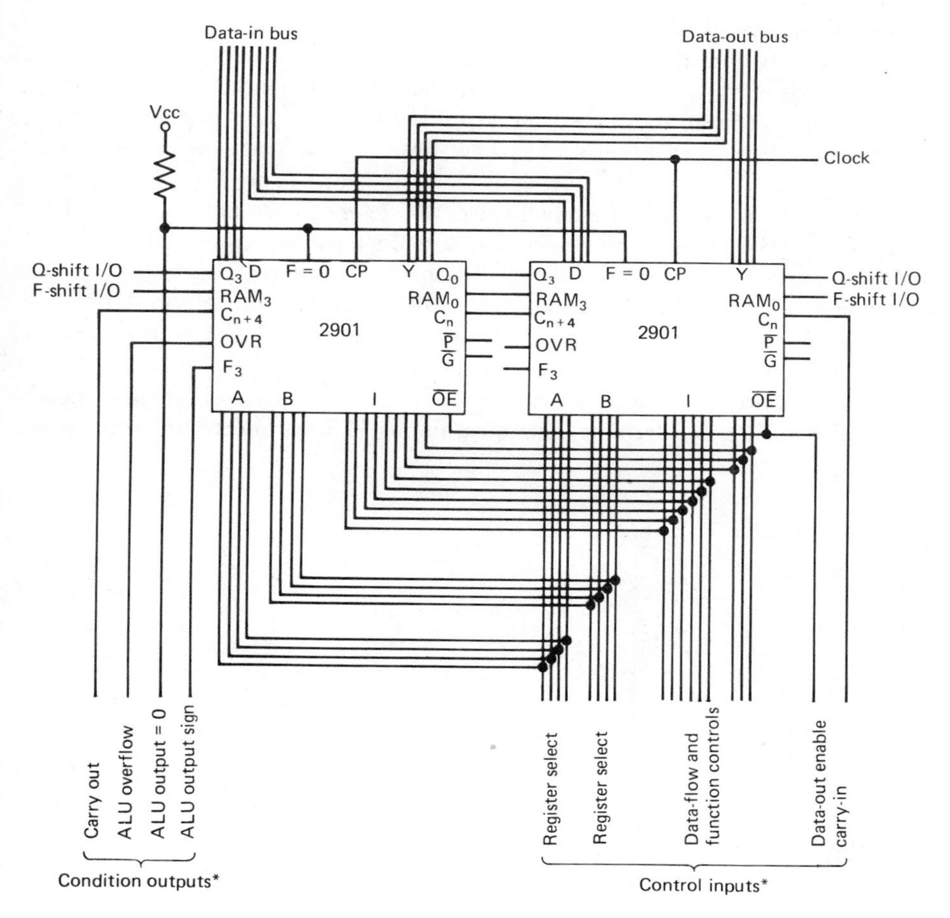

Figure 1.6. Two connected 2901 slices.

in the 2901s, a half in each 2901. An 8-bit data-in bus feeds both 2901s in parallel, and, too, the 2901s feed an 8-bit data-out bus.

The most important aspect of Figure 1.6 is the connection of the control signals and the condition outputs. Note the series/parallel connections; most of the control signals feed the 2901s in parallel, but, for a few connections, the 2901s are connected in series. In particular, the I signals are sent to the 2901s in parallel, ensuring that both perform the same data-flow, shifting, and ALU operations. On the least-significant (right) slice, the carry-in is an input from an external control source, while, on the other slice, the carry-in is connected to the carry-out of the first slice, enabling the ALUs to work as a single, ripple-carry, 8-bit ALU. Notice also the interconnection of the shifters, enabling the Q shifter and F shifter to act as two 8-bit shifters.

In most cases, the condition or status outputs are taken from only the most significant slice. For instance, the OVR and F_3 outputs on the least significant slice have no special significance for 8-bit arithmetic; thus they are unused. The F=0 output is an open-collector output, meaning that it can be wire-AND'ed, with a pull-up resistor, between slices to indicate whether the output from both ALUs (i.e., the output from the single 8-bit ALU) is zero. The carry-lookahead pins on both slices are not used, since carry-lookahead logic is not used in this simple example.

The reader is advised to study Figure 1.6 closely to see how the two 2901s behave as an 8-bit processing element, because the interconnection of slices to form devices of varying widths is the central concept of bit-slice logic.

Figure 1.7 shows the pin layout of the 2901, packaged as a 40-pin DIP (dual inline package). The only two pins that have not been mentioned to this point are V_{cc} (the supply voltage, +5V) and ground (GND).

When working with slices, one has to analyze the propagation speeds of signals from slice to slice to determine the system timing, that is, to determine the fastest

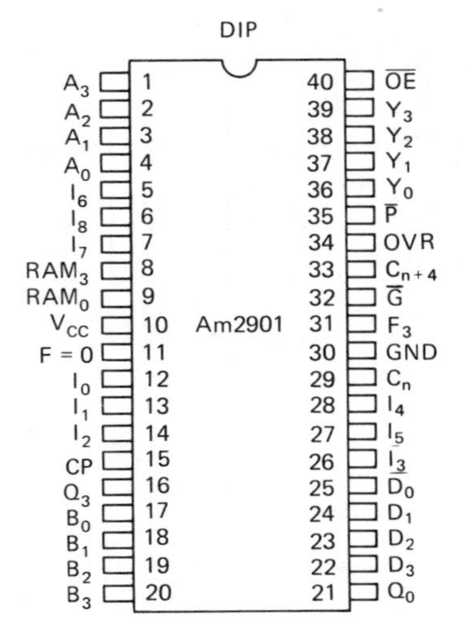

Figure 1.7. The 2901 DIP.

rate at which the slices can be driven (i.e., the cycle time, or the time between changing the set of control signals) and the time until the outputs have settled to their correct values. A detailed discussion of this is premature, but a simple analysis can be performed on Figure 1.6.

Some of the guaranteed, or worst-case, propagation times for Advanced Micro Devices' 2901A slice are listed below (typical, or expected-value, propagation times are usually about two-thirds of the guaranteed times).

From inputs A,B to output Y	– 80 ns
From inputs A,B to C_{n+4} output	– 75 ns
From inputs A,B to last status output	– 95 ns
From input C_n to last status output	– 50 ns
From input C_n to output Y	– 30 ns

The third case is the longest path in the 2901A, but it does not mean that one can drive Figure 1.6 at this rate. The reason is the ripple carry between the slices. That is, the carry-in to the left slice is not stable until t+75 ns, meaning that the outputs of this slice would not stabilize until t+75+50 ns (50 is the longest path from the C_n input to the output). Hence this system could be driven no faster than one cycle per 125 ns. (Improvements in these times have been made in newer versions of the 2901).

BIT SLICES VERSUS MICROPROCESSORS

A common source of confusion about bit-slice logic concerns the distinctions and relationships between bit-slice logic and microprocessors. For instance, common questions are: Are bit-slice devices microprocessors or members of the microprocessor family? Are microprocessors built from bit slices? Are bit-slices usable only in microprocessor applications?

There is little relationship between bit slices and microprocessors, other than the fact that both are modern LSI components. Unfortunately, the semiconductor manufacturers do not help in this regard, since they often advertise their bit-slice devices as "bit-slice microprocessor devices."

A microprocessor is an LSI device that, in the fullest sense of the word, is a stored-program computer. On the chip are an ALU, a variety of registers, a sequential-logic network for instruction decoding and control, and often a small amount of memory. A microprocessor has a fixed, and usually primitive, instruction-set architecture. Most microprocessors are constructed from metal-oxide semiconductor (MOS) devices. In short, a microprocessor is a computer.

A bit slice, however, has little in common with a microprocessor. It is not a computer, because it does not execute programs, although it may be used as a component in a computer. Unlike a microprocessor, it makes no sense to talk of the "instruction-set architecture" of a bit slice; if one were to look for an analogy, however, one might talk of its architecture as being the definition of its control signals. It is true that, in a particular design, a bit-slice device might be said to respond to a particular microinstruction architecture (if the control concept of microprogramming is used, the topic of the next chapter), but this is a characteristic of the particular design itself, not of the bit-slice device. Last, most bit slices are built from bipolar technology, a faster technology than MOS.

To answer the questions above, in no sense of the word are bit slices microprocessors. Calling a bit slice a microprocessor or a device in a microprocessor family is no more accurate than calling an ALU or a memory array a microprocessor. To the second question, microprocessors are not built from bit slices. Last, the application spectrum for bit slices is usually much different than that for microprocessors. Bit slices can be used to *build* processors, but microprocessors *are* processors. Bit slices are faster devices, meaning that they can (and are) used as components of high-speed computers. Bit slices need not be used in building only CPUs or instruction processors; they have application in the design of I/O channels, disk controllers, and so on. Finally, although the orientation of this book is the use of bit-slice logic as components in computer systems, bit slices can be used in other digital, noncomputer-based, applications, such as control systems.

SEMICONDUCTOR TECHNOLOGY

Since bit-slice devices are available in a variety of semiconductor and circuit technologies, and since they use different technologies than microprocessors, a review of the major technology families may be helpful. Using Figure 1.8 as a guide, integrated-circuit devices can be categorized in two groups: (1) devices that are built on a noninsulating silicon substrate and (2) those that are built on islands of silicon grown on an insulating substrate, such as sapphire. Although the latter technology (SOS) promises advantages in increased speeds and circuit

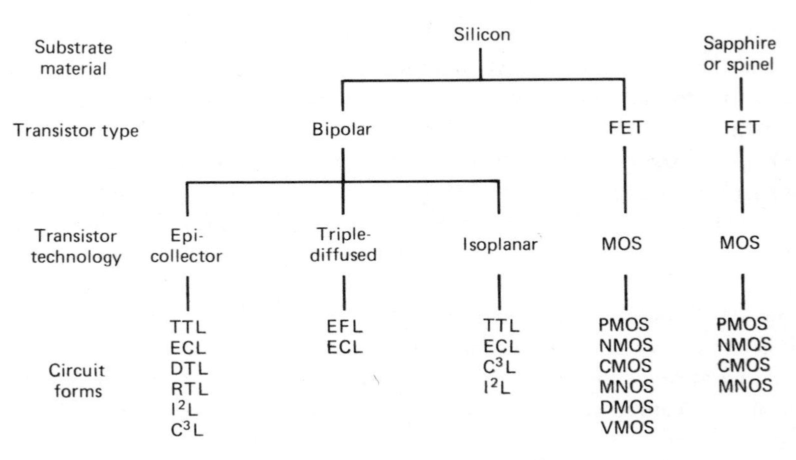

Figure 1.8. Common technology families.

densities, it has special problems (cost is one), and hence is not yet widely used. (Newer forms of semiconductor and circuit technology that have not yet made significant inroads in the product marketplace, such as magnetic bubbles, charge-coupled devices, Josephson devices, and multivalued logic, are not included in Figure 1.8.)

In the first category, one can further subdivide device types by the basic transistor type. The first type is the familiar *pnp* or *npn* junction transistor, also known as the bipolar transistor because charge carriers of both polarities (electrons and holes) are involved in their operation. The second type is the unipolar or field-effect transistor (FET), which employs a single active carrier and uses a field, or a capacitive effect, to control the movement of carriers. Since FETs are built from three materials—a metal layer, an oxide layer (silicon dioxide), and a semiconductor—they are also known as MOS devices. Most bit-slice devices are built from bipolar transistors, and the majority of microprocessors in the marketplace are built from MOSFETs.

In addition to being the technology of most microprocessors, MOS is primarily used for memories. Its main characteristics are higher density (approximately four MOS transistors can be built on the amount of silicon needed for a single bipolar transistor), low power consumption, and low speed. The three principal forms of MOS are PMOS, NMOS, and CMOS. PMOS, or *p-channel* MOS, transistors use holes as carriers. The PMOS technology was used for the original calculator and microprocessor chips, but it is infrequently used today for new devices. NMOS, or *n-channel* MOS, transistors are significantly faster, principally because the carriers are electrons, which have a greater mobility than holes. The majority of second-generation microprocessors (e.g., the Intel 8080, Motorola 6800) are NMOS devices. HMOS, or high-performance MOS, transistors are considerably smaller than normal NMOS transistors, resulting in speed and power advantages. Third-generation 16-bit microprocessors, and high-capacity memories, employ HMOS devices.

CMOS, or *complementary* MOS, devices include both PMOS and NMOS transistors on a single chip of silicon. It is a more attractive technology, since it offers advantages of slightly higher performance, extremely low power consumption, a wide operating range with respect to the power-supply voltage and ambient temperature, and high noise immunity. However, the manufacturing process is more complicated, and the packing density of transistors on the chip is reduced.

The majority of bit-slice devices are built with bipolar transistors, although some are available in CMOS and some manufacturers are known to be developing bit-slice components in other technologies, such as CMOS on sapphire. In comparison to field-effect transistors, bipolar transistors are physically larger (resulting in lower densities), significantly faster (because of lower capacitance and because the base, which resides under the emitter on the chip, can be made extremely narrow, much more so than the narrowest line that can be photolithographically resolved on the chip's surface), and consume more power. Like MOS devices, bipolar devices come in several varieties, although the bipolar varieties are distinguished by circuit design, that is, the manner in which transistors are interconnected to form logic gates. Also like the MOS devices, these forms involve tradeoffs in speed, power, density, and cost.

TTL (transistor-transistor) logic is the most widely used circuit form employing bipolar transistors. There are more types of integrated circuits in this technology than in any other technology. A TTL gate usually consists of a multiple-emitter transistor that controls a switching transistor, which in turn drives a network of output transistors. Most current processors, with the principal exceptions of microprocessors and supercomputers, are designed entirely from TTL devices. Most bit-slice devices are TTL devices, although there are many varieties of TTL, which are discussed later.

ECL (emitter-coupled, or current-mode) logic is the fastest technology available in the marketplace today; ECL logic gates can operate at subnanosecond speeds. ECL is used in most of the recent super-high-performance processors. Its drawbacks are the unavailability of a large selection of devices, larger sensitivities to temperature and voltage variations, and the need to pay special attention to power distribution, cooling, and transmission-line effects.

The third principal circuit form employing bipolar transistors is I^2L (integrated-injection logic), also called MTL (merged-transistor logic). It is a newer technology, promising increased densities and lower power consumption. The higher densities come from the elimination of resistors in the circuit design and the use of some regions of the chip as elements in more than one transistor (e.g., the collector area for a *pnp* transistor also serves as the base area of an *npn* transistor). Low power requirements stem from the elimination of resistors, the use of lower voltages, and several other factors.

Table 1.1 summarizes several characteristics of the most common MOS and bipolar families, namely typical packing densities, propagation delays, power consumption, supply voltages, and speed/power products.

Almost all current bit-slice devices are bipolar devices, namely TTL, ECL, or I^2L. Since most of the bipolar slices are TTL devices and since there are several varieties of TTL, a further exploration of the TTL family is warranted.

Table 1.2 illustrates several characteristics of the TTL subfamilies. Since the

TABLE 1.1. Typical Device Characteristics

	PMOS	NMOS	HMOS	CMOS	TTL	ECL	I^2L
Relative density (1 = densest)	3	2	1	4	5	5	2
Prop. delay per gate (ns)	>100	20–90	1	>10	2–10	.2–2	5–35
Power/gate (mW)	1–5	.1–3	1	<.1	1–19	20–60	<.5
Typical supply voltages	5 and −10	−5, 5 and 12*	5	10	5	−5 and −2	1
Typical speed/ power prod (pJ)	200	4	1	3	20	20	.5

* Most devices now require only a single +5V supply

TABLE 1.2. Characteristics of the TTL Subfamilies

Subfamily	Propagation time (ns)	Power per gate (mW)	Max. low-level output (mA)	Max. low-level input (mA)
TTL	10	10	16	−1.6
HTTL	6	22	20	−2.0
LTTL	33	1	3.6	−0.18
STTL	3	19	20	−2.0
LSTTL	10	2	8	−0.36
ASTTL	1.5	20	20	−2.0
ALSTTL	4	1	8	−0.36

output of a TTL gate draws current only when at a low, or zero, logic level, the last two columns are needed when mixing TTL subfamilies. (At a high logic level, a TTL output supplies a very small amount of current). For instance, a standard TTL device can sink enough current (16 mA) to drive up to 10 other TTL devices (each draws −1.6 mA) or drive up to 8 STTL (Schottky TTL) devices (each needs −2.0 mA). (Note that the analysis should also be performed at the high logic level, although the low levels usually dictate the maximum fanout of a device.)

An inspection of Table 1.2 shows that the STTL and LSTTL subfamilies have the best speed and power characteristics. For STTL (Schottky TTL), Schottky diodes are placed across most of the transistors, preventing the transistors from saturating and thus eliminating storage-time delays within the transistors. The resultant speed of Schottky TTL is thus very close to ECL. LSTTL (low-power Schottky TTL) devices also use Schottky diodes, but larger resistors are used in the gate structures to limit currents, and reduce power dissipation. Lower power requirements are advantageous because: (a) they reduce the cost of the power supply; (b) they reduce cooling requirements and allow increased chip counts on circuit boards; (c) the resultant lowered chip temperatures decrease failure rates; and (d) lower operating currents reduce output spiking.

Recently, several manufacturers have announced new TTL subfamilies with improved speed/power characteristics. Although the speed and power attributes of these new subfamilies vary, the last two rows in Table 1.2 contain representative characteristics.

Most bit-slice devices (e.g., the 2901) employ one of the two forms of low-power Schottky TTL. A few employ Schottky TTL, ECL, and I^2L. Some newer devices have TTL inputs and outputs, but employ ECL internally.

2

An Introduction to Microprogrammed Control

Although bit-slice devices are general building blocks and need not be used with any particular type of control logic, they are normally discussed in the context of microprogrammed control logic. In fact, many of the available bit-slice devices were designed to be used in microprogrammed control sections. That is, a set of bit slices was designed, the assumption was made that microprogramming would be used to control these slices, and additional bit slices were designed to support the development of microprogrammed control logic. Because of this close relationship between the two concepts, an understanding of microprogramming is needed to fully appreciate the nature of bit-slice logic.

In relation to other current digital-system concepts, microprogramming is a rather old concept, originated by M. Wilkes in 1951 [1]. Unfortunately, however, despite its age and widespread use, the concept has no formal concrete definition. One reason is that the term is used occasionally for other than technical reasons, leading to confusion in attempting to find a universal definition.

Another reason, which is explained in a later chapter, is that microprogramming is usually used in a binary sense (i.e., a machine is a microprogrammed machine or it is not), but the issue is not quite as clearcut as this; it is more accurate to talk of the *degree* to which a machine is microprogram controlled. That is, there are machines that everyone would agree are microprogrammed, there are machines that virtually everyone would agree are not microprogrammed, but there is also a large grey area between. For instance, one machine in this grey area is IBM's S/370 Model 135/138. There could be two valid arguments about this machine: one that it is a microprogrammed processor, the other that it is not and that in fact it does not even have the S/370 architecture, but that it is a sequential-logic-controlled machine with an unfamiliar instruction-set architecture on which a software interpreter was written to give it the outward appearance of being a member of the S/370 family.

There have been attempts to develop formal definitions of microprogramming, but none have been able to stand the test of example and counterexample. That is, given a definition of microprogramming, one can usually find a machine fitting the definition, but, at the same time, it is a machine that most people would consider to be a nonmicroprogrammed machine. Conversely, one can usually find a machine excluded by the definition that most people agree is microprogrammed. This situation has lead to several desperate definitions, such as calling microprogramming "the implementation of hopefully reasonable systems through interpretation on unreasonable machines" [2] and likening microprogramming to pornography—that it is virtually impossible to define, but you know it when you see it.

Given this situation, the safest approach is probably to avoid a definition, but this is hardly the way to begin a discussion of microprogrammed control. To begin working toward a definition, one can start with Wilkes' original motivation for the idea [3]

> Microprogramming was . . . a solution . . . to a . . . problem, namely, that of finding a systematic way of designing the control unit of an electronic digital computer.

couple this with the definition found in the first text on microprogramming [4]

> Microprogramming is a technique for designing and implementing the control function of a data processing system as a sequence of control signals, to interpret fixed or dynamically changeable data processing functions. These control signals, organized on a word basis and stored in a fixed or dynamically changeable control memory, represent the states of the signals which control the flow of information between the executing functions and the orderly transition between these signal states.

and add to that a more philosophical definition [5] in one of the many books on microprogramming which appeared in the 1975 timeframe [5–11]

> Microprogramming "corrals" key control functions into a regular structure, isolating them from the data flow. Microprogrammed control functions can be viewed as "working hypotheses," while hardwired control functions are "solid commitments." By assuming a functional perspective, the essence of microprogramming, appropriate design flexibility can be retained. As you can see, microprogramming is a design philosophy and organizational method not limited by the implementation technology.

By putting these together, one can arrive at the following working definition:

> Microprogramming is the process of producing microprograms. A microprogram is a form of stored-program logic that *explicitly* and *directly* (the emphasis is important) controls the major logic devices of a digital system (e.g., registers, ALUs, counters, busses, memory). As such, a microprogram is a substitute for a sequential-logic control network. The most common application of a microprogram, but certainly not the only application, is to give a processor a particular instruction-set architecture.

The last sentence is significant, since it implies that, in a microprogrammed processor, it is the *microprogram* that gives the processor its outward architecture, not the physical logic devices (although the physical organization of a processor is usually designed with a particular architecture in mind, that is, to allow one to produce an efficient microprogram for the architecture). For instance the fact that IBM's S/370 Model 145 is a member of the S/370 architecture family is an attribute of the microprogram residing within it. By altering its microprogram, one could give the machine a totally different architecture, for example, give it the architecture of a DEC PDP-11 or an Intel 8080 (albeit an expensive 8080).

To clarify a point surrounding the use of the prefix "micro," note that microprogramming is *not* the programming of microprocessors. A microprocessor may or may not be designed as a microprogrammed machine (most are not), but the act of writing programs for a microprocessor (e.g., 8080, Z80, 6800) is programming, not microprogramming.

Another way of explaining the nature of a microprogram is to view a computing system as a hierarchy, as shown in Figure 2.1. At the top of the hierarchy sits the end user, and the bottom or center sit the basic logic circuits, such as ANDs, ORs, and flipflops. (If one were to carry the hierarchy to the extreme, one might define the bottom as basic molecular materials, such as silicon, aluminum, and gold). The logic circuits define an interface level of boolean functions. Atop this interface sit logic devices (e.g., registers, adders, shifters, memories); these are intercon-

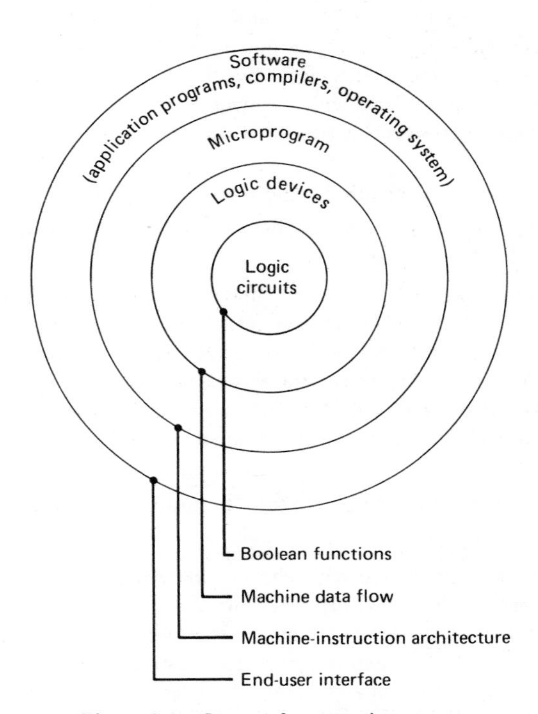

Figure 2.1. Layers of a computing system.

nected with busses to form a higher interface level: the machine data flow, or the register-transfer level.

At the next level one finds this elusive concept of microprogramming. The microprogram is designed to form a higher interface known as the machine-instruction repertoire, or the computer architecture. (In a nonmicroprogrammed machine this level still exists, but the transformation is performed by a sequential-logic control network.) The highest level is the programming level, where one finds application programs, compilers, utility programs, and operating systems, which interact to provide an end-user interface, such as a query language used by a librarian or a set of terminal functions used by a banking teller. Hence we see that, in a microprogrammed computer, the role of the microprogram is to create an abstraction for use by programs, namely an abstraction consisting of a set of machine instructions, data types, and data-addressing schemes. The microprogram creates this abstraction from the physical logic devices in the system.

One last way to view the subject of microprogramming is to pose a commonly asked question: Is microprogramming programming, or is it engineering? Like everything else related to the subject, the question is not an easy one to answer, because microprogramming has elements of both disciplines. One can get a feeling for its relative proximity to programming and engineering by examining the concepts with which a microprogrammer is typically concerned. In particular, the microprogrammer usually must understand the characteristics of the physical logic devices in the system (e.g., registers, counters, ALUs, memory arrays, latches, shift registers). The interconnections among these devices (i.e., the data flow) are a key concern of the microprogrammer. For instance, he or she must have answers to such questions as: Can I route the contents of this register directly into that register, or do I have to route the data through the ALU? Also, the microprogrammer is concerned with circuit timings to some degree. One must know when particular events will occur, the access and cycle times of memories, and so on. In contrast (fortunately), none of these are concerns of programmers, even those coding in machine language.

At the risk of overgeneralizing, the objectives of the microprogrammer are normally different from those of a programmer, and thus the tradeoffs are different. The microprogrammer's primary objective is usually speed. Contrasted with this, programmers are not (or should not be) as interested in squeezing the last cycle out of their programs as they are in other objectives, such as program understandability and extensibility.

One last clue to the answer can be obtained by examining the educational background of today's microprogrammers. Using one company that performs a considerable amount of microprogramming as an example (IBM), most people who perform "real" microprogramming (i.e., microprogramming in the true technical sense) have engineering titles and engineering degrees.

On the other hand, since a microprogram is a form of stored program logic, it must have attributes in common with software programming. It does, and the microprogrammer can learn much from the programmer (the subject of Chapter 9). In summary, microprogramming requires both engineering and programming knowledge (hence, the motivation for the derivation of the synonym of microprogram, *firmware,* in recognition that it is somewhere between hardware and software).

A HYPOTHETICAL MICROPROGRAMMED MACHINE

Because of the widespread confusion about the subject of microprogramming, the most effective approach for introducing the subject is in a bottom-up fashion: begin with the details of designing a microprogrammed machine, and then, once the mystique of microprogramming has been lifted, introduce the general concepts, subtleties, and advanced concepts. This section will describe the data flow of a simple hypothetical processor, illustrate a method of controlling the processor (developing a microprogrammed-control solution), describe some optimizations of the design (some of necessity, others for reasons of cost), and illustrate the writing of a simple microprogram for the design.

Rather than starting completely from scratch, Figure 2.2 will be the starting point. Figure 2.2 represents the initial data-flow design of the processor. The processor is constructed from only eight active components: four registers (I, A, B, and D), one full binary adder, a shifter, a two-input multiplexer, and a memory array. These components have been tied together with a variety of busses. Also, two registers with a fixed content of zero have been added. Of course, the processor designer and microprogrammer are very concerned with the width of the components and data paths, but their width in this example is not important (assume a width of 16 bits if you wish).

For the sake of discussion, most of the registers have three-state outputs. Examine register I. It has two control inputs: S and E. When the value on the S input rises from low to high (or 0 to 1) value, the value on the X bus will be stored in the I register. During the time that control-input E is in the high (1) state, the value in the I register is gated onto its output bus (the bus feeding the left input of the adder).

Register A has two outputs. One, a three-state output, feeds the adder bus and is controlled by A's E input control. The other output of A, not a three-state output, always feeds the addressing logic in the memory array. The B and D registers have two three-state outputs; thus each has two output-enable controls. The D register is fed by a multiplexer. If M is 0, the X-bus contents flow through the multiplexer; otherwise, the value on the memory-data bus flows through the multiplexer. Since a memory word can flow to or from the D register, the D register might be called the *memory data register.* Similarly, A could be called the *memory address register.*

Setting the memory SL input to 1 causes a memory word to be selected (the word selected is the one whose address is in the A register). Setting R/W to 1 causes the word's value to be gated to the data bus; setting R/W to 0 causes the value on the data bus to be stored into the word. We will assume here that the

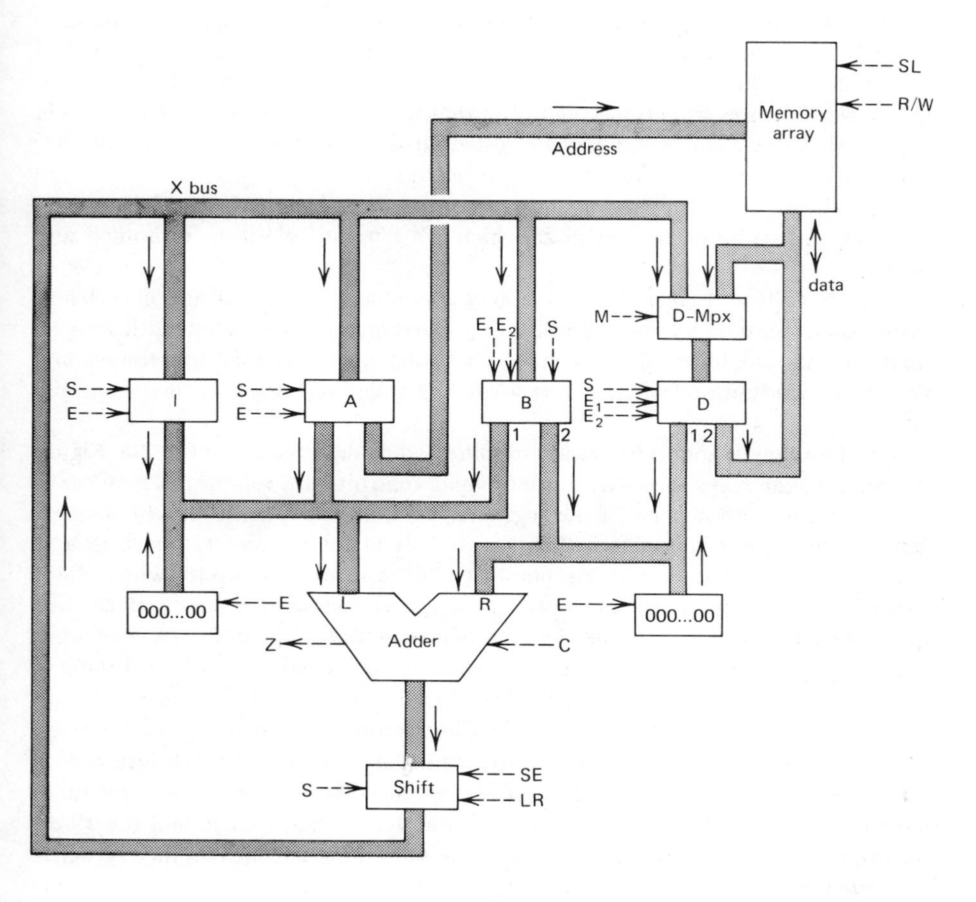

Control key	
S:	Set register from input
E:	Enable register output
M:	Select input 1 or 2
SL:	Initiate memory operation
R/W:	Indicates memory read or write
SE-LR:	Shift right, left, or none (two lines)
C:	Adder carry-in
Z:	High if adder output is zero

Figure 2.2. Initial data-flow design.

speed of the memory array is the same as the registers, although this is usually not the case in real systems.

Note that the adder output passes through a shift register before moving onto the X bus. When the shifter's S input rises from 0 to 1, the adder output is stored in the shifter, shifted left or right by 1 bit, and gated onto the X bus. The SE input, when it rises from 0 to 1, indicates that the value in the shift register should be shifted left or right by 1 bit (the injected bit is assumed to be 0). The LR control specifies whether the shift is left (0) or right (1).

The adder has a 1-bit carry input, meaning the adder performs the function L+R+C. It has an output signal Z, which is a 1 if and only if the computed sum is zero.

Designs such as Figure 2.2 are largely a product of intuition and experience, rather than based on any pre-established engineering priciples. That is, there is no textbook that would specify that, for a particular instruction-set architecture and particular cost and performance objectives, the design of Figure 2.2 is the optimal one.

The data flow is somewhat easier to follow when depicted as Figure 2.3. Figure 2.3 replaces the register set and output-enable controls with conceptual switches at the input and output ports of the registers. Each switch is controlled by a single control line. For instance, switch 9 corresponds to the set control for register I, and switch 1 corresponds to its output-enable control. An input switch closes when its control line rises from 0 to 1 and remains closed for only a short time (just long enough to ensure that the data will be stored in the register). An output switch closes when its control line becomes a 1, and remains closed until the line drops to zero. All of the controls have been assigned numbers from 1 to 19.

The next step is to make this processing section do something useful (e.g., perhaps we are trying to design a processor having the DEC PDP-8 instruction-set architecture). The obvious thought is that, to make Figure 2.3 do something, we have to be able to set the 19 control lines. To do this, we can add the 19-bit control register shown in Figure 2.4. For the sake of clarity, bit 1 in the register is connected to switch 1, bit 2 to switch 2, and so on. By placing a value in this register, we can specify all 19 control signals and thus cause the processing section to do something.

The introduction of the control register, however, is only the beginning. The next question is: Place a value into the control register *from where*? A second consideration is that having a *single* control value is not likely to be sufficient; what we really need is a *set of control values that vary over time,* so that a preplanned sequence of control values are gated into the control register at periodic intervals.

To accomplish this, one is faced with two basic alternatives. One could devise a complex random-logic network consisting of logic gates, flipflops, and counters. The function of this network would be to deliver a new value to the control register at periodic intervals. If we take this approach, the machine would be classified as a nonmicroprogrammed, *sequential-logic-controlled* (or *hardwired*) machine. Since the topic at hand is microprogrammed control, this alternative is

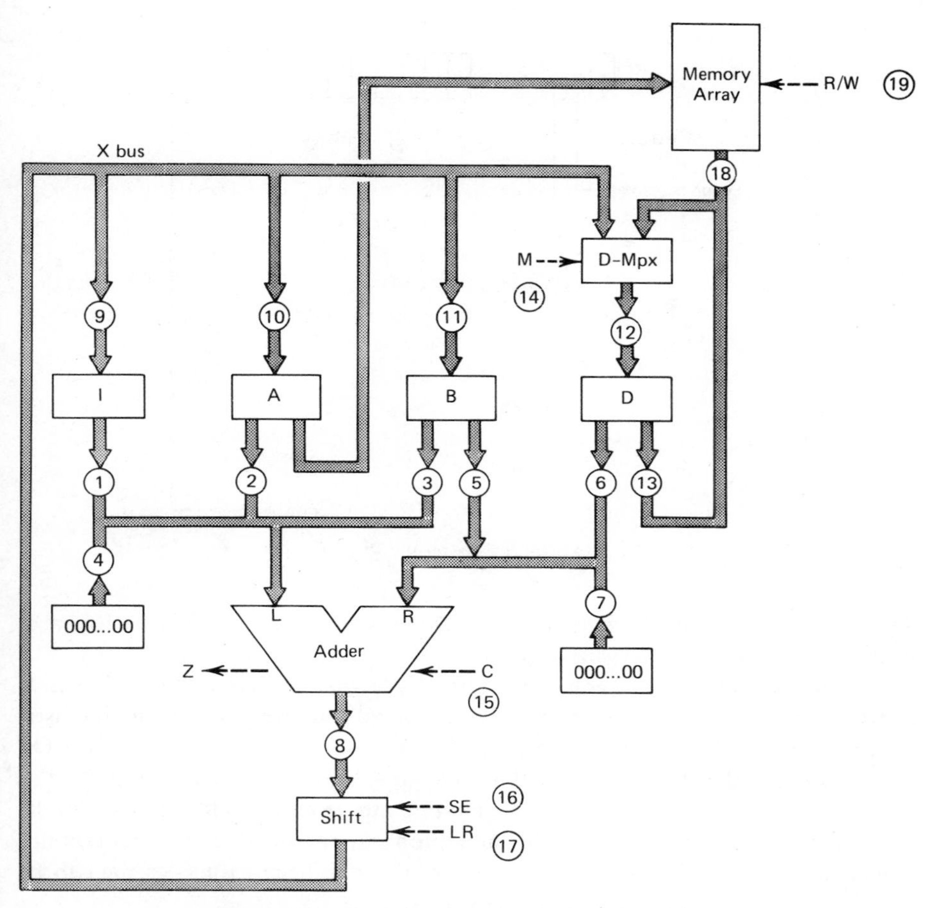

Figure 2.3. Equivalent data-flow design with switches.

not pursued here, although a later section in this chapter evaluates the advantages and disadvantages of both approaches.

The other approach is the following. Since we have to deliver a sequence of 19-bit values to the control register, why not store these values in another memory array and have the memory array periodically deliver the next set of 19 bits to the control register? This leads us to the control-section design in Figure 2.5. What we have is a 19-bit control register with its own set and output-enable controls, a 19-bit-wide memory with an output-enable control, and a counter. Whenever the INC line rises (from 0 to 1), the counter adds one to its current value.

Since the word *periodic* has been used, a clock must be introduced. Initially, have the clock simply drive the set and enable lines in Figure 2.5, such that each time the clock fires, a new word is addressed in the control memory, the word is read into the control register, and the control register feeds the 19 control points. Perhaps one inverts the clock pulse and feeds this to the INC line on the counter.

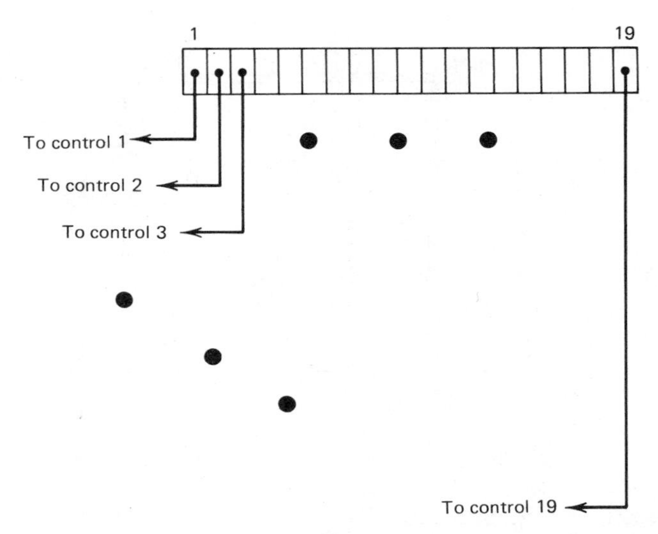

Figure 2.4. Control register for the 19 control points.

The end result is that each time the clock fires, the *next* word in the control memory controls the processing section.

Determining the clock frequency is usually a matter that requires extensive analysis (the matter is discussed in later chapters), but we will take a simplistic approach here. Of course, one wants the clock to run as rapidly as possible. On the other hand, the clock has to be slow enough to allow state changes specified in the control register to have sufficient time to occur (e.g., the adder is given sufficient time to add, the set-up, hold, and propagation times of the registers are observed). Let us define the clock speed to allow sufficient time for data to be

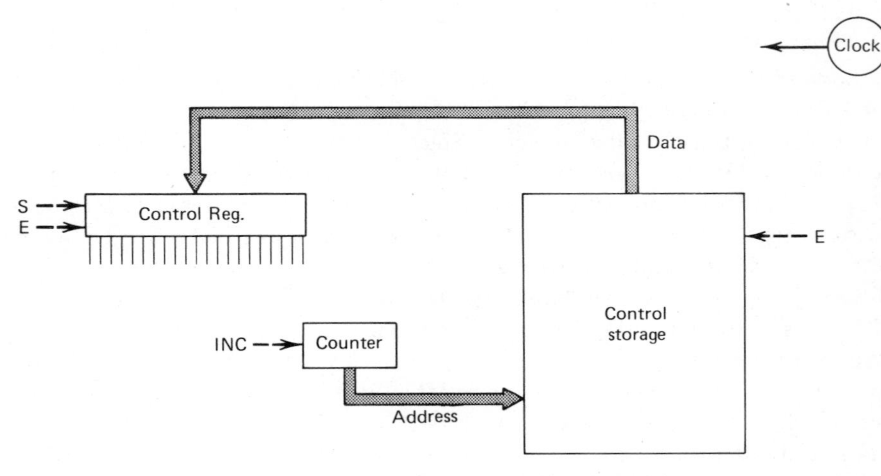

Figure 2.5. Initial microprogrammed control section.

gated out of the registers, through the adder and shifter, and back into a register. By examining the worst-case times for the devices along this path, as well as the time for the control word to be addressed and sent to the nineteen control points, we will make the assumption that the clock frequency is 5 MHz. That is, the clock period is 200 nanoseconds, during which the clock output is high for approximately 100 nanoseconds and low for approximately 100 nanoseconds.

At this point, a few definitions can be made. The clock period can be called the *machine cycle time,* which is the interval between successive sets of control signals being fed to the processing section. Each string or word of nineteen bits in the control memory is called a *microinstruction*. Hence a microinstruction is a set of control signals or state specifications that controls the data flow and device functions in the processing section for the duration of one machine cycle. The collection of stored microinstructions can be termed a *microprogram*. Hence, returning to the definitions of microprogramming, one should now see that it is "a systematic way of designing the control unit," that it "corrals key control functions into a regular structure," and that a microprogram is a form of stored-program logic that explicitly and directly controls the major logic devices of a processing unit. For the hypothetical processor in its current form, Table 2.1 defines the meaning of each bit in the microinstruction.

Continuing with the definitions, the memory containing the microprogram is normally called the *control storage,* and the counter in Figure 2.5 might be termed a very primitive *microprogram sequencer*. Depending on the technology used, control storage is sometimes called a *ROS* or *ROM* (read-only memory), *PROM* (programmable read-only memory, a memory that can be "programmed" only once, usually by using a high voltage to burn out fused links), *EPROM* or *EAPROM* (a PROM that can be altered more than once by first resetting it with high-intensity ultraviolet light or by a high current and then using a high voltage to store charges), or *WCS* (writable control storage, a RAM). Be careful to note, however, that most of these terms describe types of storage devices, not their use. For instance, a PROM can be used for purposes other than control storage (e.g., to hold software programs, such as an operating system for a microprocessor, in a nonvolatile medium); thus, "programming" a PROM does not necessarily imply that one is microprogramming.

Timing

The astute reader has probably recognized at this point that the design will not behave correctly as it stands; in particular, it is fraught with timing problems. To cite one, notice what would happen if one were to write a microinstruction to set A to the sum of A and B. That is, the task is to gate the A and B registers to the adder inputs, set the carry input to zero, gate the adder output to the shifter, perform no shift, and gate the shifter output (the X bus) into the A register. Presumably, following Table 2.1, the microinstruction would be coded as

0100100101000000000

TABLE 2.1. Definition of the Microinstruction

Bit	Function if Bit is Set
1	I → Adder L-bus
2	A → Adder L-bus
3	B → Adder L-bus
4	0 → Adder L-bus
5	B → Adder R-bus
6	D → Adder R-bus
7	0 → Adder R-bus
8	Adder → Shifter
9	X-bus → I
10	X-bus → A
11	X-bus → B
12	D-Mpx → D
13	D → Memory bus
14	Memory bus → D-Mpx (Otherwise, X-bus → D-mpx)
15	Adder carry-in
16	Shift
17	Right shift (Otherwise, left shift)
18	Initiate memory operation
19	Memory read (Otherwise, memory write)

When the control signals from this microinstruction reach Figure 2.3, the following actions will occur:

1. First, the current contents of the X bus (shift register) will move into register A, since switch 10 closes on the rising edge of the pulse on control line 10. At the same time, the output of the adder (probably unpredictable at this point, since no inputs to it have been activated yet) is gated into the shifter. Hence the actual value that moves into the A register is somewhat unpredictable, since it is dependent on the set-up and hold time requirements for the register and the physical wire length of the X bus. In any event, this is not what should happen.
2. Switches 2 and 5 are now activated and remain closed for approximately 100 ns. By the time that the sum propagates to the output of the adder, switch 8 is open (it was edge activated) and the sum goes nowhere. Actually, this is no great loss, since the sum is not A+B, but B plus the unpredictable value of A from point one above.

What we have is a *race condition* (register A is being modified before its contents have been gated into the adder) and several synchronization problems (e.g., the switch to the shifter is closed and reopened before the output of the adder has time to stabilize).

To illustrate another problem, suppose that two successive microinstructions had the form

xxxxxxxx1xxxxxxxxxx
xxxxxxxx1xxxxxxxxxx

In other words, we are examining two microinstructions that gate the X bus into the I register. Recall that switch 9 only closes momentarily on a rising control pulse. However, the control signal in both microinstructions is a 1 and there is nothing to guarantee that it will drop to 0 "between microinstructions," meaning that the second microinstruction might not cause switch 9 to close.

The control section (Figure 2.5) also has timing problems. The most obvious is that the control register is set at the time that control storage is enabled. The result is that the microinstruction never reaches the control register. Another problem is that the control register is set and enabled for output at the same time, resulting in spurious control signals.

A step toward a solution is the realization that one must somehow sequence different sets of events within the machine cycle, taking account of race conditions and device set-up, hold, and propagation times. The solution requires a detailed analysis of each device, but to illustrate the general idea we will develop a simplistic solution.

The solution is to subdivide the machine cycle into a sequence of subcycles. The subcycles can be identified as the following sequence:

1. Control storage is enabled, and the current microinstruction is read.
2. The control register is set, storing the current microinstruction.
3. Output of the control register is enabled, and control storage is disabled.
4. Output-enable-controls 1 to 7 and 13, and memory-select (18) are set if so indicated in the current microinstruction.
5. Switch 8 is set if so indicated in the current microinstruction, thus gating the ALU output into the shifter. A rising pulse edge is applied to the INC control of the counter.
6. Switches 1 to 7, 13, and 18 are disabled. Switch 16 is set if so indicated in the current microinstruction, thus causing a shift to occur.
7. Switches 9 to 12 are set if so indicated in the current microinstruction, thus storing the shifter (X-bus) output.

Controls 14, 15, 17, and 19 are passive signals (they do not initiate action; they specify information about an action that is triggered by another signal); hence they do not require special timings as long as we can ensure that they have their proper values when their corresponding control signals are triggered.

To create this sequence within a machine cycle, one needs clock pulses within the major 200-ns clock period. One way to do this is to create other pulses with a 200-ns period but offset from the major clock pulse. This can be accomplished by building a sequence of delay circuits, running the clock into the delay units, and then using the output of the delay units as intermediate timing signals. A delay unit can be something as simple as a sequence of pairs of inverters.

This idea is implemented in Figure 2.6. Waveform T_1 is the output of the clock. Waveform T_2 has the same frequency but has been delayed by 50 ns. Two other waveforms, T_3 and T_4, have been generated in a similar fashion. By ANDing and ORing these signals together, one can generate a large variety of timing signals. By writing logic equations for the control points in the system, one can design a small amount of logic to create the seven-step sequence listed earlier.

As an illustration of a few of the derived sequences, we determined that the initial actions are to enable control storage, then set the control register from the control-storage output (after delaying for sufficient time to allow the microinstruction in control storage to be accessed), and then enable the control-register output and disable the control-storage read. The first can be accomplished by feeding T_1 AND $\overline{T}_2$ to the E control on control storage. Assuming that the control storage has a worst-case access time of about 40 ns, the control-register set control is connected to T_2, meaning that the control register will be filled on the rising edge of T_2. Since we want the control register to feed the controls in the processing section from this time

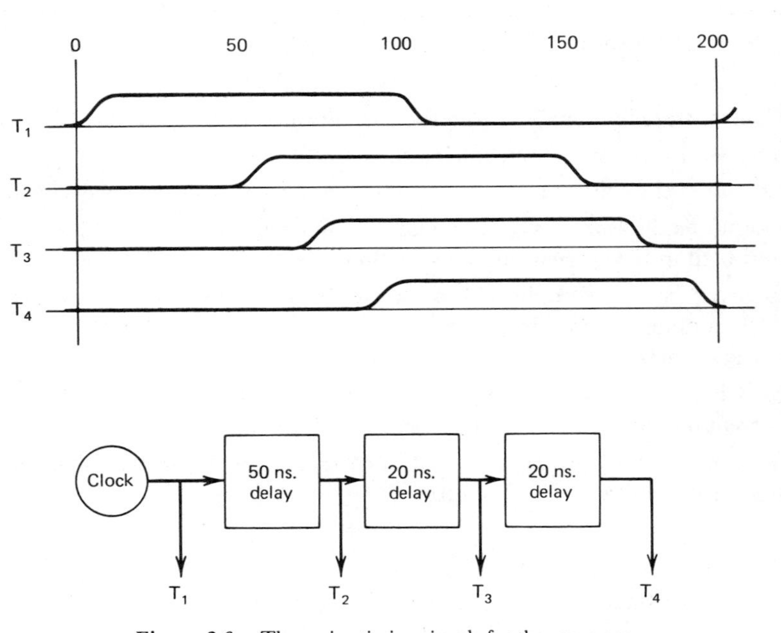

Figure 2.6. The major timing signals for the processor.

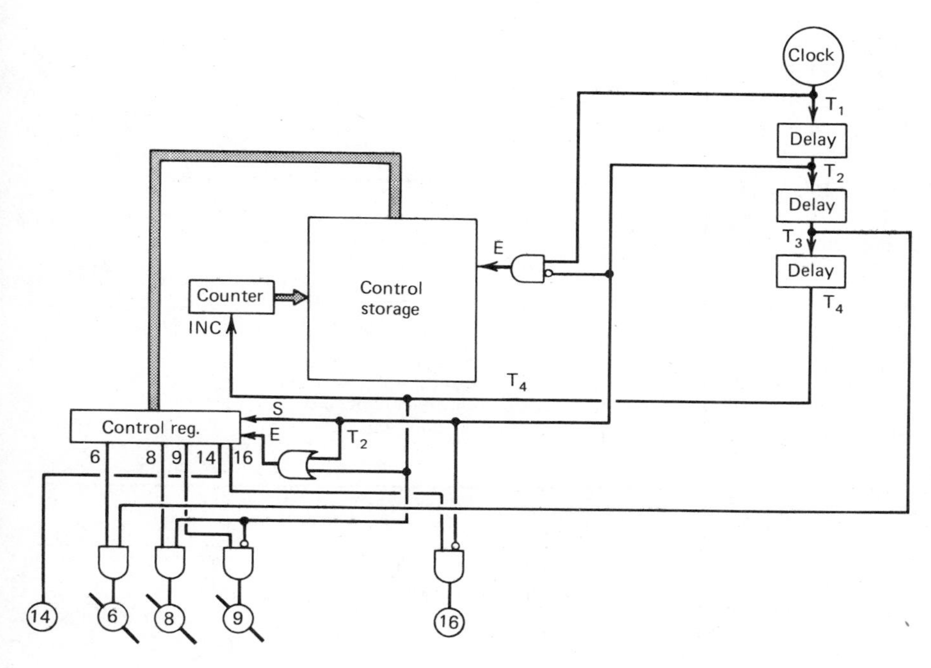

Figure 2.7. Partial interconnection of the timing signals.

until the end of the machine cycle, we can feed the output-enable control of the control register from T_2 OR T_4.

Another of the requirements was to enable switches 1 to 7, 13, and 18 early in the cycle (after the control register has started delivering its output), if the current microinstruction so indicates, and disable them later. We will use clock pulse T_3. For instance, rather than connecting control-register output 6 directly to switch 6, we will AND control-register output 6 and T_3, and feed this to switch 6. Switch 8 can be set a while later (allowing enough time for values to propagate out of the registers and through the adder); thus switch 8 is fed by microinstruction bit 8 AND T_4. The shift operation must be delayed until the shift register has been filled. Hence, the shift control (16) is controlled by the falling edge of T_2, causing the shift to occur (if the current microinstruction so specifies) about 60 ns after switch 8 is activated. Switches 9 to 12 are triggered late in the cycle; each is fed by the corresponding microinstruction bit ANDed with $\overline{T}_4$. Thus, registers I, A, B, and/or D are set from the X bus on the falling edge of T_4, providing that the corresponding bit in the microinstruction is a 1. A few representative interconnections to accomplish this sequencing are shown in Figure 2.7.

The timing introduced by the delayed clock pulses and the added logic gates is summarized in Figure 2.8. One can now verify that our original troublesome microinstruction, namely

010010010100000000

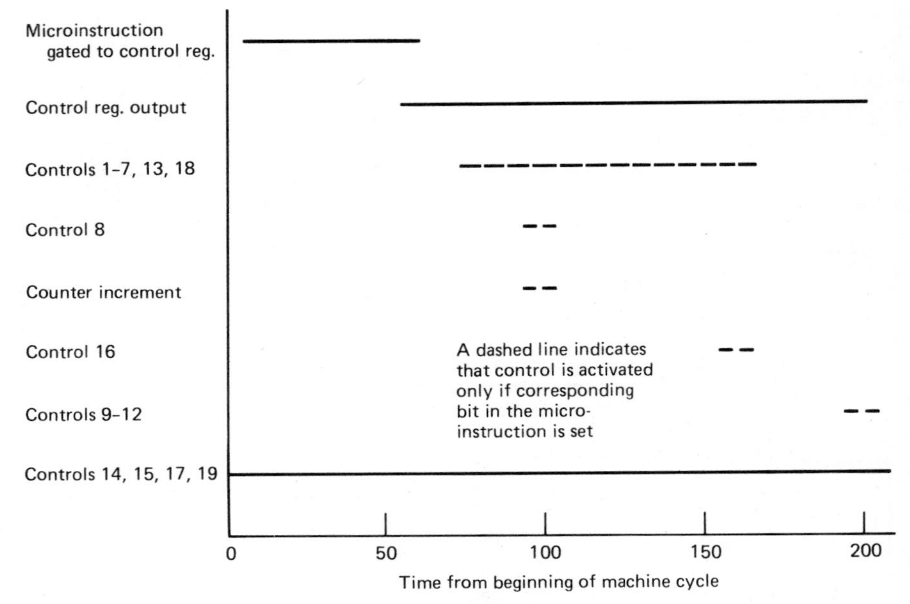

Figure 2.8. Processor timings within a machine cycle.

will now do what it is supposed to do, that is, set the A register to the sum of the contents of the A and B registers.

A Miniature Microprogram

At this point, it may be instructive to write a small microprogram. Suppose the problem at hand is the following:

> Register A contains the address of a main-storage word. Add the value of this word to four times the value in the B register. Put the result in the main-storage word and increment the A register by one.

Using Table 2.1, the reader may wish to attempt this on his or her own before reading further.

This problem can be solved in three microinstructions by taking advantage of the parallelism in the data flow. In the first microinstruction, a read is initiated from main storage into the D register, and the B register is multiplied by 4 by adding it to itself and shifting the sum left. In the second microinstruction, the main-storage value is added to the new value in the B register and placed in the D register. In the third microinstruction, the result is written to main storage and the A register is incremented. The microinstructions are

```
0010 1001 0011 0101 011
0010 0101 0001 0000 000
0100 0011 0100 1010 010
```

Optimizing the Microinstruction

Continuing with the design of the processor, an issue that might be considered is whether a 19-bit control word is needed, that is, whether we can accomplish the same with fewer bits. The 19-bit control word gives us 2^{19}, or 524,288, possible unique microinstructions. If a large number of these turn out to be useless, the control word contains considerable redundacy, and presumably one can reduce the size of the microinstruction.

The obvious motivation for this line of thinking is a reduction in the amount of control storage needed, which would lead to lower chip counts, lower packaging costs, and lower power supply and cooling requirements. Another motivation is eliminating the ability to write microinstructions that are unpredictable, useless, or nonsensical.

The basic methodology is to examine the processor for combinations of control bits that are related to one another (in the sense of being mutually exclusive or redundant), and, when a situation is found, combine and encode them. The most obvious example is bits 1 to 4. These bits control the input to the left side of the adder (I, A, B, or zero). Since we would expect at most one of bits 1 to 4 to be set in a given microinstruction (setting more than one of them would lead to unpredictable results), they can be encoded into fewer than four bits. There are actually five states to consider—just switch 1 set, just 2 set, just 3 set, just 4 set, and none set—but since the latter state appears to serve no use, we can encode them into two bits. The two bits are arbitrarily defined as

00 - Zeros → adder L-bus
01 - I → adder L-bus
10 - A → adder L-bus
11 - B → adder L-bus

To accomplish this, we need decoding logic between the control register and the switches in the data flow. As an illustration, assume that these two bits are bits 1 and 2 in the microinstruction (bits 3 and 4 can be eliminated or used for another purpose). Earlier, switch 1 (output of the I register) was fed from an AND fed by control-register bit 1 and clock pulse T_3. Now, switch 1 must be set according to the equation

$$\overline{b_1} \wedge b_2 \wedge T_3$$

This is accomplished by the logic in Figure 2.9. Similar decoding logic is added for switches 2 to 4. Hence, we have eliminated 2/19 of control storage, but at the expense of additional decoding logic. There is a tradeoff here, of course, in terms of circuits and speed; the decoding logic represents additional circuits and possibly additional logic delays. In terms of circuit and chip counts, the cost of the decoding logic is amortized across all the words of control storage. Thus it is usually a good tradeoff, if control storage is large, but perhaps an unwise choice if we foresee a requirement for only a small number of control-storage words.

The time penalty of the decoding logic can usually be absorbed at no cost. In all systems, there is usually a single critical path through the logic that dictates the

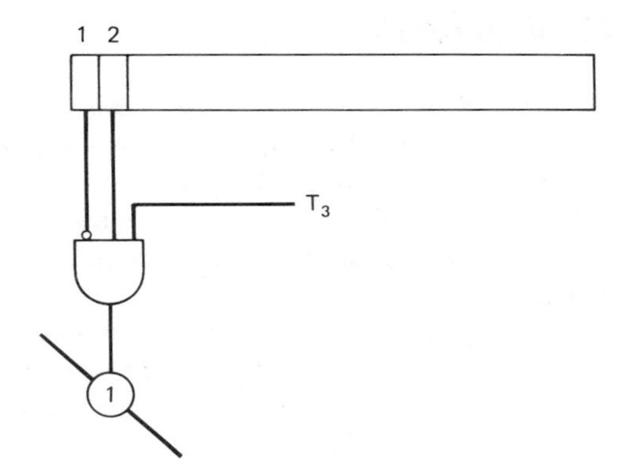

Figure 2.9. Example of a microorder decoder.

maximum clock speed. In most cases, the logic path containing the decoding logic would not be the critical path, and its use does not affect the cycle time. Of course, if it is used on the critical path, or if its use makes this path the new critical path, the designer is faced with a tradeoff.

Notice that bits 1 and 2 in the microinstruction have a new property: neither bit has a meaning when viewed alone; they mean something only when viewed together. In other words, bits 1 and 2 control the input to the left side of the adder. Such a field in a microinstruction is called a *microorder*. A microorder is a set of one or more bits in a microinstruction that, when viewed together, control a fragment of the data path.

Referring back to Figure 2.3, there are other encodings that can be performed. Switches 5 to 7 can be encoded into two bits, leaving one combination of the two bits unused. Note that the X bus can be gated to four destinations: I, A, B, and/or D. Although these are not mutually exclusive in the sense that switches 1 to 4 or 5 to 7 were (e.g., one may want to gate the X bus into I and B simultaneously), experience might tell us that the flexibility of gating the X bus into more than one register simultaneously is not needed. If so, there are five states of the X-bus destination: just I, just A, just B, just D, and nowhere. Since the "nowhere" state is significant here (we do not want to force ourselves into the situation of having to modify one of the four registers every cycle), three bits are needed. They might be encoded as

000 - no operation
001 - set switch 9
010 - set switch 10
011 - set switch 11
100 - set switch 12 and set multiplexer input (14) to zero
101–111 - unused

The unused states could be used, if desired, to allow the X bus to be gated into a few selected combinations of the four registers.

Assuming that the shift register is intended only as a temporary holder of the output of the adder for each machine cycle, one can see that a bit to control switch 8 (the shifter input) is not needed. When we are using the adder in a microinstruction, bit 8 should always be set. When we are not using the adder, there is no harm in setting switch 8, since the 3-bit X-bus microorder above allows us to gate the X bus nowhere. Hence this control bit can be eliminated; switch 8 will be fed by just clock pulse T_4.

Other redundancies one can spot are switches 12 and 13, in relation to bits 18 and 19 (the memory controls). When 18 and 19 are set (indicating a read operation), switch 12 should be closed, switch 13 should be open, and the D-mpx input (14) should be 1. When bit 18 is a 1 and 19 is a zero (write operation), switch 13 should be closed. Hence, control bits 12 to 14 are not needed, since the state of controls 12 to 14 can be derived (decoded) from bits 18 and 19. (Note that the X-bus microorder also controls points 12 and 14.)

A revised definition of the microinstruction is presented in Table 2.2. This encoding has reduced the total control-storage requirement by 37% and has greatly reduced the possibility of coding meaningless or erroneous microinstructions.

TABLE 2.2. Encoded Microinstruction Definition

Microorder	Bits	Setting	Function
Adder left input	1–2	00	Zeros
		01	I
		10	A
		11	B
Adder right input	3–4	00	Zeros
		01	B
		10	D
		11	Unused
Adder carry-in	5	0	Carry-in = 0
		1	Carry-in = 1
Adder-output shift	6–7	00	No shift
		01	Shift left
		10	Shift right
		11	Unused
X-bus destination	8–10	000	None
		001	I
		010	A
		011	B
		100	D
		101–111	Unused
Memory operation	11–12	00	No operation
		01	Read into D
		10	Write from D
		11	Unused

Note that one could encode even further. For instance, since fields 6–7 and 8–10 contain unused settings, they could be encoded together. One can see that if the X-bus destination is set to 000, the setting of the shift microorder is irrelevant. Hence, there are $1 + 4 \times 3$ meaningful states of these two microorders, meaning that, if they were combined, four bits would be needed, saving another bit.

Extensions

Although the processor might have some utility in its current form, it still has two major weaknesses, one of which has a simple solution. The only arithmetic operation that can be performed is addition. It may be true that, if one wants to perform such functions between two registers as subtraction, AND, and OR, one could devise microprogrammed algorithms to do so via addition, but it is likely that they would be inefficient. An obvious improvement is the replacement of the adder with an arithmetic-and-logic unit. Assume the ALU performs the four functions

$$L + \overline{R} + C$$
$$L + R + C$$
$$L \wedge R$$
$$L \vee R$$

Two bits will be added to the microinstruction, indicating which function the ALU should perform during this machine cycle.

The other weakness concerns the control-storage sequencing mechanism in Figure 2.5. One problem is the limited time of operation the machine appears to have, because of the sequencing through control storage. If control storage contains 1024 words, in 1024 x 200 ns, or 205 microseconds, the machine hits the end of control storage, exhausting the supply of microinstructions. This implies that some type of *branching capability* is needed, that is, the ability of a microinstruction to set the sequencing counter to something other than the address of the next control-storage word.

A second problem is the absence of decision-making ability in the microprogram. Given that microprograms are a form of stored-program logic, one wishes to embody algorithms in microprograms, and most algorithms require decision-making capability, such as the conditional-branch (or jump) instruction in machine instruction sets and the IF-THEN-ELSE construct in programming languages. In other words, what is needed is the ability to test one or more conditions in the processing section and have these conditions influence the address of the next microinstruction to be selected.

Both problems express a need for a conditional-branching capability within control storage. A wide variety of solutions exist, but here we will explore only one simple approach. Two fields will be added to the microinstruction. Assuming that control storage contains 1024 words, the first field will contain 10 bits. Its purpose is to contain an address of a control-storage word (i.e., an address of a microinstruction). The second field contains two bits. If it is set to 00, it indicates that the microinstruction for the next cycle should be fetched from the current control-storage address plus one. If 01, the microinstruction for the next cycle is fetched from the location specified by the 10-bit branch-address field (an unconditional branch). If 10, it indicates a conditional branch. If the ALU output in the current cycle is zero, the next microinstruction is fetched from the location specified by the 10-bit branch address; otherwise it is fetched from the current control-storage address plus one.

Given this simple branching mechanism, one further extension comes to mind. First, the 10-bit branch-address field in the microinstruction is an expensive resource, because it represents 38% of the control-storage space (the current microinstruction width has grown to 26 bits). Second, other than using the two registers of zero and the ALU carry-in, there is no way to inject a constant value into the processing section (e.g., we currently have a problem if we want to add eight to the value in the A register). Both problems can be overcome by giving the 10-bit branch-address field a dual use. A bus will run from this field in the control register to the right side of the ALU. A switch will be placed in this bus, and since the ALU-right-input microorder (bits 3 and 4) has an unused value (11), this value will close this switch. Hence, if a microinstruction is not performing a branch, the branch-address field can be used to gate a constant value into the ALU. Such a field is often called an *emit* field, because it contains a value to be emitted into the processing section.

The current design of the control section of the processor is shown in Figure 2.10 (clock signals are not shown). Bits 15 and 16 in the microinstruction are the branch-control microorder, and bits 17 to 26 represent the branch-address/emit

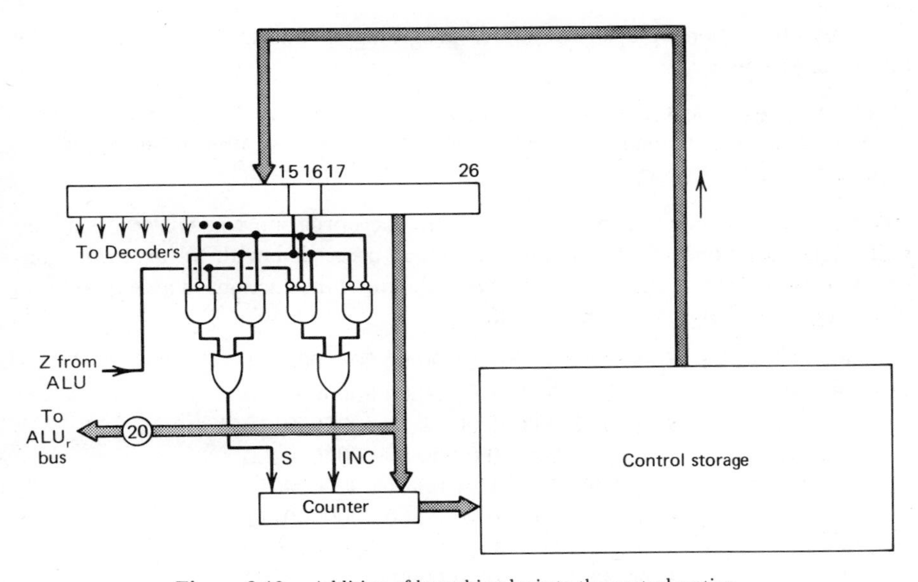

Figure 2.10. Addition of branching logic to the control section.

microorder. Z is a line from the processing section having the value 1, if the ALU output is zero.

The final definition of the microinstruction for the system is illustrated in Table 2.3.

Microprogramming the Processor

The last step in the development of the hypothetical processor is the microprogramming of the control section. For instance, if we were designing it to act as a CPU with an instruction-set architecture of 16-bit instructions where the first eight bits represent an operation code, the microprogram would have the following general organization. It would select a particular main-storage word (whose address might have been kept in one of the registers) and read its value into the D register. The value would represent a 16-bit machine instruction (or portion thereof, if the data-path width is less than 16 bits). The microprogram would then examine the first eight bits and eventually branch to a unique point in the microprogram. (Our hypothetical processor cannot do this "256-way branch" efficiently; special branching mechanisms usually exist in real processors to perform this branch efficiently.) At this point, control has been transferred to a point in the microprogram corresponding to a particular machine instruction (i.e., we have "decoded" its operation code). At each of these points, a microprogrammed algorithm would be written to simulate (or "emulate" or "interpret") the machine instruction (e.g., algorithms to perform the operations of multiply, move, jump-to-subroutine, etc., machine instructions). At the end of each of these algorithms, the microprogram would increment its "current machine instruction counter" and branch to the starting point (i.e., begin fetching and decoding the next machine instruction).

As an illustration of a smaller microprogram for this machine, consider the following problem

> The A register contains the address of a main-storage word. Add together the values of the eight words beginning at this location and place the sum at main-storage location 7.

The microinstructions are shown below. The microprogram was written as a loop with eight iterations. Each microinstruction is preceeded with an arbitrary control-storage address assigned to it. Beneath the microinstructions is an explanation of the function of each microinstruction.

```
1: 00 11 0 00 001 01 00 00 0000001001
2: 00 00 0 00 011 00 00 00 0000000000
3: 01 00 0 00 001 00 01 10 0000000110
4: 10 00 1 00 010 00 00 00 0000000000
5: 11 10 0 00 011 01 00 01 0000000011
6: 00 11 0 00 010 00 00 00 0000000111
7: 00 01 0 00 100 00 00 00 0000000000
8: 00 00 0 00 000 10 00 00 0000000000
```

TABLE 2.3. Final Microinstruction Definition

Microorder	Bits	Setting	Function
ALU left input	1–2	00	Zeros
		01	I
		10	A
		11	B
ALU right input	3–4	00	Zeros
		01	B
		10	D
		11	Emit (microinstruction bits 17–26)
ALU carry-in	5	0	Carry-in = 0
		1	Carry-in = 1
ALU output shift	6–7	00	No shift
		01	Shift left
		10	Shift right
		11	Unused
X-bus destination	8–10	000	None
		001	I
		010	A
		011	B
		100	D
		101–111	Unused
Memory operation	11–12	00	No operation
		01	Read into D
		10	Write from D
		11	Unused
ALU function	13–14	00	Add (L+R+C, C = microinstruction bit 5)
		01	Subtract ($L+\overline{R}+C$)
		10	OR ($L \vee R$)
		11	AND ($L \wedge R$)
Branch control	15–16	00	Next address = current address + 1
		01	Next address = microinstruction bits 17–26
		10	If ALU output = 0, next address = microinstruction bits 17–26, else next address = current address + 1
		11	Unused
Branch address and emit	17–26		

1. Read the first word into D. Store the value 9 in I.
2. Zero the B register (to be used to hold the sum).
3. I=I−1. If the result is zero, branch to location 6.
4. A=A+1.
5. B=B+D. Read the next word into D. Branch to location 3.
6. A=7.
7. D=B.
8. Write D into main storage.

Before leaving the hypothetical example, one clarification should be made. It is that microprogrammers do not normally write microprograms as strings of 0s and 1s as shown above. If the amount of microprogramming to be done is not trivial, one usually invents a symbolic microprogramming language and writes a software assembler program that executes on an existing system. The function of the assembler is to translate microinstructions expressed in the symbolic language to the strings of 0s and 1s. The symbolic languages are often crude (compared to programming languages); for instance a crude, positional, symbolic representation of the microinstruction at location 5 in the example above might be

B,D,0,NOSHIFT,B,READ,ADD,BRANCH,3

or a keyword-oriented alternative with default values for missing microorders, such as

ALUL=B ALUR=D DEST=B MEMORY=READ OP=ADD
BTYPE=BRANCH ADDR=3

or a more-flexible form with symbolic labels on microinstructions, such as

B=B+D READ GOTO LOOPHEAD

OBSERVATIONS

The discussion of the hypothetical processor contained a simplifying assumption that should be discussed at this point, particularly because it is an unrealistic assumption and its elimination is an important concern of the microprogrammer. The issue is the assumption that the memory operates at the same speed as the remainder of the processing section, that is, that the memory read-cycle and write-cycle times are less than the machine cycle time. In most systems, this is not the case; the memory cycle time is greater than the machine cycle time, typically two to four times greater. In other words, we assumed that, if a microinstruction specified that a word is to be read from storage into the D register, the operation would finish by the start of the next machine cycle. A more realistic assumption would be that the data does not move into the D register until two cycles later.

Given the fact that the memory interface is heavily used, and given the microprogrammer's usual objective of speed, one can see the implications of this delay on the microprogram. In fact, the microprogrammer views memory in much the same way that a programmer views an input/output device. Since the microprogrammer does not want to waste valuable processor cycles waiting for a memory request to be completed, the microprogram is usually written so that there is useful work to perform in the microinstructions active between the time a memory request is initiated and the time that it is expected to be complete. Where the programmer employs the techniques of I/O buffering and I/O overlap (e.g., reading a record from a tape device into memory before the record is needed by the program), the microprogrammer employs analogous techniques (e.g., reading a data word from memory into the processor in advance of the microinstruction that needs it, and instruction lookahead—reading the next machine instruction from memory during the time that the current machine instruction is being processed).

At this point, one should see considerable differences between programming and microprogramming. Microprogramming is different in two significant respects: parallelism (a microinstruction can potentially cause many parallel events to occur) and a higher degree of asynchronous operations (e.g., memory cycles). Where a programmer thinks of a program in terms of sequential operations (because of the definition of most programming languages and computer architectures), the microprogrammer, to achieve maximum efficiency, thinks in terms of parallel operations. This often has a significant effect on the design of algorithms, since, when designing a microprogrammed algorithm, one normally tries to use as many parallel functions as possible in each machine cycle.

ADVANTAGES OF MICROPROGRAMMED CONTROL

Earlier in the chapter, two control alternatives were mentioned, although only one, microprogrammed control, was discussed in detail. The other alternative is *random-logic, sequential-logic,* or *hardwired* control. That is, one could take the entire microprogrammed control mechanism (control storage, decoding logic, the sequencing mechanism, and the microprogram itself) and design a logic circuit to perform the same function. The logic circuit, designed from combinatorial logic devices, flipflops, counters, and so on, would periodically send signals to the control points of the processing section. It would also receive feedback from the processing section (e.g., whether the ALU output was zero), which would influence the subsequent control signals. Since one finds systems with both control mechanisms (although the microprogrammed alternative has begun to dominate, particularly in the design of processors), the advantages and disadvantages of each should be discussed. It should be noted that *either* control alternative can be used with bit-slice logic, although bit-slice logic was designed with microprogrammed control in mind.

Table 2.4. lists the major advantages and disadvantages of microprogrammed control compared with the sequential-logic-network alternative. The first advantage is not listed first because it is considered to be the most important; it is listed first because it was the original motivation for the invention of the microprogramming concept. That is, a microprogram gives the designer a structured and regular method of specifying the control logic of a digital system, as opposed to the ad hoc and irregular form of most sequential-logic control networks.

The second advantage is considered to be the most important one, and it surfaces in many contexts. If one is attempting to extend the function of a processor (e.g., add a new machine instruction), the task is considerably easier if the processor is microprogrammed. The change entails adding new microinstructions in control storage and possibly modifying some existing microinstructions. If the processor is controlled by a logic network, the change entails altering and extending the circuitry, a considerably harder task. In other words, the change to the microprogrammed machine requires no physical changes (with the exception, if the control storage is a ROM or PROM, of plugging in a new memory), but the change to the hardwired machine requires logic design and physical changes, possibly even requiring the redesign of chips and board layouts.

This advantage also applies in other contexts. If a design error is found (e.g., a multiplication algorithm is found to be incorrect), the error is much easier to correct if the algorithm is embodied in a microprogram rather than a logic network. If a design change is encountered during the development process (e.g., the architect redefines the representation of floating-point numbers), one can more readily respond to the change if the system is microprogrammed. The same holds for design improvements, such as the discovery of a faster algorithm to perform division of decimal numbers. If the processor must emulate a second architecture (i.e., it must execute programs written for a foreign instruction-set architecture), one can accomplish this by writing another microprogram for the processor. A sequential-logic-controlled processor would require the development of a new logic network.

The third advantage is one of cost, but notice the contradiction in Table 2.4, since the advantage column claims "cheaper," but the disadvantage column states "more costly . . ." The explanation is that the sequential-logic approach appears

TABLE 2.4. Microprogrammed Control Versus Random Logic

Advantages	Disadvantages
1. A more orderly and uniform way of design	1. More costly for simple machines
2. Ease of change	2. Sometimes slower
3. Cheaper	3. Designs are easier to copy
4. More suited to the LSI environment	4. Users want to make microprogram changes
5. Better diagnostic capability	
6. Higher system reliability	

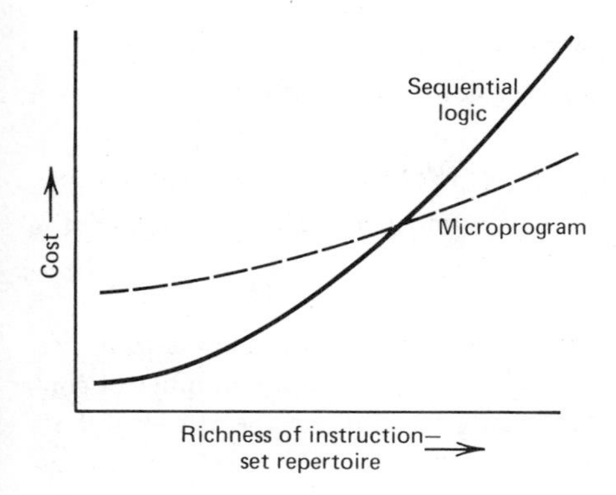

Figure 2.11. Cost relationship of both control alternatives.

to be cheaper for "simple" processors, but the microprogrammed approach is cheaper for all others.

This is illustrated by the curves in Figure 2.11, plotting the cost of the control mechanism versus the complexity or richness of the processor's instruction-set architecture. For a simple processor having a small set of primitive machine instructions, the logic network is small and simple with a low cost. However, as the complexity of the architecture increases (more, and more-powerful, instructions), the cost of the logic network increases rapidly, because its usual unstructured and irregular form means that its complexity grows more rapidly than the richness of the instruction set. Indeed, random-logic control networks in processors having many complex machine instructions are enormously complicated.

For a simple processor with the small, primitive instruction repertoire, the microprogrammed alternative is usually more expensive, since one is faced with the cost of control storage, the control-storage sequencing mechanism, and the decoding logic for microinstructions. However, these largely represent a one-time cost. As the complexity of the processor increases, one needs more control storage and additional microinstructions. Therefore, the cost rises, but not nearly as rapidly as with the logic network. Furthermore, since control storage is regular and random logic is not, the logic-network approach requires more circuit packages, leading to higher layout and packaging costs.

Figure 2.11 shows a crossover point, beyond which the microprogrammed alternative is cheaper. Unfortunately, no one has empirically derived these curves, since doing so would require the design of random-logic and microprogrammed control units across a range of instruction-set repertoires. However, indications are that the crossover point is reached quickly, meaning that the microprogrammed alternative has a lower development and manufacturing cost for all but very simple processors.

The fourth advantage implies that microprogrammed control is a better choice when employing large-scale integration. When one is analyzing a system design to find sections that might be integrated onto a single chip, the following criteria must be met

1. The section must contain a large number of circuits.
2. The section must have a limited number of connections to other sections (in recognition of pin-count limitations).
3. The circuitry must exhibit few, if any, changes over the lifetime of the design.
4. One must be able to produce and use a large number of the chips (to make the economics of LSI viable).

The control unit, be it microprogrammed or a logic network, appears to fit criteria 1 and 2. It represents a lot of circuitry and has a manageable number of connections to the outside world (i.e., the processing section, or the ALU and its associated registers and other devices). The logic network fails in relation to criteria 3, however. Each time a change is required (e.g., an extension, design change, or design correction), the chip would have to be redesigned and remanufactured, leading to economic disaster. However, since the microprogrammed approach has "programmable" control logic, the majority of changes would entail modifications and additions to the control-storage contents, resulting in no need to redesign or remanufacture the chip(s), particularly if control storage is a RAM or PROM.

The microprogrammed approach also has a large edge in relation to the fourth criterion. One is likely to develop an LSI chip for control storage and another for the microinstruction sequencing logic. In this case, these chips are likely to be usable in other designs (in fact, there are LSI microinstruction-sequencer bit-slice chips, the subject of Chapter 4). The probability of being able to use a random-logic-network control chip in other designs is virtually zero.

Another advantage is enhanced diagnostic capabilities. Consider the task of writing diagnostic programs—programs that service personnel might use to pinpoint a failure when a processor is behaving improperly. In a nonmicroprogrammed machine, the only choice is to write software diagnostic programs. Unfortunately, most of the possible failures will render the processor unable to execute programs. For instance, if the ALU, the memory interface, a register, or other critical component has failed, the processor will not be able to even fetch and execute the first instruction of the diagnostic program.

In a microprogrammed machine, one has another alternative: writing the diagnostics as microprograms, either by dynamically loading them into control storage or having them be permanently resident in control storage. The diagnostic microprograms can be written so that they "gingerly feel their way throughout the system." That is, the diagnostics need not rely on the majority of the processor's logic devices being operable at the start. Thus, the "microdiagnostics" have a better chance of running than do software diagnostics and, because of their closer relationship with the system, they can more accurately pinpoint the failing component.

The last advantage of the microprogrammed approach is reliability; a microprogrammed system should be more reliable than a nonmicroprogrammed system. In part, this stems from advantage 1—the fact that microprogramming is a more orderly and uniform approach to the design of the control section, implying that one should expect fewer design errors. Furthermore, in a microprogrammed control section, the majority of the gates are memory cells (control storage). This represents regular circuitry, which can be packaged in higher densities on a chip than irregular circuitry (i.e., random control logic). Given this and given the earlier discussion explaining why the microprogrammed alternative is more suitable for LSI designs, one would expect a microprogrammed control section to contain considerably fewer chips than a random-logic network. Since system reliability is closely correlated to chip and pin counts, one would expect a lower failure rate in the microprogrammed system.

To be fair, microprogrammed control also has disadvantages. Table 2.4 lists the disadvantages, although the consensus seems to be that the advantages greatly outweigh the disadvantages.

The first disadvantage has already been discussed. The second disadvantage, performance, stems from two considerations:

1. Some time is wasted fetching microinstructions from control storage, meaning that the processing section sits idle for some fraction of each machine cycle.
2. Because of the fixed-length machine cycle, some time may be wasted if the operations specified by particular microinstructions are completed before the end of the machine cycle. In other words, the machine cycle time is based on worst-case calculations. One looks for the longest possible path through the processing section that can be specified by a microinstruction; this dictates the machine cycle time. However, many microinstructions may evoke shorter paths of signal flow through the processor, meaning that the processor sits idle during the later parts of the machine cycles in which these microinstructions are in control.

The early belief was that microprogrammed control required a sacrifice in speed. Indeed, this was evident in the initial family of processors in IBM's S/360 family (the first widespread use of microprogramming), where the slower models were microprogrammed and the high-speed models were controlled by sequential logic. However, a later high-speed addition to the family, the Model 85, was a microprogrammed processor, and high-performance members of the S/370 family, such as the Models 168 and 3033, employ microprogrammed control.

Today, microprogrammed control is not considered to require a significant sacrifice in speed. Consideration 1 is largely solved by a technique known as *microinstruction pipelining* (discussed in Chapter 5). Consideration 2 is usually found to result in only an insignificant penalty in time, and can be eliminated, if necessary, by techniques discussed in Chapter 5.

The third disadvantage is that microprogrammed machines are easier to copy or re-engineer than sequential-logic-controlled machines. In other words, if company *A* has a system *X* and company *B* wishes to produce a competitive system *X*, *B* will have an easier time if *X* is microprogrammed, since the system's control logic and algorithms will be easier to comprehend. As an illustration, an article in *Business Week,* discussing the efforts of a company attempting to produce processors with the same architecture as IBM's S/370, made the observation: "Ippolito's design group capitalized on the fact that computers became much easier to copy when a design trend known as microprogramming caught on a few years ago." [12].

The fourth disadvantage may or may not be a disadvantage, depending on one's point of view. Consider a family of microprogrammed processors where each model has the same instruction-set architecture but different price/performance characteristics. Consider the owner of one of these models who has spent considerable time modifying and improving the manufacturer's operating-system software and now, hearing about the subject of microprogramming and thinking of it as just another level of software, decides that modifying the microprogram is desirable (e.g., he has heard rumors that the performance of the system can be improved significantly by taking small, heavily used, pieces of application or system programs and replacing them with additional microcode). The disadvantages from the manufacturer's point of view are that the customer is now dealing with the system at an unarchitected and undocumented level, and the customer's programming is now dependent on that particular model of the family (impeding efforts to sell him, in the future, a faster and more costly model), since, although the models in the family are identical at the instruction-set architecture level, they are rarely similar at the microprogramming level. Of course, if the processor models were sequential-logic controlled, the situation is much less likely to arise.

In summary, although there are pro's and con's concerning the use of microprogrammed control, the advantages usually greatly outweigh the disadvantages. The majority of processor types today, as well as other complex digital systems, such as I/O device controllers, use some form of microprogrammed control.

INFORMATION SOURCES

As mentioned earlier, bit-slice logic is closely allied with the subject of microprogramming. Thus, the use of bit-slice logic requires an understanding of microprogram design techniques. To obtain further information on the subject, one can consult the references at the end of the chapter. Also, Chapters 5, 8, and 9 of this book discuss additional aspects of the subject. The Association of Computing Machinery (ACM) has a special-interest group on microprogramming that publishes a quarterly publication, *SIGMICRO Newsletter,* and a yearly conference proceedings. *EURMICRO Journal* is a publication devoted to microprogramming and microprocessors, and the IEEE publications *IEEE Transactions on Computers* and *Computer* occasionally have special issues devoted to the subject of microprogramming.

REFERENCES

1. M. V. Wilkes, "The Best Way to Design an Automatic Calculating Machine," *Manchester University Inaugural Conference Proceedings*. Manchester, England, 1951, pp. 16–18.
2. R. F. Rosin, "The Significance of Microprogramming," *SIGMICRO Newsletter,* **4**(4), 24–39 (1974).
3. M. V. Wilkes, "Ten Years and More of Microprogramming," *SIGMICRO Newsletter,* **8**(4), 11–13 (1977).
4. S. S. Husson, *Microprogramming: Principles and Practices*. Englewood Cliffs, N.J.: Prentice-Hall, 1970.
5. J. M. Galey and R. L. Kleir, Eds., *Microprogramming: A Tutorial on the Queen Mary*. New York: IEEE, 1975.
6. G. G. Boulaye, *Microprogramming*. New York: Wiley, 1975.
7. A. K. Agrawala and T. G. Rauscher, *Foundations of Microprogramming: Architecture, Software, and Applications*. New York: Academic, 1975.
8. C. Boon, Ed., *Microprogramming and System Architecture, Infotech State of the Art Report 23*. Berkshire, England: Infotech, 1975.
9. A. B. Salisbury, *Microprogrammable Computer Architecture*. New York: American Elsevier, 1976.
10. N. Sondak and E. Malloch, Eds., *Microprogramming*. New York: Artech House, 1977.
11. H. Katzan, Jr., *Microprogramming Primer*. New York: McGraw-Hill, 1977.
12. "Control Data's Bid to Break IBM's Grip," *Business Week,* **2431,** 55 (May 10, 1976).

3

ALU/Register Slices

To date, the majority of bit-slice devices have fallen into two categories: ALU/register slices, discussed in this chapter, and microprogram sequencer slices, discussed in Chapter 4. A few remaining varieties of slices are illustrated in Chapter 6.

In the discussions of the ALU/register slices, each slice is illustrated with two diagrams: a diagram of its physical layout and a diagram of its external connections. The same notation will be used for each slice to ease the transition from slice to slice. However, because of this, these diagrams differ from those used by the manufacturers of the slices. Hence, the manufacturers' diagrams are also included to ease the transition to the manufacturer's detailed specifications.

Each slice is discussed in a consistent manner. The order of discussion is its physical organization, external connections, controls, representative timing information, and, usually, one or more short examples of its use. None of the slices is discussed in full detail; for information on their switching waveforms and electrical characteristics, one should consult the manufacturers' specifications. The last section of the chapter discusses the relative advantages and disadvantages of each slice and contains several tables summarizing their characteristics.

THE 2901 SLICE

The 2901 ALU/register slice, introduced in Chapter 1, is the most widely used slice today. Among other things, it has been used as the heart of the processing section in several new CPUs. It was first produced by Advanced Micro Devices, Inc., as part of their large Am2900 family of bit-slice devices and related support chips [1–2]. The 2901 device is now second-sourced by a large number of semiconductor manufacturers. The Am2900 family includes two ALU/register slices, several microprogram sequencer slices and support chips, a memory-interface slice, a status and shift-control device, bus transceivers, a priority-interrupt unit, RAMs and PROMs, registers, multiplexers, and others. Other designations, such as the 2901A and 2901B, represent faster versions of the 2901.

The organization of the 2901 is shown in Figure 3.1. All data paths are four bits wide. One key element is the 16-word RAM forming a bank of 16 4-bit registers. It is a 2-port RAM, meaning that two words (registers) can be selected simultaneously. The device has a pair of 4-bit control signals that select the two registers. The register currently selected by one pair of lines is designated the A register; its value appears on the A bus. The other control lines select the register designated as the B register, whose value appears on the B bus. As shown, a value may be stored in the register array. If this is done, the value is stored in the B register. The A and B registers may be the same, in which case the same value appears on the A and B busses simultaneously.

The A and B busses feed two latches. When the clock input to the slice is high, the selected registers are enabled onto the A and B busses and pass through the latches. When the clock input is low, the latches hold their prior inputs, thus eliminating the potential race condition when gating a value into the B register.

The 4-bit ALU can perform three arithmetic (base-two) and five logical operations. The R port of the ALU is fed from a multiplexer, allowing one to gate the A register, the D bus (an external bus coming into the 2901), or zeros into the R

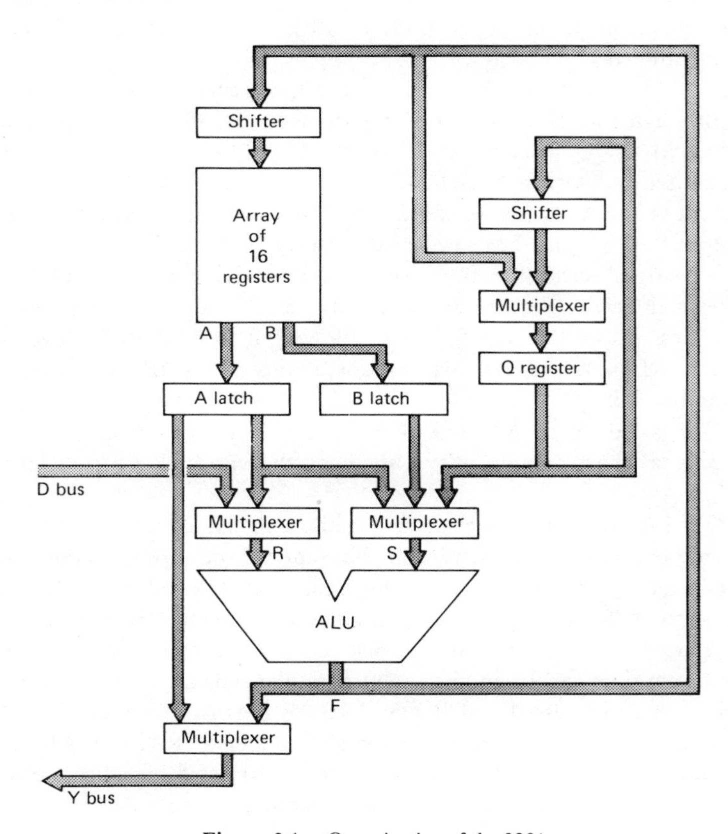

Figure 3.1. Organization of the 2901.

port. Likewise, the S port of the ALU is fed by a multiplexer, allowing one to gate the A register, B register, Q register, or zeros into the S port. These multiplexers and the characteristics of the register array allow one to perform such operations as

$$R3 = R2 + R3 + 0/1 \qquad R3 = D + Q + 0/1$$
$$R3 = R3 + R3 + 0/1 \qquad R3 = D + 0 + 0/1$$
$$R3 = D + R2 + 0/1 \qquad R3 = 0 + R2 + 0/1$$

but not

$$R4 = R2 + R3 + 0/1 \qquad R3 = Q + Q + 0/1$$

Certain other operations, such as

$$R3 = R2 + R2 + 0/1$$

are prohibited because of the way the control signals were encoded. (The designation 0/1 above represents the carry-in to the operation. Unless otherwise noted, this occurs on all arithmetic operations in all bit slices and will usually not be explicitly mentioned.)

The chip also contains another register, the Q register. Q can be used for any purpose, although the motivation for its existence is for use in multiplication and division algorithms. Q can be gated into the S port of the ALU and/or gated through a shifter and back into itself.

The output of the ALU can be gated to a variety of destinations. The chip has a 3-state output bus (Y); this bus can be fed with the ALU output (the F bus) or the value of the register selected as the A register. The ALU output can also be gated into the register array (the register currently selected as the B register). When gated so, it first passes through a shifter, allowing it to be shifted left or right by one bit. The ALU output can also be gated into the Q register, passing first through another shifter.

Figure 3.2 is Advanced Micro Devices' diagram of the 2901's organization. Figure 3.3 is a more detailed view, showing the external control signals and busses.

Figure 3.4 illustrates the external connections to the 2901. The 2901 has 40 pins; the only two pins not shown in the diagram are the supply voltage (5V) and ground. The 24 input pins consist of the input bus (D) and 20 control signals. These should be self-explanatory except for the nine I signals, which are discussed in detail later. The 10 output pins consist of the output bus (Y) and six status indicators. Four pins can be inputs or outputs, depending on the function being performed. As was discussed in Chapter 1, these pins run from 2901 to 2901 to interconnect the two shifters. On the most-significant slice, RAM_3 and Q_3 are used to inject a 0 or 1 when shifting right, or to test the ejected bit when shifting left. The RAM_0 and Q_0 pins are used in a similar manner on the least-significant slice.

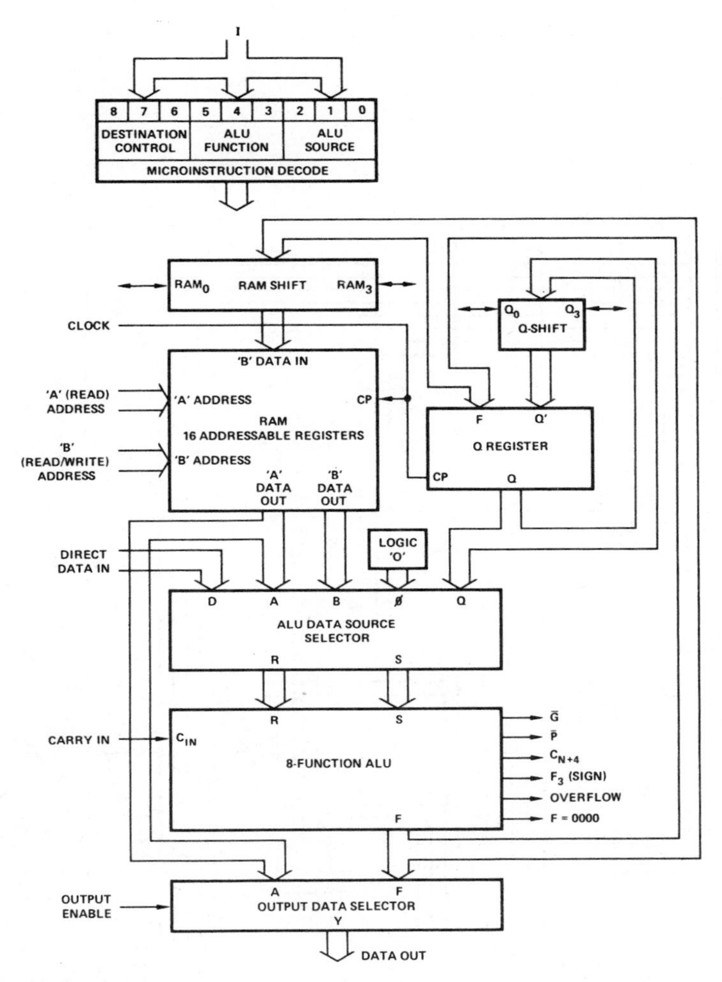

Figure 3.2. Organization of the 2901 (Copyright 1978 Advanced Micro Devices, Inc. Reproduced with permission of copyright owner.).

These pins can also be used, by external logic, to interconnect both shifters to form a double-length shifter for multiplication and division algorithms.

2901 Functions

The nine I inputs control the ALU function, the shifters, and the routing of data. Three of the I lines control the multiplexers feeding the ALU, as shown in the upper left box in Figure 3.5.

A second group of three I lines specify the function of the ALU. When perform-

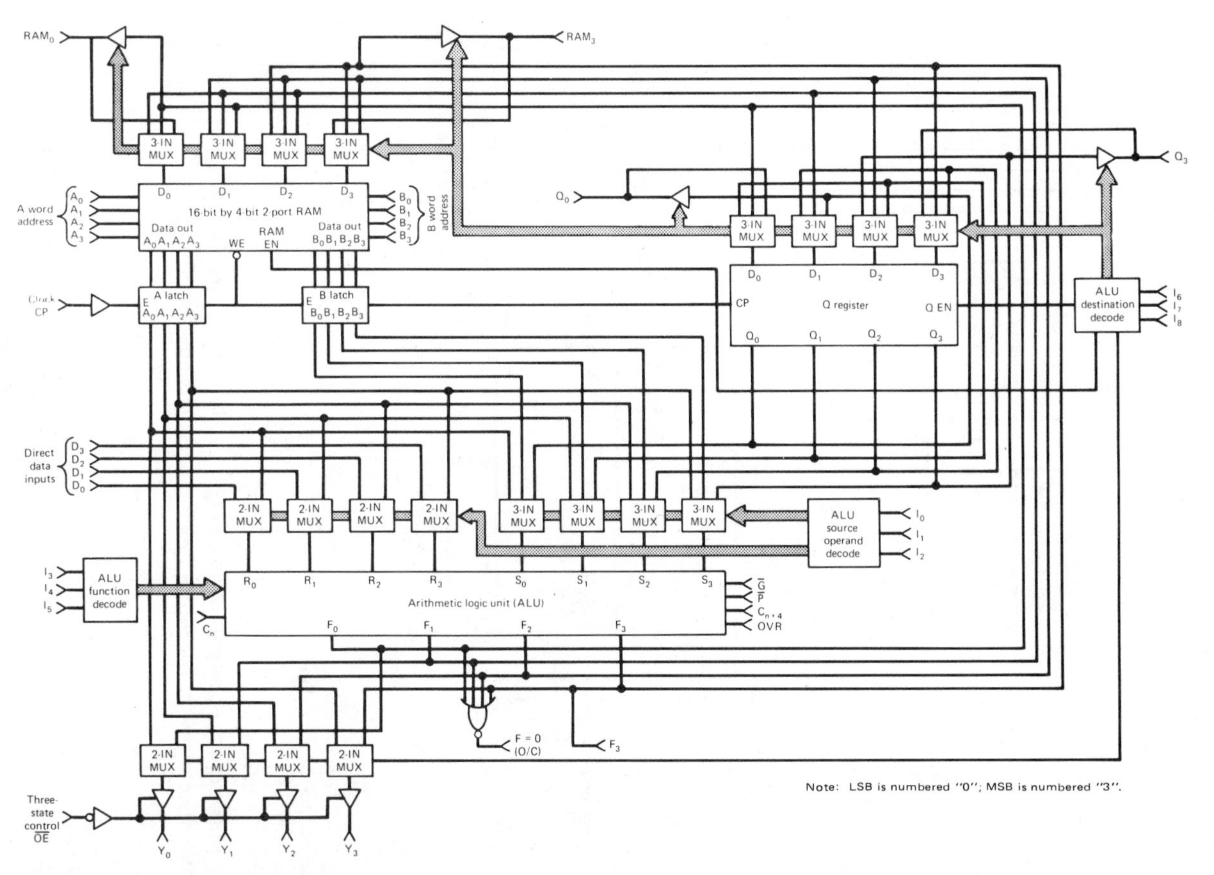

Note: LSB is numbered "0"; MSB is numbered "3".

Figure 3.3. Detailed view of the 2901 (Copyright 1978 Advanced Micro Devices, Inc. Reproduced with permission of copyright owner.).

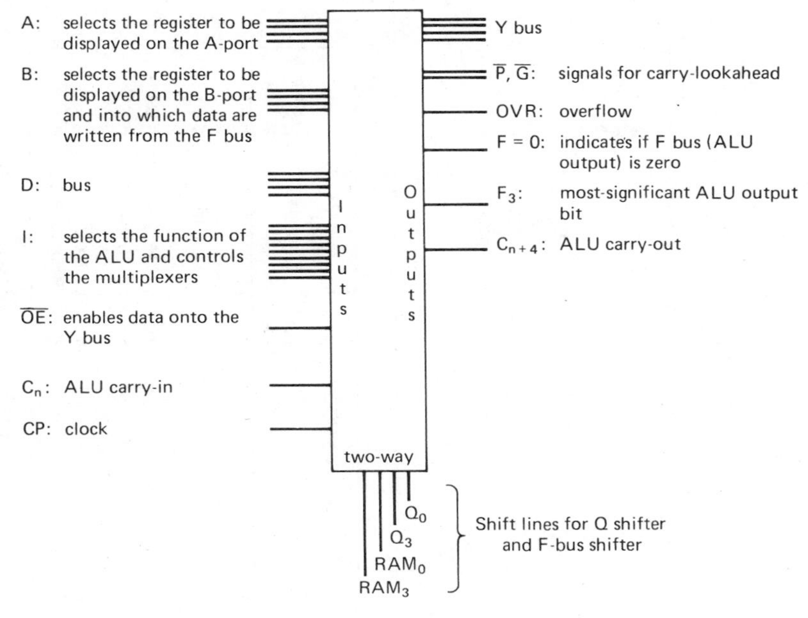

Figure 3.4. 2901 external connections.

ing an addition or subtraction, the C_n pin is an input. Thus, by feeding a 0 or 1 to the C_n input of the least-significant slice, the available arithmetic functions are

$$\begin{aligned} &R + S \\ &R + S + 1 \\ &S - R - 1 \\ &S - R \\ &R - S - 1 \\ &R - S \end{aligned}$$

In Figure 3.5, such expressions as S–R mean the addition of S to the one's complement of R. Hence, if the C_n input is set to 0, "S-R" means S-R-1; if C_n is set to 1, S-R means S-R.

The remaining three I lines, as shown in Figure 3.5, control the two shifters, the Q-register multiplexer, and the Y-bus multiplexer. In the diagram, X is a "don't care" condition, F/2 and Q/2 represent a right (down) shift, and 2F and 2Q represent a left (up) shift. Figure 3.5 also specifies the state of the four shifter pins under each of the eight combinations of control values. For instance, setting the last three I inputs to 111 (assuming positive logic, that is, 1=high, 0=low) specifies that the value of the F bus is shifted left and loaded into the register designated by the B lines, the Q register is not altered, and the F bus is gated to the Y bus. Under these conditions, the RAM_0 pin is an input and feeds the low-order bit of the value to be placed in the B register, and RAM_3 is an output containing the value of the high-order bit on the F bus (the "shifted-out" bit).

MICRO CODE				ALU SOURCE OPERANDS	
I_2	I_1	I_0	Octal Code	R	S
L	L	L	0	A	Q
L	L	H	1	A	B
L	H	L	2	O	Q
L	H	H	3	O	B
H	L	L	4	O	A
H	L	H	5	D	A
H	H	L	6	D	Q
H	H	H	7	D	O

ALU Source Operand Control.

MICRO CODE				ALU Function	Symbol
I_5	I_4	I_3	Octal Code		
L	L	L	0	R Plus S	$R + S$
L	L	H	1	S Minus R	$S - R$
L	H	L	2	R Minus S	$R - S$
L	H	H	3	R OR S	$R \vee S$
H	L	L	4	R AND S	$R \wedge S$
H	L	H	5	$\overline{R}$ AND S	$\overline{R} \wedge S$
H	H	L	6	R EX-OR S	$R \forall S$
H	H	H	7	R EX-NOR S	$\overline{R \forall S}$

ALU Function Control.

MICRO CODE				RAM FUNCTION		Q-REG. FUNCTION		Y OUTPUT	RAM SHIFTER		Q SHIFTER	
I_8	I_7	I_6	Octal Code	Shift	Load	Shift	Load		RAM_0	RAM_3	Q_0	Q_3
L	L	L	0	X	NONE	NONE	$F \to Q$	F	X	X	X	X
L	L	H	1	X	NONE	X	NONE	F	X	X	X	X
L	H	L	2	NONE	$F \to B$	X	NONE	A	X	X	X	X
L	H	H	3	NONE	$F \to B$	X	NONE	F	X	X	X	X
H	L	L	4	DOWN	$F/2 \to B$	DOWN	$Q/2 \to Q$	F	F_0	IN_3	Q_0	IN_3
H	L	H	5	DOWN	$F/2 \to B$	X	NONE	F	F_0	IN_3	Q_0	X
H	H	L	6	UP	$2F \to B$	UP	$2Q \to Q$	F	IN_0	F_3	IN_0	Q_3
H	H	H	7	UP	$2F \to B$	X	NONE	F	IN_0	F_3	X	Q_3

X = Don't care. Electrically, the shift pin is a TTL input internally connected to a three-state output which is in the high-impedance state.
B = Register Addressed by B inputs.
Up is toward MSB, Down is toward LSB.

ALU Destination Control.

Figure 3.5. 2901 I control signals (Copyright 1978 Advanced Micro Devices, Inc. Reproduced with permission of copyright owner.).

Also, although the Q shifter is not being used, the Q_3 pin is an output, containing the high-order bit of the Q register, and the Q_0 pin is in a high-impedence state.

Given an understanding of the control signals of the 2901, one can now visualize the format of a microinstruction in a design incorporating 2901s. Although the designer is free to choose any microinstruction format and representation, a typical format might be that of Figure 3.6. As was explained in Chapter 1, if multiple slices are cascaded, most of the control signals are fed to all slices in parallel. The sixth microorder, the ALU carry input, would feed the C_n pin of the least-significant slice. If the design did not require the flexibility of explicitly feeding a 0 or 1 carry input to the processing section, this bit might be eliminated and replaced with logic that decodes the ALU-function microorder to set C_n to 0 if an addition is indicated, or a 1 if a subtraction is indicated.

The next microorder controls the output-enable line of the slices. Normally, one would not expect to see an explicit microorder for this, as other microorders might specify the destination of the Y bus (i.e., a register or other device external to the 2901), and the decoding of these microorders would control the output-enable line.

The sample microinstruction has two 1-bit microorders that send a bit into the vacated position whenever a shift is done. For example, the Q-shift-input microorder would be gated to the Q_3 pin of the most-significant slice and the Q_0 pin of the least-significant slice, the one to which it is currently gated being determined, perhaps, by some decoding logic that examines the fifth microorder to determine which type of shift is being done.

Another item of interest is the status generated by an ALU/register slice. These outputs are used to interconnect cascaded slices, as well as for use by the microprogram sequencing logic to direct conditional branching. At any one time, the 2901 has eight status outputs. Depending on the type of shift being done, two of the four bidirectional shift lines indicate the value being ejected. These four pins are usually used to interconnect the shifters in cascaded slices, although the RAM_3 and Q_3 pins on the most-significant slice, and the RAM_0 and Q_0 pins on the least-significant slice, can be used to test the value of the ejected bit.

The C_{n+4} pin, as illustrated in Chapter 1, contains the carry-out bit of the ALU operation and is used to interconnect cascaded slices when using ripple carry.

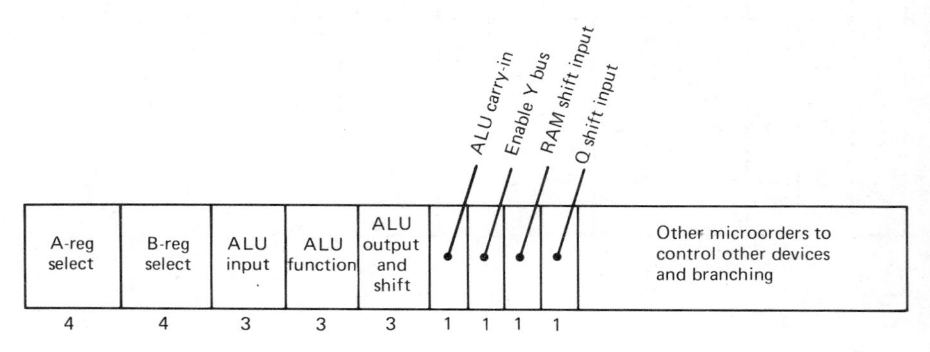

Figure 3.6. Typical microinstruction format in a 2901 design.

When using external carry-lookahead logic to increase the speed of cascaded slices, the C_{n+4} pin is normally not used.

The F_3 output contains the high-order bit on the F bus, the high-order bit of the ALU result. With cascaded 2901s, usually only the F_3 pin of the most-significant slice is used; if two's-complement arithmetic is being performed, it indicates the sign of the result.

The F=0 open-collector output indicates whether the ALU result is zero. By tying the F=0 pins on each cascaded 2901 together, one has a signal indicating whether the total result is zero. The OVR (overflow) pin indicates whether the result of a two's-complement operation in the ALU has overflowed into the sign bit.

The P and G outputs represent carry propagate and generate signals from the ALU. They are used (as will be illustrated shortly) to drive external carry-lookahead logic, allowing a set of cascaded 2901s to be driven faster than if the slower ripple-carry technique is used. Figure 3.7 defines the state of these and two other outputs as a function of the current ALU operation.

2901 Timing

The clock input to the 2901 controls the register array, the shifters, the Q register, and the A and B latches to the ALU. When the clock input is high, the latches are open (transparent) and pass the values of the registers currently selected as the A and B registers. When the clock is low, the latches close and retain their values. For Advanced Micro Devices' Am2901A, the minimum amount of time that the clock must be held low is 30 ns, the minimum high time is 30 ns, and the minimum clock period is 100 ns. (All are guaranteed times.)

Figure 3.8 is an oversimplified view of the timing of the 2901. Notice that the control inputs must be stabilized at their required states at the beginning of the cycle. These times are called *set-up times* and, in most cases, are expressed relative to the rising edge, or low-to-high transition, of the clock input. As an example, the guaranteed, or worst-case, set-up times for the Am2901A are shown in Figure 3.9 (typical set-up times are approximately 40% of the guaranteed times). For instance, the I signals from the current microinstruction must be present at the Am2901's pins at least 80 ns before the rising edge of the clock pulse.

Another timing consideration is propagation delays—the time from when an input signal is established to when a particular output is stable. Figure 3.10 lists guaranteed propagation delays for the 2901, although it is a summary and one should consult the manufacturers' literature for exact information. Figure 3.10 lists times for three versions of the 2901: the Am2901A from Advanced Micro Devices and the IDM2901A and IDM2901A-1 from National Semiconductor. Advanced Micro Devices has more recently produced another version—the AM2901B—with approximately the same speed as the IDM2901A, although complete timing specifications were not available at the time that this book was written. Similarly, National Semiconductor is developing a faster version of their 2901, the IDM2901A-2.

LOGIC FUNCTIONS FOR G, P, C_{n+4}, AND OVR

The four signals G, P, C_{n+4}, and OVR are designed to indicate carry and overflow conditions when the Am2901A is in the add or subtract mode. The table below indicates the logic equations for these four signals for each of the eight ALU functions. The R and S inputs are the two inputs selected according to Figure 2.

Definitions (+ = OR)

$P_0 = R_0 + S_0$ $\quad G_0 = R_0S_0$

$P_1 = R_1 + S_1$ $\quad G_1 = R_1S_1$

$P_2 = R_2 + S_2$ $\quad G_2 = R_2S_2$

$P_3 = R_3 + S_3$ $\quad G_3 = R_3S_3$

$C_4 = G_3 + P_3G_2 + P_3P_2G_1 + P_3P_2P_1G_0 + P_3P_2P_1P_0C_n$

$C_3 = G_2 + P_2G_1 + P_2P_1G_0 + P_2P_1P_0C_n$

I_{543}	Function	$\overline{P}$	$\overline{G}$	C_{n+4}	OVR
0	R + S	$\overline{P_3P_2P_1P_0}$	$\overline{G_3 + P_3G_2 + P_3P_2G_1 + P_3P_2P_1G_0}$	C_4	$C_3 \forall C_4$
1	S − R	← Same as R + S equations, but substitute $\overline{R_i}$ for R_i in definitions →			
2	R − S	← Same as R + S equations, but substitute $\overline{S_i}$ for S_i in definitions →			
3	$R \vee S$	LOW	$P_3P_2P_1P_0$	$\overline{P_3P_2P_1P_0} + C_n$	$\overline{P_3P_2P_1P_0} + C_n$
4	$R \wedge S$	LOW	$\overline{G_3 + G_2 + G_1 + G_0}$	$G_3 + G_2 + G_1 + G_0 + C_n$	$G_3 + G_2 + G_1 + G_0 + C_n$
5	$\overline{R} \wedge S$	LOW	← Same as $R \wedge S$ equations, but substitute $\overline{R_i}$ for R_i in definitions →		
6	$R \forall S$	← Same as $\overline{R \forall S}$, but substitute $\overline{R_i}$ for R_i in definitions →			
7	$\overline{R \forall S}$	$G_3 + G_2 + G_1 + G_0$	$G_3 + P_3G_2 + P_3P_2G_1 + P_3P_2P_1P_0$	$\frac{\overline{G_3 + P_3G_2 + P_3P_2G_1}}{+ P_3P_2P_1P_0 (G_0 + \overline{C_n})}$	See note

Note: $[\overline{P}_2+\overline{G}_2\overline{P}_1+\overline{G}_2\overline{G}_1\overline{P}_0+\overline{G}_2\overline{G}_1\overline{G}_0C_n] \forall [\overline{P}_3+\overline{G}_3\overline{P}_2+\overline{G}_3\overline{G}_2\overline{P}_1+\overline{G}_3\overline{G}_2\overline{G}_1\overline{P}_0+\overline{G}_3\overline{G}_2\overline{G}_1\overline{G}_0C_n]$ + = OR

Figure 3.7. Definition of 2901 status outputs (Copyright 1978 Advanced Micro Devices, Inc. Reproduced with permission of copyright owner.).

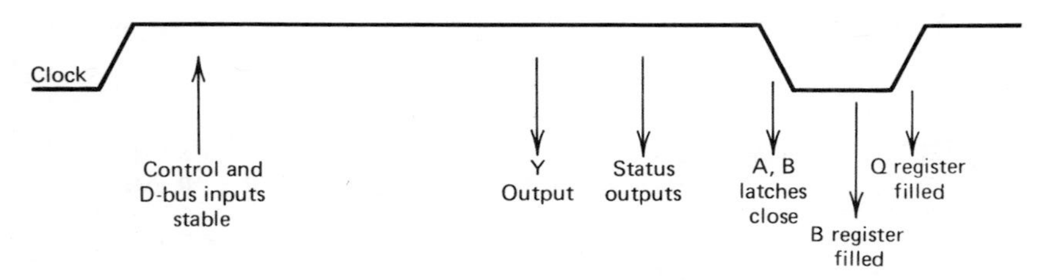

Figure 3.8. Simplified view of 2901 timing.

Figure 3.10 points out a frequent source of confusion, namely the last row, the propagation times from the rising clock edge. At first glance, these values appear to be inconsistent with Figure 3.8. Figure 3.8 shows the outputs appearing before the rising edge, while Figure 3.10 might imply that they are not available until 50 to 72 ns after the rising clock edge.

To resolve this, recall that the outputs should be independent of the rising clock edge; all that this clock edge does is enable a write into the Q register. However, these outputs *are* dependent on the previous rising edge, since the Q register must stabilize in the previous cycle before it is available for use in the current cycle. Hence these propagation delays (those in the last row) are measured from the *previous* rising clock edge. Normally these delays are not found to be on the critical speed path; therefore they are often neglected.

The presentation of timing information in this book for comparative purposes is a problem, since manufacturers do not specify device speeds in an identical fashion. Times are usually specified in one or more of the following categories: (1) typical at 25°C, (2) guaranteed (maximum for propagation times, minimum for set-up and hold times) at 25°C, (3) typical over the operating range (usually 0°C to 70°C), and (4) guaranteed over the operating range. However, not all the devices' timings are specified in the same category. Since most, but not all, devices are specified in terms of guaranteed times over the operating range (the worst case), and, since this category is usually used by system designers when analyzing

From input	Minimum set-up time to clock rise
A, B	100*
D-bus	70
C_n	55
I	80*
RAM_0, RAM_3, Q_0, Q_3	25

* - the maximum of 1) this time and 2) the clock low time + 30 ns.

Figure 3.9. Representative Am2901A guaranteed set-up times.

From input \ To output	Y	F_3	C_{n+4}	$\overline{G}, \overline{P}$	F=0	OVR	Shift outputs RAM_0 RAM_3	Q_0 Q_3
	80	80	75	65	87	85	95	—
A, B	65	65	65	60	70	65	70	—
	50	50	50	45	44	60	55	—
	45	45	45	35	57	55	65	—
D	40	40	40	35	55	45	50	—
	32	32	32	30	33	40	35	—
	30	30	20	—	47	30	50	—
C_n	30	30	20	—	40	30	35	
	25	22	16	—	28	25	35	
	55	55	55	50	67	65	75	30
I	55	50	50	45	60	50	60	30
	40	40	40	32	38	45	45	30
Clock	60	60	60	50	72	70	80	30
	60	60	60	50	60	55	60	30
	50	45	45	40	48	51	55	30

Notes: 1. All times are worst-case and are in nanoseconds.
2. In each entry, the first time is for the AM2901A, the second for the IDM2901A, and the third for the IDM2901-A.

Figure 3.10. Representative 2901 guaranteed propagation delays.

system times and clock speeds, all times presented hereafter in the book are in this category unless otherwise noted.

Comparisons of set-up times and propagation delays among the varieties of 2901 slices is not a simple task. One reason is that a large number of parameters are specified, but, in most designs, the parameters do not have equal significance, because only a few of them fall on the design's critical timing path. Needless to say, the comparison of timing specifications of *different* slices is even more difficult. A chart at the end of this chapter compares, over sample design situations, the speeds of different 2901 versions as well as the other ALU/register slices discussed in this chapter.

Cascading the 2901

Figure 3.11 shows four 2901 slices cascaded to form a 16-bit processing section, although most of the control inputs are not shown. The two shifters are interconnected with carry-lookahead logic (e.g., 74S182 or Am2902) to form a 16-bit ALU. As shown, most of the conditions are taken from the most-significant slice.

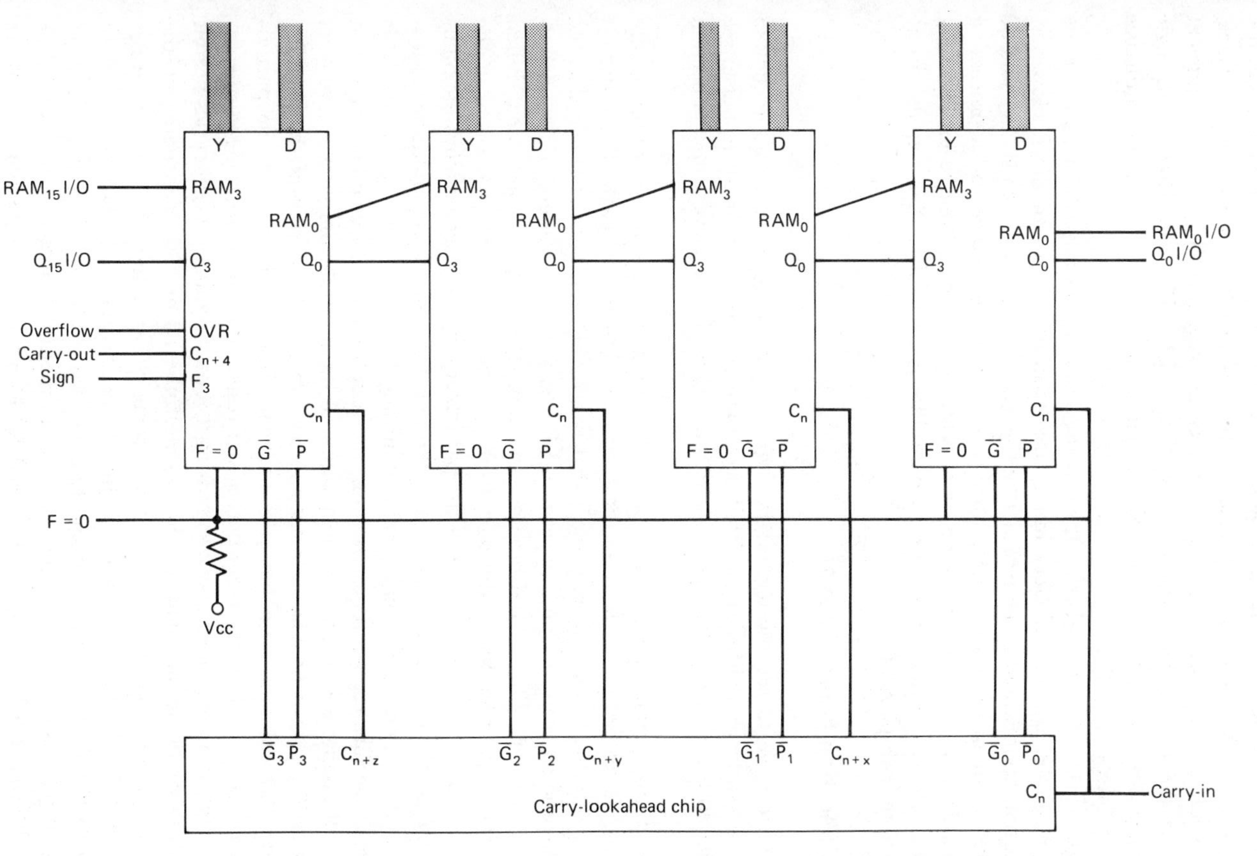

Figure 3.11. 16-bit processing section built from four 2901 slices.

The carry-lookahead generator is optional, but usually desirable when the number of cascaded slices is three or more. If it were not present, one would use ripple carry, where the C_{n+4} signal of a slice feeds the C_n input of the next slice. This would require one to slow down the clock cycle such that there is sufficient time for the carries to ripple from ALU to ALU. The carry-lookahead logic generates the carry inputs considerably faster. Most books on logic design discuss this concept.

Figure 3.11 does not illustrate the control inputs. Normally, the signals for the I, A, B, Y-bus enable, and clock inputs are fed in parallel to all slices. However, in unusual applications, this need not be the case; for instance, two slices could be directed to do an addition and the other two to do an OR. If the 16-bit processing section of Figure 3.11 must also do byte (8-bit) operations on the 16-bit registers, one might devise a microinstruction where different sets of control signals can be directed to both pairs of 2901s.

Multiplication Using the 2901

It is instructive to see the use of the 2901 in performing multiplication, not so much to illustrate multiplication itself, since the process is well known, but to see another illustration of 2901 interconnections, as well as a clever interconnection trick.

The multiplication algorithm to be used is the add and shift algorithm, where one cycle is needed for each bit in the multiplier. In each cycle, the low-order bit of the multiplier register is examined. If it is a 1, the multiplicand is added to the partial-product register (initially set to 0); if it is a 0, zero is added to the partial-product register. The partial-product and multiplier registers are then shifted right, the high-order bit of the partial-product register is set to the sign of the addition result, the low-order bit of the partial-product register becomes the high-order bit of the multiplier register, and the low-order bit of the multiplier is discarded. Assuming two's-complement multiplication is being performed, this process is performed n-1 times (n being the number of multiplier bits). The last cycle is the same except that, if the addition of the multiplicand would have been performed, a subtraction is performed instead to adjust the sign of the product. The product contains n+n bits, the most-significant half being in the partial-product register and the least-significant half residing in the multiplier register.

The algorithm described above sounds as if it may require several cycles per bit (e.g., because of the decision to add the multiplicand or zero, depending on the low-order multiplier bit), but, employing some clever design, only one cycle per bit is needed (i.e., multiplying two 16-bit values requires 16 2901 clock cycles). Figure 3.12 illustrates the configuration of two 2901 slices to multiply 8-bit values. If we wish to dedicate the two slices to multiplication, the slices would be physically connected as shown in Figure 3.12. More likely, one would want to use the 2901s for other purposes as well. In this case, a bit in the microinstruction might specify "configure the 2901s for multiplication," and the interconnections in Figure 3.12 would be made by additional multiplexer circuits.

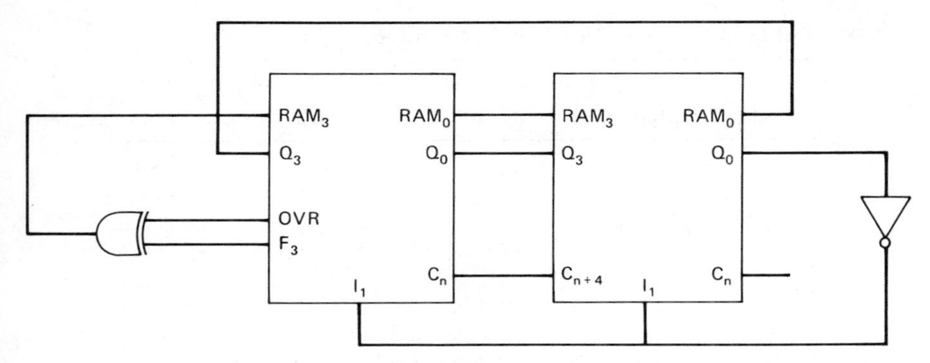

Figure 3.12. Two 2901 slices connected for multiplication.

In relation to the algorithm described above, the B register will serve as the partial-product register and Q is the multiplier register. Notice the interconnection of the RAM and Q shifters; this gives one a double-length shifter needed by the algorithm. Also notice the unusual connection of the Q_0 pin to the I_1 control pin. The I_1 pin will not be controlled by the microinstruction; it is controlled by the low-order bit in Q.

As initial conditions, assume that the multiplicand is in register 1, register 2 is zero and is specified as the B register, and the multiplier is in the Q register. The two sets of control signals (microinstructions) needed are:

1. A=0001 B=0010 I_{876}=100 I_{543}=000 I_{210}=0x1 C_n=0
2. A=0001 B=0010 I_{876}=100 I_{543}=001 I_{210}=0x1 C_n=1

The first microinstruction gates either A,B or 0,B to the ALU, depending on whether Q_0 (the low-order bit in the Q register) is 1 or 0, respectively. The ALU inputs are added and the result is on the F bus. The F bus is shifted right before being moved into B. Also, the Q register is shifted right. The bit shifted out of the F bus becomes the high-order bit of the Q register, and the high-order bit of the B register is filled with F_3 exclusive-OR OVR. This is the sign of the partial product. The F_3 output will represent the sign of the partial product unless an overflow occurs, in which case F_3 represents the complement of the sign (hence the need for the exclusive-OR operation).

The first microinstruction is repeated six more times (e.g., by designing sequencing logic that allows one to iteratively execute one or more microinstructions for a specified number of iterations). The second microinstruction is used in the eighth cycle. Its only difference is the setting of the carry input to 1 and the setting of the ALU function. Since Q_0 still feeds the I_1 control input, the ALU function is B-A (if Q_0 is 1) or B-0 (if Q_0 is 0). After this cycle, the double-length product is found in registers B (register 2) and Q, the high-order half in B and the low-order half in Q.

As an example, assume that we wish to multiply +3 by −13. Register 1

TABLE 3.1. Eight-bit Multiplication Operation

Cycle	ALU Operation	F-bus	R2	Q
			00000000	11110011
1	R1 + R2	00000011	00000001	11111001
2	R1 + R2	00000100	00000010	01111100
3	0 + R2	00000010	00000001	00111110
4	0 + R2	00000001	00000000	10011111
5	R1 + R2	00000011	00000001	11001111
6	R1 + R2	00000100	00000010	01100111
7	R1 + R2	00000101	00000010	10110011
8	R2 − R1	11111111	11111111	11011001

contains 00000011 (+3), register 2 is set to 0, and Q contains 11110011. Table 3.1 illustrates the eight cycles. The 16-bit product in register 2 and Q is 1111111111011001, which is −39 in two's-complement representation.

2901 Varieties

As mentioned earlier, the 2901 is to ALU/register bit slices what the 8080 is to 8-bit microprocessors. As a result, many varieties of the 2901 are now available. In addition to there being several second-source manufacturers, Advanced Micro Devices has produced two newer versions (Am2901A and Am2901B) and has another version, the Am2901C, under development. Each is available in a wide number of packages, including hermetically sealed DIPs, plastic DIPs, flat packages, and dice. In addition, National Semiconductor produces two fast versions of the 2901 that employ ECL technology internally.

The newer versions of the 2901, following semiconductor trends, are both faster and cheaper. For instance, the Am2901A is approximately 50% faster than the Am2901, and the Am2901B is over twice as fast as the Am2901. 2901 slices sold for $30 in 1975; in 1979 they could be obtained for under $10.

Given these varieties of 2901 slices, all of which have identical organizations and external connections but differ greatly in speed, it is instructive to see how these speed improvements were achieved. The original Am2901 was a low-power Schottky device with a die size of 33,100 square mils. The faster Am2901A is considerably smaller, having a die size of 19,700 square mils. It, like the Am2901, is a low-power Schottky device. However, the speed improvements stem from the elimination of a few gates from the critical-speed paths (e.g., A and B inputs to the carry-lookahead outputs) and from better distribution of the supply-voltage and ground lines on the chip. The faster Am2901B is still a low-power Schottky device, but with a die size of 15,000 square mils. Projection masking is used to reduce the sizes of transistors and conductors on the chip. The speed improvement results from the lower capacitance of the smaller transistors.

National Semiconductor took a different approach with their 2901 versions. The IDM2901A uses internal ECL circuitry to achieve fast switching speeds, with translation logic at all input and output connections to achieve TTL levels externally. ECL to TTL translators are normally very power consuming, but improved circuit designs in the IDM2901A have apparently overcome this. The higher speed IDM2901A-1 is actually not a different component; it is simply a screened IDM2901A, that is, an IDM2901A that has been tested to the set-up and propagation-speed specifications of the IDM2901A-1.

THE 3002 SLICE

The 3002 ALU/Register slice [3–7] was one of the first bit-slice devices. It was initially produced by Intel Corp., but like the 2901, is now produced by second-source manufacturers.

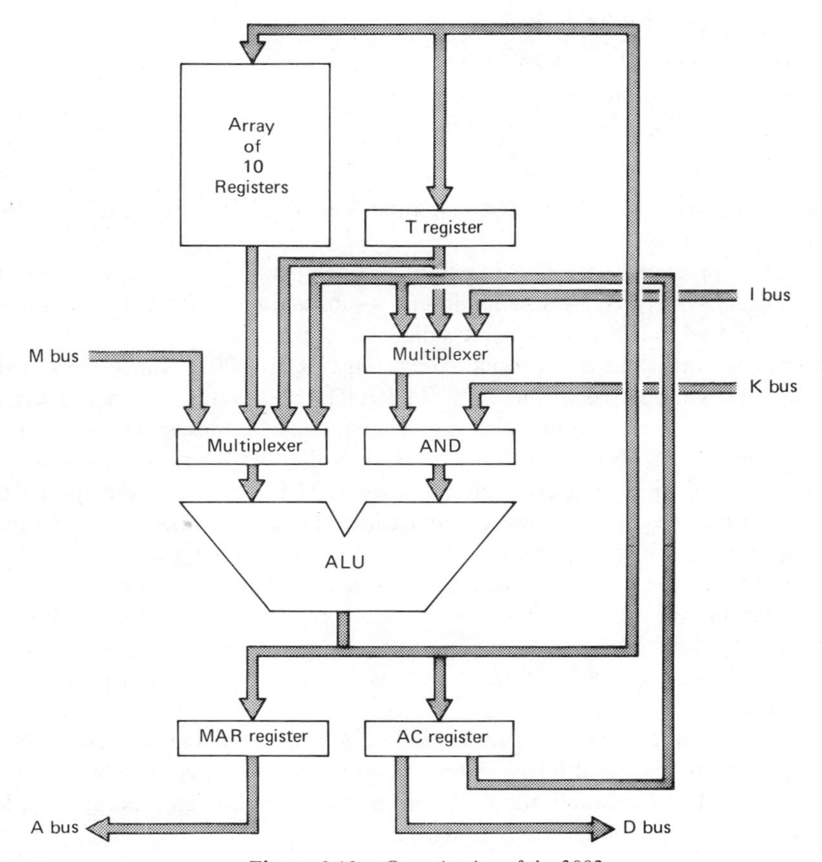

Figure 3.13. Organization of the 3002.

The organization of the 3002 is shown in Figure 3.13. Unlike the 2901, the 3002 is a 2-bit slice in a 28-pin package. The 3002 contains 13 registers, ten of which reside in an array. Unlike the 2901, the register array in the 3002 has only one port, meaning that only one register may be used in any particular cycle. The AC (accumulator) and T registers are more flexible than the registers in the array in that they can be gated to more places and more operations can be applied to them. The 13th register is the one labeled MAR.

Notice that the 3002 has five external busses, where the 2901 has only two. Two of the three input busses, M and I, can be gated into the left and right sides of the ALU, respectively. The K input bus serves a special purpose which will be discussed later; as shown, K feeds an AND circuit at the right side of the ALU, meaning that it is used in every ALU operation. The A and D busses are three-state output busses. Although the five busses can be used for any purpose, their intended uses in a processor are

M = data bus from memory
I = data bus from I/O devices
K = mask data from microinstruction
A = address bus to memory
D = data bus to memory

Hence, the MAR and AC registers are intended to function as memory-address and memory-data registers.

The ALU performs only five functions, namely, addition, right shift, AND, OR, and exclusive-NOR. Noticeably absent is subtraction, which will be discussed later. Figure 3.14 in the manufacturer's diagram of the 3002.

Figure 3.15 illustrates the external connections to the 3002. Notice the absence of register-select inputs, as in the 2901. The 3002, to its credit, has more external busses. However, these require pins on the chip, and the designers' goal was to limit the number of pins to 28. This decision had several serious consequences. First, the control lines to select registers, specify ALU functions, and specify the ALU inputs and destination were crammed into the seven F lines. The resultant high degree of encoding (1) makes the 3002 difficult to understand, (2) leads to inconsistencies in the way different operations can be used in conjunction with different registers, (3) causes certain common operations (e.g., subtration) to be absent, and (4) places severe restrictions on the combinations of registers that can be used in a given cycle. Second, the 3002 has very few status-output pins, a limitation discussed later.

Most of the connections in Figure 3.15 should be self-explanatory, except for F and K, which are discussed later. There are only two shift pins. The 3002 has no independent shifters, although the ALU can perform a shift-right operation (a left shift can be done by adding a value to itself).

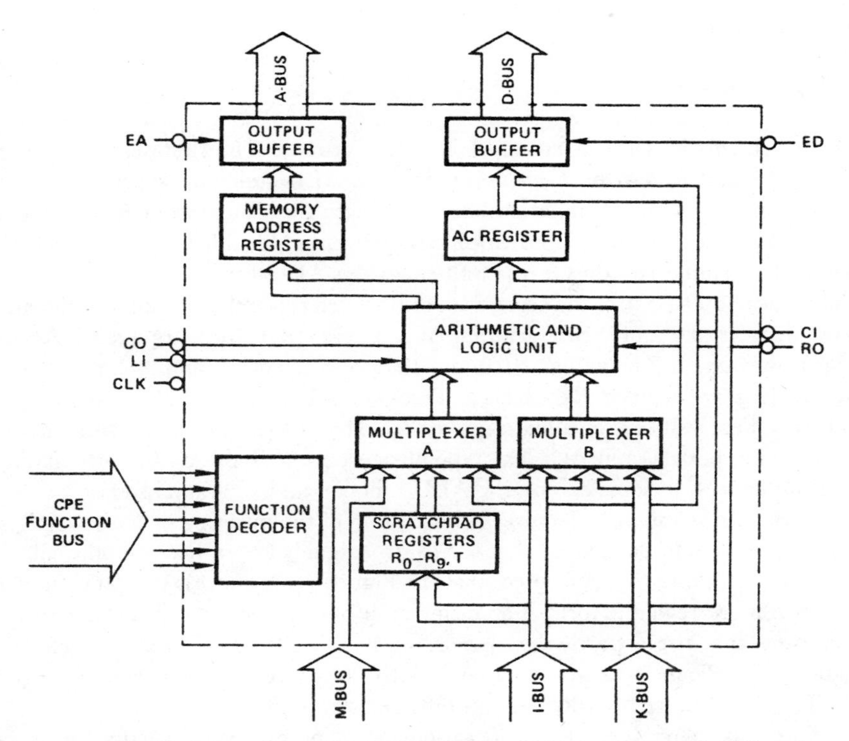

Figure 3.14. Organization of the 3002 (Reproduced with permission of Intel Corp.).

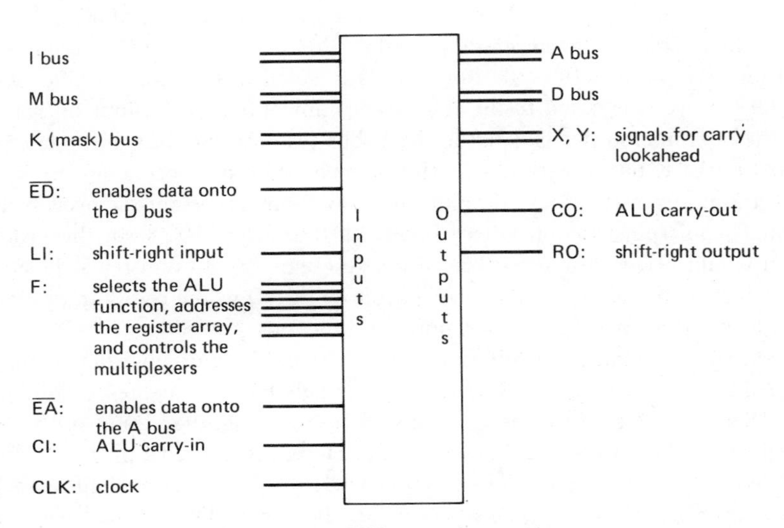

Figure 3.15. 3002 external connections.

3002 Functions

The F control lines are partitioned into two groups. The first set, consisting of three lines and called the *F-group,* selects the ALU function, specifies the ALU right source, and, in some situations, influences the decoding of the second set. The other set, consisting of four lines, designates the ALU left source and destination and, in some circumstances, influences the ALU function and right source. The values 0000-1001 of the second set of lines reference the registers in the array, 1100 denotes T, and 1101 denotes AC. These 12 values are called *R-group* (register-group) *I.* The values 1010 and 1011 also denote T and AC, respectively; these values are termed *R-group II.* The values 1110 and 1111 again specify T and AC; these values are called *R-group III.* Hence there are three different ways of specifying both T and AC. The type of reference to T or AC (i.e., the R-group used) influences, to some extent, the ALU function and righthand source.

Figure 3.16 is the manufacturer's description of the meaning of the seven F control lines. It should be obvious that the control signals are rather complicated.

Figure 3.17, and the associated notes in Figure 3.18, express the same information in a more readable form. For instance, if one is interested in determining all operations that use a particular bus or register, Figure 3.17 is more useful than Figure 3.16. Please note that, in Figure 3.16, R_n represents registers 0-9 plus AC and T, but in Figure 3.17 R_n denotes only registers 0-9.

Before analyzing the 3002's operations, it is necessary to discuss the K bus, since the K bus is always an input to the ALU. In most situations, the value on the K bus is ANDed with the value in the AC register, and this result becomes the righthand input to the ALU. The K bus is useful when manipulating individual bits in a register. Figure 3.19 illustrates three different forms that a microinstruction might have in a system employing the 3002. The first three microorders are gated directly to all 3002s, the fourth (LI) is gated to the most-significant 3002, and the fifth (CI) is gated to the least-significant 3002. In the first microinstruction, the next microorder is gated to the K bus (each pair of bits in the microorder is gated to the K bus of each 3002); thus *n* is equal to the processing-section width (twice the number of 3002s). For instance, assuming an eight-bit processing section and a microinstruction where F=0101110 and K=11000000, the processing section would OR the first two bits of the AC register into register 5. Hence such a microinstruction allows one to apply a bit mask (K) to the righthand part of the ALU, leading to powerful bit-manipulation functions.

When such masking operations are not needed, the microinstruction might be designed as in Figure 3.19b. Here there is a single bit that is gated to all K inputs in parallel. Hence this bit operates as a switch on the righthand port of the ALU.

A third idea is expressed in Figure 3.19c. Here, assume a 16-bit processing section. One KC bit in the microinstruction feeds the K inputs on half of the 3002 slices; the other bit feeds the K inputs on the other slices. This gives one the ability to mask the high- or low-order half of the source to the righthand port of the ALU, useful when the processing section is 16 bits wide but one needs to deal with byte (8-bit) quantities.

F-GROUP	R-GROUP	MICRO-FUNCTION	
	I	$R_n + (AC \wedge K) + CI \rightarrow R_n, AC$	
0	II	$M + (AC \wedge K) + CI \rightarrow AT$	
	III	$AT_L \wedge \overline{(I_L \wedge K_L)} \rightarrow RO \cdot LI \vee [I_H \wedge K_H) \wedge AT_H] \rightarrow AT_H$	
		$[AT_L \wedge (I_L \wedge K_L)] \vee [AT_H \wedge (I_H \wedge K_H)] \rightarrow AT_L$	
	I	$K \vee R_n \rightarrow MAR$	$R_n + CL + K \rightarrow R_n$
1	II	$K \vee M \rightarrow MAR$	$M + CI + K \rightarrow AT$
	III	$(\overline{AT} \vee K) + (AT \wedge K) + CI \rightarrow AT$	
	I	$(AC \wedge K) - 1 + CI \rightarrow R_n$ (see Note 1)	
2	II	$(AC \wedge K) - 1 + CI \rightarrow AT$	
	III	$(I \wedge K) - 1 + CI \rightarrow AT$	
	I	$R_n + (AC \wedge K) + CI \rightarrow R_n$	
3	II	$M + (AC \wedge K) + CI \rightarrow AT$	
	III	$AT + (I \wedge K) + CI \rightarrow AT$	
	I	$CI \vee (R_n \wedge AC \wedge K) \rightarrow CO$	$R_n \wedge (AC \wedge K) \rightarrow R_n$
4	II	$CI \vee (M \wedge AC \wedge K) \rightarrow CO$	$M \wedge (AC \wedge K) \rightarrow AT$
	III	$CI \vee (AT \wedge I \wedge K) \rightarrow CO$	$AT \wedge (1 \wedge K) \rightarrow AT$
	I	$CI \vee (R_n \wedge K) \rightarrow CO$	$K \wedge R_n \rightarrow R_n$
5	II	$CI \vee (M \wedge K) \rightarrow CO$	$K \wedge M \rightarrow AT$
	III	$CI \vee (AT \wedge K) \rightarrow CO$	$K \wedge AT \rightarrow AT$
	I	$CI \vee (AC \wedge K) \rightarrow CO$	$R_n \vee (AC \wedge K) \rightarrow R_n$
6	II	$CI \vee (AC \wedge K) \rightarrow CO$	$M \vee (AC \wedge K) \rightarrow AT$
	III	$CI \vee (I \wedge K) \rightarrow CO$	$AT \vee (I \wedge K) \rightarrow AT$
	I	$CI \vee (R_n \wedge AC \wedge K) \rightarrow CO$	$R_n \overline{\oplus} (AC \wedge K) \rightarrow R_n$
7	II	$CI \vee (M \wedge AC \wedge K) \rightarrow CO$	$M \overline{\oplus} (AC \wedge K) \rightarrow AT$
	III	$CI \vee (AT \wedge I \wedge K) \rightarrow CO$	$AT \overline{\oplus} (I \wedge K) \rightarrow AT$

Note: 1. 2's complement arithmetic adds 111...11 to perform subtraction of 000...01.

SYMBOL	MEANING
I, K, M	Data on the I, K, and M busses, respectively
R_n	Contents of register n (R-Group I)
AC	Contents of the accumulator
AT	Contents of AC or T, as specified
CI	Data on the carry input
CO	Data on the carry output
L, H	As subscripts, designate low and high order bit, respectively
$+$	2's complement addition
$-$	2's complement subtraction
$\wedge$	Logical AND
$\vee$	Logical OR
$\overline{\oplus}$	Exclusive-NOR
$\rightarrow$	Deposit into

Figure 3.16. Definition of F control inputs (Reproduced with permission of Intel Corp.).

		ALU left source							ALU right source					ALU destination					
$F_{4\text{-}6}$	$F_{0\text{-}3}$	R_n	T	AC	M	11	$\overline{AC} \vee K$	$\overline{T} \vee K$	$AC \wedge K$	$I \wedge K$	$T \wedge K$	K	ALU function	R_n	T	AC	MAR	CO output note	Comments
0	0–9	X							X				+	X		X		1	Whenever the ALU function is +,
0	12		X						X				+		X	X		1	CI (ALU carry-in) is a third
0	13			X					X				+			X		1	input to the ALU.
0	10				X				X				+		X			1	
0	11				X				X				+			X		1	
0	14		X							X					X			1	See note 4.
0	15			X						X						X		1	
1	0–9	X										X	$\vee$				X	1	Each of these performs two
		X										X	+	X					functions. MAR is loaded with
1	12		X									X	$\vee$				X	1	K or-ed with a source, and K is
			X									X	+		X				added to a register.
1	13			X								X	$\vee$				X	1	
				X								X	+			X			

1	10				X							X	∨				X	1	
					X							X	+		X				
1	11				X							X	∨				X	1	
					X							X	+			X			
1	14							X			X		+		X			1	
1	15						X		X				+			X		1	
2	0–9					X			X				+	X				1	Adding value to 11...11 has
2	12					X			X				+		X			1	the effect of subtracting 0 or 1
2	13					X			X				+			X		1	from the value, depending on
2	10					X			X				+		X			1	the value of CI.
2	11					X			X				+			X		1	
2	14					X				X			+		X			1	
2	15					X				X			+			X		1	
3	0–9	X							X				+	X				1	
3	12		X						X				+		X			1	
3	13			X					X				+			X		1	These are identical to the cor-
3	10				X				X				+		X			1	responding $F_{4\text{-}6} = 0$ functions.
3	11				X				X				+			X		1	
3	14		X							X			+		X			1	
3	15			X						X			+			X		1	

Figure 3.17. Definition of F control inputs.

		ALU left source							ALU right source					ALU destination					
$F_{4\text{-}6}$	$F_{0\text{-}3}$	R_n	T	AC	M	11	$\overline{AC} \vee K$	$\overline{T} \vee K$	$AC \wedge K$	$I \wedge K$	$T \wedge K$	K	ALU function	R_n	T	AC	MAR	CO output note	Comments
4	0–9	X							X				$\wedge$	X				2	
4	12		X						X				$\wedge$		X			2	
4	13			X					X				$\wedge$			X		2	
4	10				X				X				$\wedge$		X			2	
4	11				X				X				$\wedge$			X		2	
4	14		X							X			$\wedge$		X			2	
4	15			X						X			$\wedge$			X		2	
5	0–9	X										X	$\wedge$	X				2	
5	12		X									X	$\wedge$		X			2	
5	13			X								X	$\wedge$			X		2	
5	10				X							X	$\wedge$		X			2	
5	11				X							X	$\wedge$			X		2	
5	14		X									X	$\wedge$		X			2	Identical to $F_{0-3} = 10$ and 11.
5	15			X								X	$\wedge$			X		2	

6	0–9	X				X		$\vee$	X			3	
6	12		X			X		$\vee$		X		3	
6	13			X		X		$\vee$			X	3	
6	10				X	X		$\vee$		X		3	
6	11				X	X		$\vee$			X	3	
6	14		X				X	$\vee$		X		3	
6	15			X			X	$\vee$			X	3	
7	0–9	X				X		$\overline{\veebar}$	X			2	ALU function is exclusive-NOR.
7	12		X			X		$\overline{\veebar}$		X		2	
7	13			X		X		$\overline{\veebar}$			X	2	
7	10				X	X		$\overline{\veebar}$		X		2	
7	11				X	X		$\overline{\veebar}$			X	2	
7	14		X				X	$\overline{\veebar}$		X		2	
7	15			X			X	$\overline{\veebar}$			X	2	

Figure 3.17. *(Continued)*

No.	Note
1	CO is the carry-out of the addition operation.
2	CO $\leftarrow$ CI $\vee$ (L $\wedge$ R), where L and R represent the ALU left and right sources. CI is ORed with the word-wise OR of L $\wedge$ R. If CI is set to 0 on the least-significant slice, CO on the most-significant slice can be used to test for a zero result.
3	CO $\leftarrow$ CI $\vee$ R. Similar to note 2, CO can be used to test the ALU right source for zero.
4	This operation defies description. However, if K is set to 00, it performs a right shift of AC or T. That is, RO is set to the rightmost bit of the register, the rightmost bit of the register is set to the leftmost bit, and the leftmost bit is set to the value of LI.

Figure 3.18. Notes to Figure 3.17.

If we assume the design of Figure 3.19b, one can simplify the list of 3002 functions in Figure 3.17. For instance, if KC is set to 1, the functions corresponding to F-group 3 in Figure 3.17 are

$$
\begin{aligned}
&\mathrm{R} + \mathrm{AC} \rightarrow \mathrm{R}\\
&\mathrm{T} + \mathrm{AC} \rightarrow \mathrm{T}\\
&\mathrm{AC} + \mathrm{AC} \rightarrow \mathrm{AC}\\
&\mathrm{M} + \mathrm{AC} \rightarrow \mathrm{T}\\
&\mathrm{M} + \mathrm{AC} \rightarrow \mathrm{AC}\\
&\mathrm{T} + \mathrm{I} \rightarrow \mathrm{T}\\
&\mathrm{AC} + \mathrm{I} \rightarrow \mathrm{AC}
\end{aligned}
$$

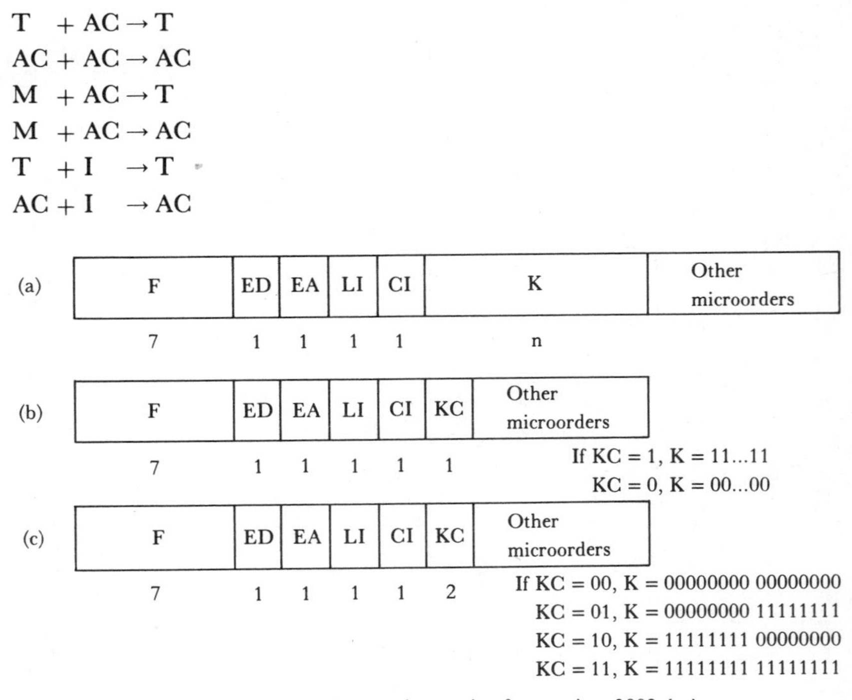

Figure 3.19. Sample microinstruction formats in a 3002 design.

If KC=0, the right ALU input is zero, allowing the left source to be copied to the destination. CI can be used to increment this value by 1.

Reviewing Figure 3.17, one sees that a few functions in F-group 0 have two destinations, the functions in F-group 1 can increment or decrement a register (by setting K=00, CI=1 or K=11, CI=0) while simultaneously moving a value into MAR, and that CO (ALU carry-out), during a logical operation, can be used as a zero/nonzero test.

Also by reviewing Figure 3.17, one sees a large difference in the flexibility of the registers. AC is most flexible, since it can appear as a left source, right source, and destination, and it is used by more control functions than any other register. T is somewhat less flexible, the registers in the register array are considerably less flexible, and MAR is least flexible; MAR cannot be used as a source, and one cannot store an arithmetic result into it.

As mentioned earlier, there are common operations that cannot be performed in a straightforward way in the 3002. One of these is subtraction. Subtraction is normally done within an ALU by complementing the subtrahend and adding it to the minuend. In Figure 3.17, one sees that, in one circumstance, the complement of AC OR K is an ALU source. However, the other source in this situation is AC AND K, meaning that this cannot be used to represent subtraction; the best that can be done is to use this to form the two's complement of the value in AC. Hence, performing an operation such as AC=R4−R5 requires at least three cycles: (1) transfer R5 to AC, (2) form the two's complement of AC, and (3) moving the sum R4+AC to AC.

Unfortunately, this operation is more difficult than indicated, since these three steps cannot be performed in three cycles. Step 1 can be performed by setting F to 0101000 (0,5 in the chart) and K to zeros. Step two is performed by setting F to 1111001 (1,15 in the chart), K to zeros, and CI to 1. However, notice that there is no F value that will set AC to R4+AC. The closest one is F=0100000 (first row in the chart), but this stores the sum in R4 as well as AC. Assuming we do not want R4's value to be altered (i.e., we wish to compute AC=R4−R5, not AC,R4=R4−R5), the following operations are needed: (1) transfer R4 to AC, (2) transfer AC to a working register (e.g., R9), a register whose contents may be altered, (3) transfer R5 to AC, (4) form the two's-complement of AC, and (5) AC,R9 = R9+AC.

This small example points out several problems in the 3002:

1. Subtraction is awkward. (The 3002 does allow one, however, to directly decrement AC and T by 1.)
2. One cannot add the value of an R register to that of AC without altering the value in R.
3. One cannot directly move one register in the array to another.
4. Furthermore, one can observe that the ways in which different registers and input busses can be used together in operations are severely limited.
5. The number of distinct ALU functions is small. Although some literature on the 3002 refers to "over 40 arithmetic and logical operations," this number is

obtained by counting such operations as AC=M+AC and AC=AC+AC as two different ALU functions.

Another operation that cannot be performed in a straightforward manner is comparison. Here one normally subtracts one value from another, routes the result nowhere, and tests the sign of the ALU output. First, in the 3002, there are no ALU-output-sign and ALU-result-is-zero outputs. One can test the carry-output signal to determine a less-than or greater-than relationship, but there is no easy way to test for equality. A larger problem is the absence of a function that routes the ALU result nowhere. If one wishes to accomplish this, a technique called *conditional clocking* can be used. One feeds the clock inputs of the 3002 slices from an AND gate. The inputs to this AND gate are the system clock and an added 1-bit microorder in the microinstruction. If this bit in the current microinstruction is 0, the clock signal to the 3002 slices is disabled, thus allowing one to use the 3002 in the current machine cycle without storing into a register (since all register loads are triggered by the rising clock edge).

Using the 3002 in a Processor

Although literature on the 3002 slice discusses its use in the processing section of a CPU, the 3002 is not well suited to such applications. Some of the reasons are the problems discussed earlier, for example, the difficulty in doing subtraction and the lack of a two-port register array, making register-to-register operations awkward.

Another significant problem is the absence of overflow detection in the ALU. All instruction-set architectures specify the presence of an overflow condition when performing arithmetic operations, but there is no way to detect this condition in the 3002. When using two's-complement arithmetic, overflow is the condition where the carry-in to the high-order bit (the sign) is unequal to the carry-out of the ALU. The 3002 has only an ALU carry-out signal, meaning that there is no way to detect overflow.

A further problem is multiplication. The 2901 slice, given its two shifters and 2-port memory, could multiply at a rate of one cycle per multiplier bit, given a small amount of external logic and a microinstruction sequencing design that allows one microinstruction to be repeated *n* times. The 3002, assuming one wants a double-length product, the case in the 2901 example, requires an average of 4.5 cycles per multiplier bit. Assuming that AC is initially zero, T holds the multiplicand, R1 holds the multiplier, and R0 is available as a working register, each iteration consists of the following cycles:

1. Examine the low-order bit of T. If the bit is 0, branch to step 5.
2. Store AC in R0.
3. Transfer R1 to AC.
4. AC ← AC + R0 (also modifies R0).
5. Shift AC right.
6. Shift T right, inserting the ejected bit from step 4 into the high-order bit of T.

One also needs a flipflop to transfer the RO output from step 5 to the LI input in step 6. The product is found in AC,T.

Given the difficulty of using the 3002 in a processor, the obvious question is: what is the 3002 suited for? To answer this, one must go back to the original characteristics of the 3002. Its major distinctions over the 2901 slice are its larger number of external busses and its bit-masking (K bus) facility. These features might imply that the 3002 is more suited to control applications. One such application is an I/O-device controller, for example, a microprogrammed disk controller. Here, one needs several busses (e.g., a control bus to the device, control and address busses to the main processor), and a bit-manipulation facility is useful in establishing and testing control signals. The lack of overflow detection is not a problem, since the controller deals with system-generated values (e.g., addresses, byte counts), where overflows should not occur. Furthermore, subtraction, multiplication, and division are usually not needed in an I/O controller. As an example, Intel Corp. uses the 3002 slice in their SBC 202 diskette controller.

3002 Timing

The data presented here is for the N3002 from Signetics Corporation, which is faster than Intel's 3002.

The minimum clock period (cycle time) of the N3002 is 70 ns. The F, CI, and LI control inputs, and the I, K, and M busses must be set up before the leading (falling) clock edge. When the clock input is low, the ALU result is stored in the destination register(s).

Figure 3.20 displays the worst-case set-up and propagation times in a format similar to that used for the 2901. Comparing Figure 3.20 to Figures 3.9 and 3.10, the N3002 appears, on paper, to be somewhat faster than the 2901A. It is, subject to the following two considerations. First, the 3002 has a 2-bit ALU where the 2901 has a 4-bit ALU. Hence, a more accurate comparison would be two cascaded 3002 slices to one 2901, adding in the extra time needed to do a carry lookahead between the 3002s. Second, the 3002 is faster *only in those applications for which it is suited.* A processor built from 3002 slices might have a smaller

From input	Minimum set-up time to trailing clock edge
F	65
I, K, M	57
CI, LI	38

From input \ To output	X, Y, RO	CO	A, D
	Max. propagation times (ns).		
F	41	52	—
I, K, M	33	44	—
CI	—	20	—
EA, ED	—	—	20
CLK	48	56	23

Figure 3.20. Representative N3002 guaranteed times.

cycle time than one built from 2901s, but the latter processor is likely, on a macroscopic basis, to be considerably faster because the limitations of the 3002 (e.g., no subtraction, slower multiplication and division, single-port register array, no overflow detection) might require considerably more machine cycles to accomplish the same amount of work.

Cascading the 3002

Figure 3.21 illustrates four cascaded 3002 slices in an 8-bit microprogrammed processor. This diagram represents a section of logic from the Intel SBC 202 diskette controller. The controller consists of two printed-circuit boards: the channel board and the interface board. The channel board fetches and decodes I/O commands from the system memory, controls the interface board, and monitors device status and errors. The interface board controls the mechanical operation of the disk drives and transfers data between the drives and the channel board. Figure 3.21 represents the heart of the channel board.

The 3002 slices are connected in a ripple-carry arrangement. The carry-out line of the most-significant slice and the shift-out line of the least-significant slice are wire-NORed to provide a single condition that can be tested by the microprogram

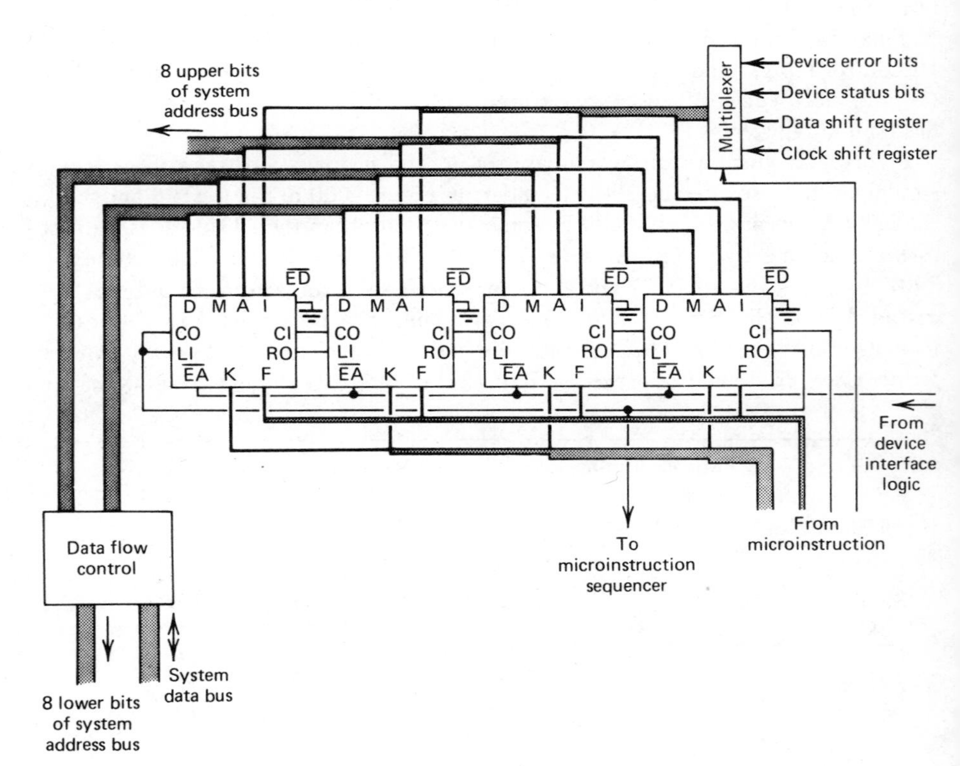

Figure 3.21. Cascaded 3002 slices in a diskette controller.

sequencing logic. The RO output from the least-significant slice is connected to the LI input on the most-significant slice, allowing the device to perform a rotating shift. The microinstruction, among other things, controls the 3002 functions, the carry-in signal, the K bus, and a multiplexer of conditions and data from the interface board. The microinstruction contains an 8-bit mask-bit field which can be gated onto the K bus, allowing the testing of combinations of status and error bits on the I bus. The size of the microprogram is approximately 500 32-bit microinstructions.

The A output bus can be gated onto the system address bus. Enabling of the A bus (EA) is under the control of the interface board. The D output bus is gated, via external logic, to the system address or data busses. The D bus is always enabled ($\overline{ED}$, which is active low, is grounded). This external logic can also gate the system data bus to the M bus of the 3002 slices.

Note the advantages of the 3002 slice over the 2901 in this type of environment. The greater number of external busses provided by the 3002 allow it to be connected in parallel to a greater number of external devices. The ANDing of the K bus gives one considerable flexibility in testing combinations of bit values. The drawbacks discussed earlier do not present problems in this application.

THE MC10800 SLICE

The MC10800 processor slice is not only the fastest slice in this chapter, but it is also one of the most functionally flexible slices. It is a member of Motorola's M10800 family of ECL bit slices and other LSI devices [8,9].

The organization of the MC10800 is shown in Figure 3.22. All data paths are four bits wide. In contrast to the 2901, which contains a large register array and few internal data paths, the MC10800 has many internal data paths and no register array. (It has no register array because the amount of logic in Figure 3.22, when this circuit was designed, taxed the density of ECL logic; if a register array is needed, it must reside on additional chips.)

The ALU is relatively simple, although it can perform BCD (binary coded decimal) addition and subtraction. However, a rich set of arithmetic and logical operations is available because of the complex input networks to the ALU. One port (X) of the ALU is fed from a complementer, which can form the one's, nine's or no complement of its input. The complementer is independently controlled, meaning that it can be used when the ALU is performing logical, as well as arithmetic, operations. The complementer is fed from a mask circuit, similar to the K-bus mask facility in the 3002. The mask can pass either of its two inputs, or pass the AND or OR of its two inputs. Again, the mask is independently controlled, meaning that it can be used in conjunction with arithmetic or logical operations in the ALU.

The Y port of the ALU has similar flexibility. The Y port can be fed with the output of the multiplexer, the output of the multiplexer ORed with the accumulator register, 0000, 0010 (+2), or 1110 (–2).

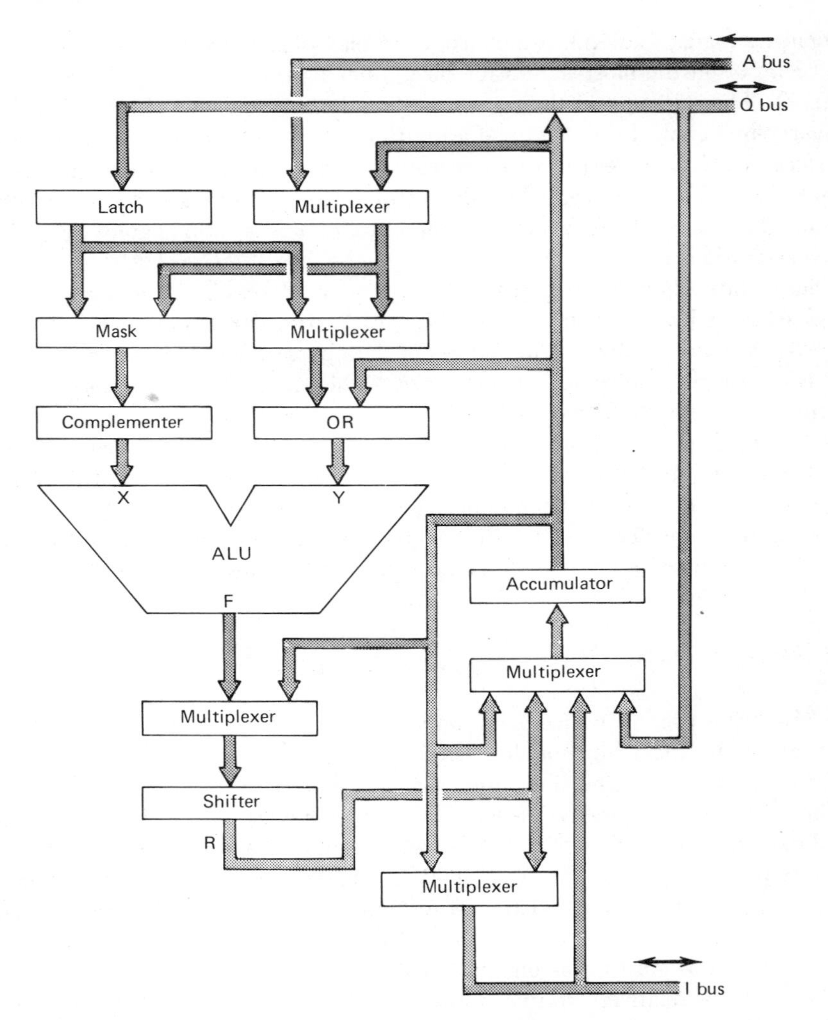

Figure 3.22. Organization of the MC10800.

The output of the ALU can be gated to a shifter, which can perform a left shift, logical right shift, or arithmetic right shift. One destination of the shifter is the accumulator, the lone register on the chip. As shown in Figure 3.22, the accumulator has a large number of possible input sources and a large variety of destinations.

The MC10800 has three 4-bit data busses. The A bus is an input bus, but Q and I are bidirectional. The A and Q busses feed the ALU input multiplexers. The Q bus, as well as the I bus, can be gated directly into the accumulator. The accumulator can be gated to the Q and I busses, and the shifter output can be gated onto the I bus. Among other things, these busses would be used to connect the MC10800 to an external register array.

Figure 3.23 is the manufacturer's diagram of the organization of the MC10800.

The external connections to the MC10800 are illustrated in Figure 3.24. The chip has 48 pins in a quad inline package (four rows of pins). The only pins not shown are two grounds, two to a –5.2V supply, and two to a –2.0V supply. As would be expected, because of the large number of internal data paths, the chip has a large number of control inputs; the 17 AS pins control the routing of data and the ALU function.

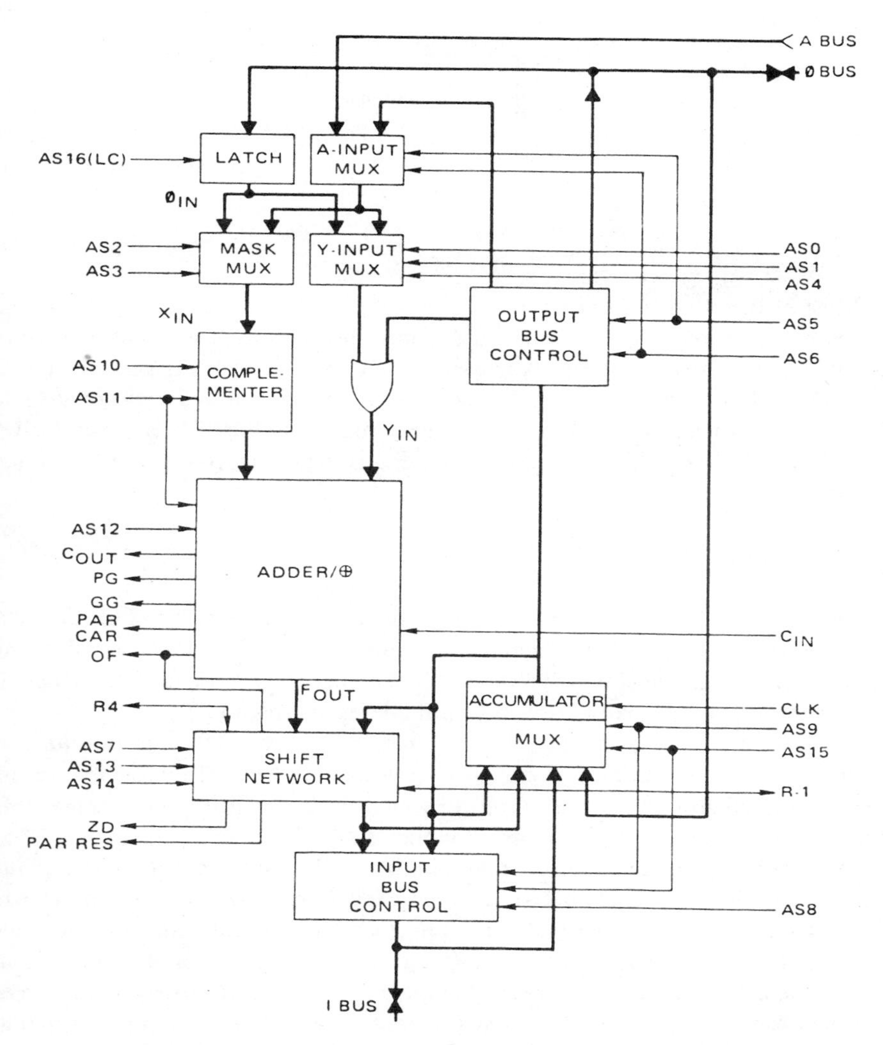

Figure 3.23. Organization of the MC10800 (Reproduced with permission of Motorola Inc.).

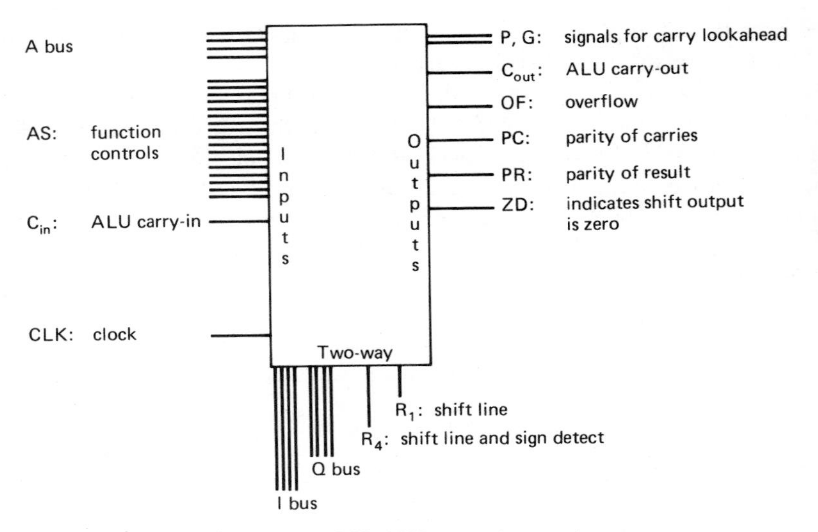

Figure 3.24. MC10800 external connections.

Most of the status output lines are self-explanatory. The OF signal detects an arithmetic overflow when a binary operation is in progress. It is also connected to the shifter and is raised during a left shift if the shift changes the sign bit (i.e., if the value of the high-order bit changes). Two parity outputs are provided to perform external parity checking and/or generate parity bits. PR is the exclusive-OR of the four result bits from the shifter. PC is the exclusive-OR of the internal carries in the ALU.

MC10800 Functions

As mentioned above, the MC10800 has 17 function-control inputs. This provides one with an impressive 131,072 functions, although, as might be expected, many of these are meaningless or redundant. The number of meaningful unique operations, however, is approximately 5000, still an impressive number.

The definition of the 17 controls is cleaner than in the 3002, but not quite as clean as in the 2901 (meaning that there is a small amount of interaction among the control subgroups). Rather than present a bit-by-bit definition, Figure 3.25 summarizes the meaning of the control inputs.

As shown, five of the control inputs influence the input to the Y port of the ALU. In addition to being able to gate the Q and A busses and the accumulator to the Y port, two ORed values and four chip-generated constants can be inputs. The 0010 and 1110 inputs present an interesting consideration. They imply that, unlike the 2901 and 3002, one cannot simply decode the microinstruction and feed all the slices the same control inputs. For instance, if one did this with four cascaded MC10800s while specifying the +2 (0010) Y input, the Y input across the 16-bit ALU would be +2222. Hence, to perform an addition of +2, the designer must provide additional decoding logic to cause the Y input of the least-significant slice to be +2 and the Y inputs of all other slices to be 0000. In the case of a –2 function, the controls to the least-significant slice would specify a Y input of –2, and the controls to the other slices would specify a Y input of 1111. This is the first slice discussed in which the control inputs sent to all slices are not identical. The additional logic required for this is eliminated in some of the subsequent slices discussed in this chapter by *relative position control,* additional pins that are permanently wired by the designer to "tell" the slice that it is the least-significant slice, most-significant slice, or neither.

The discussion above describes the intended use of the +2 and –2 inputs (and, of course, the carry input to the least-significant slice extends this to increment by +3 and decrement by –1). However, for instance, if the application in which the MC10800 was used required a frequent addition of 32, the control inputs could be

ALU Y INPUT (AS_0, AS_1, AS_{4-6})
0000
1111
Q
A
ACC
0010 (+2)
1110 (–2)
A ∨ ACC
Q ∨ ACC

ALU COMPLEMENTER INPUT (AS_{2-3}, AS_{5-6})
A ∧ Q
ACC ∧ Q
Q
A
ACC
A ∨ Q
ACC ∨ Q

ALU FUNCTION (AS_{10-12})
+ (binary)
+ (BCD)
⊬ (exclusive or)

COMPLEMENTER (AS_{10-11})
Invert
No complement
9's complement (For BCD subtraction)

SHIFT SOURCE (AS_7)
ACC
F (ALU)

SHIFT TYPE (AS_{13-14})
None
Left
Right logical
Right arithmetic

ACC SOURCE ($AS_{9,15}$)
R (shift)
Q
I
ACC

I-BUS SOURCE (AS_{8-9}, AS_{15})
R (shift)
ACC
Disable

Q-BUS SOURCE (AS_{5-6})
Q
ACC ∧ Q

Q LATCH (AS_{16})
Enabled
Latched

Figure 3.25. MC10800 functions.

varied such that the +2 Y input is specified for only the slice next to the least-significant slice (i.e., add 0020).

The inputs to the other port of the ALU (via the complementer) are also shown in Figure 3.25. Many of these values involve an AND or OR operation performed by the masking device in Figure 3.22. Hence, the Q bus can serve a purpose similar to that of the K bus in the 3002 slice. Most, but not all, of these inputs can appear with any of the Y inputs (the exceptions are the cases where both are fed from the multiplexer of the A bus and accumulator register as shown in Figure 3.22).

These inputs pass through a complementer before reaching the ALU's X port. The complementer can pass its input unchanged, pass the one's complement (logical inversion) of its input (useful for logical operations, or, taken with a C_{in} input of 1, for two's-complement subtraction), or pass the nine's complement of its input (for BCD subtraction).

The ALU is rather simple, performing only addition and the exclusive-OR operation. However, by use of the mask and OR networks, the complementer, and the flexibility in gating inputs to the ALU, a large number of arithmetic and logical functions are possible. For instance, one can perform the operation $\bar{A}$ OR Q by specifying A as the ALU Y-port input, A AND Q as the input to the complementer, inversion as the complementer operation, and exclusive-OR as the ALU operation.

The MC10800 is one of the few slices that can perform BCD arithmetic. BCD arithmetic operations are usually slower because they require two additions (the second being 6's correction), but, according to the manufacturer, the inclusion of BCD arithmetic did not compromise the chip's speed. Apparently, no detection of invalid BCD data is performed. (The OF, or overflow, status output is not used during BCD operations, meaning that it could have been used to signal an invalid BCD digit, but Motorola either did not think of this, chose not to do so, or could not squeeze the additional gates needed onto the chip.)

Figure 3.25 also shows that either the ALU result or the accumulator can be routed through the shifter. If the accumulator is selected, the ALU output is gated nowhere. Being able to do this with the ALU output is useful, because it allows one to use the ALU to compare two values (using a subtraction) without having to modify a register. The 2901 also has this property, but the 3002 does not.

The shifter performs four operations and has two connection pins: R_4 and R_1. In addition to logical shifts, the shifter can perform an arithmetic right shift. If this operation is specified, pin R_1 contains the ejected bit and pin R_4 is not used; instead, the injected bit is equal to the high-order bit. This has the effect of retaining the sign (high-order) bit in a right shift. Again, this is a situation in which the same function signals would not be sent to all cascaded slices. To perform an arithmetic right shift, this function would be specified for the most-significant slice and logical right shift would be specified for all other slices.

There is a single left shift operation. However, the difference between an arithmetic and logical left shift is the ability to detect a change in the sign of the result. The MC10800 allows this by indicating this situation on the OF (overflow) pin. Hence the OF pin is driven by both the ALU and the shifter.

When doing a shift, the R_1 and R_4 pins are used to connect the shifters in cascaded slices. When no shift is specified, R_4 outputs the high-order bit, meaning that R_4 on the most-significant slice can be used to detect the sign of the ALU result.

Figure 3.25 shows that the I bus can be used as an input to feed the accumulator (the accumulator can also be fed from the shifter, the Q bus, and itself). The I bus can also be used as an output, fed from the shifter or the accumulator. The Q bus is primarily an input bus, but, when specified by control lines AS5 and AS6, the accumulator value can be output on the Q bus. In this situation, the accumulator is ANDed with the data on the bus.

The Q bus passes through a latch before entering the ALU network. When control line AS16 is 0, the data ripples through the latch. When it is 1, the latch holds the value on the Q bus. Hence this control signal behaves like a clock signal controlling the latch.

MC10800 Timing

The clock input to the MC10800 controls the accumulator; on the rising clock edge, data is stored in the accumulator. Representative set-up times and propagation delays are shown in Figure 3.26. If one compares these times to those of the Signetics N3002 in Figure 3.20, the ECL MC10800 does not look significantly faster. (A good column to compare is the propagation delays to the carry-lookahead outputs.) However, when one considers that (1) the MC10800 is specified over a wider temperature range (up to 85°C), (2) the MC10800, with its mask, multiplexers, and shifter, has longer data paths, and (3) the MC10800 is functionally much richer, the MC10800 looks considerably faster. The minimum clock

From input	Minimum set-up time to clock rise (ns)
A, Q busses	38
I bus	7
AS	36
C_{in}	19
R_1, R_4	8

From input \ To output	Maximum propagation times (ns) I	P, G	C_{out}	OF, ZD R_1, R_4	Q
A, Q	49	27	28	44	—
C_{in}	25	—	9	19	—
AS	64	38	39	60	17
R_1, R_4	13	—	—	—	—
CLK	58	43	45	49	13

Note: Times are over the range −30°C to +85°C

Figure 3.26. Representative MC10800 guaranteed times.

cycle time is not specified by the manufacturer, but it is estimated to be approximately 75 ns.

THE SBP0401A SLICE

The SBP0401A is a 4-bit ALU/register slice produced by Texas Instruments, Inc. [10]. Unlike the previous slices discussed, the SBP0401A is not a member of a family of bit-slice components. This slice employs I^2L technology, although its input and output signals are TTL compatible.

The organization of the slice is shown in Figure 3.27. The chip contains a single-port array of eight registers. As shown in the diagram, the eighth register (also called PC, or program counter) has special significance in that a few things can be done with it that cannot be done with the other seven registers.

The ALU performs a wide variety of arithmetic and logical functions. The A port of the ALU can be fed from the register array or an external input bus (DIP). The B port of the ALU can be fed from the DIP bus, the WR ("working") register, or the XWR ("extended working") register, two additional registers on the slice with motivations similar to those of the Q register in the 2901 slice. The

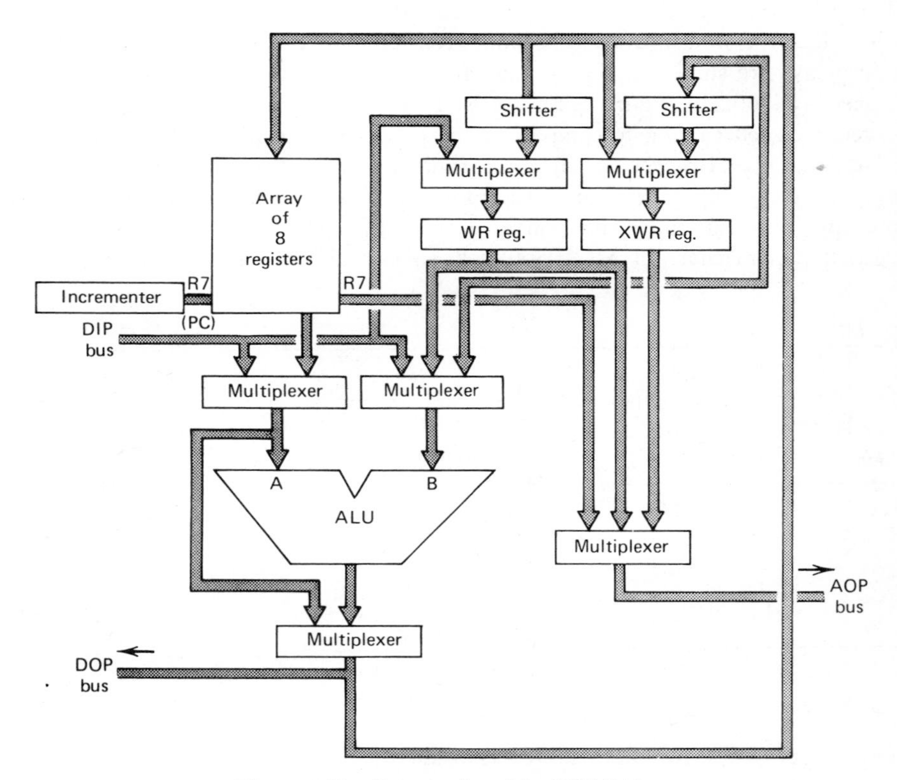

Figure 3.27. Organization of the SBP0401A.

WR and XWR registers can be interconnected to perform a wide variety of double-length logical and arithmetic shifts and circulates (circular shifts).

The output of the ALU can be gated to an output bus (DOP), the register array, WR, or XWR. When gated to WR, the value passes through a shifter. The slice contains a second output bus, AOP. In most cases, WR is gated to the AOP bus. In a limited number of situations, XWR is gated to the AOP bus. A single control line allows one to override this and gate register 7 (PC) to the AOP bus.

Register 7 (PC) is not contained in the register array; it is a counter. PC is intended (but not required) to be the program, or instruction, counter in a processor, that is, the register containing the main-storage address of the next machine instruction to be executed. Under microprogram control (i.e., under the control of specific input signals to the slice), PC can be independently incremented by 1 or 2. That is, PC can be incremented while the ALU is being used for another purpose.

The manufacturer's diagram of the SBP0401A is shown in Figure 3.28. Actually, there are two slices with the same organization: the SBP0400A and SBP0401A. The only difference is that the SBP0400A contains an additional register used to buffer the nine control signals. This is of use in designs incorporating *microinstruction lookahead* or *pipelining* (discussed in Chapter 5).

Figure 3.29 illustrates the external connections of the SBP0401A. The chip has 40 pins; the only two pins not shown are the current input and ground. The ALU function, multiplexer control, and register selection is performed by nine control-signal inputs, labeled the S, D, and OP groups. In most cases, one cannot relate specific control lines or groups of lines to specific functions; most of them are highly encoded. However, when D=00, the three S lines designate a register in the register array and the four OP lines specify which of 16 ALU functions is to be performed. PCCIN specifies whether the PC register is to be incremented during the current cycle, and another input, ENINCBY2, specifies whether the increment is 1 or 2. PCP, if set, causes PC to be gated to the AOP bus.

The two pins labeled POS represent a recent organizational advance in bit slices. Such pins will be seen on subsequent slices in this chapter. Their purpose is *relative position control.* Quite simply, they are used to tell each slice whether it is configured as the most-significant (highest-order) slice (MSS), least-significant (lowest-order) slice (LSS), or intermediate slice (IS). Relative position control has two motivations:

1. Economy of pins, since some of the pins can have different definitions, depending on the position of the slice (the reason for the three "see text" notes in Figure 3.29).
2. Simplification of microinstruction decoding (e.g., the discussion of the MC10800 slice mentioned several situations where one would not send the same function-control inputs to all slices).

Table 3.2 defines the POS settings and the interpretations of the three dual-use pins under each position setting. Notice that there are two settings to specify that

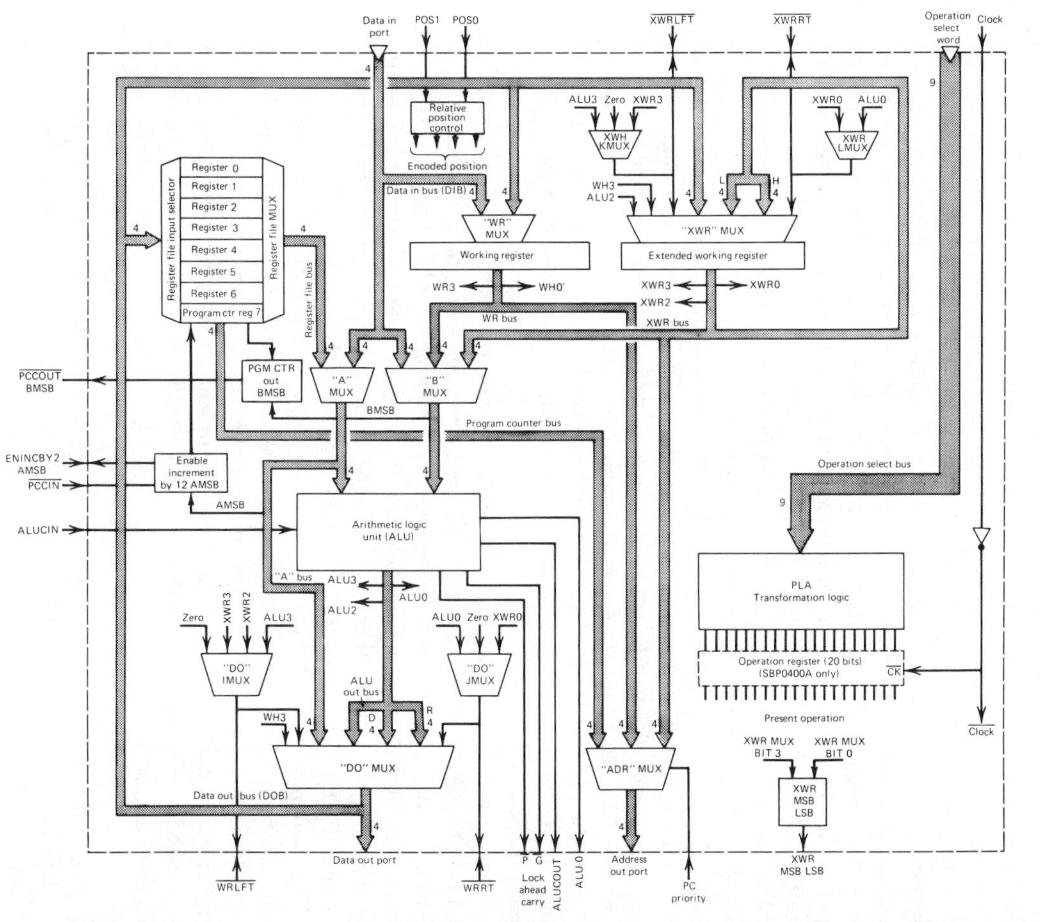

Figure 3.28. Organization of the SBP0401A (Copyright 1977 Texas Instruments Inc. Reproduced with permission.).

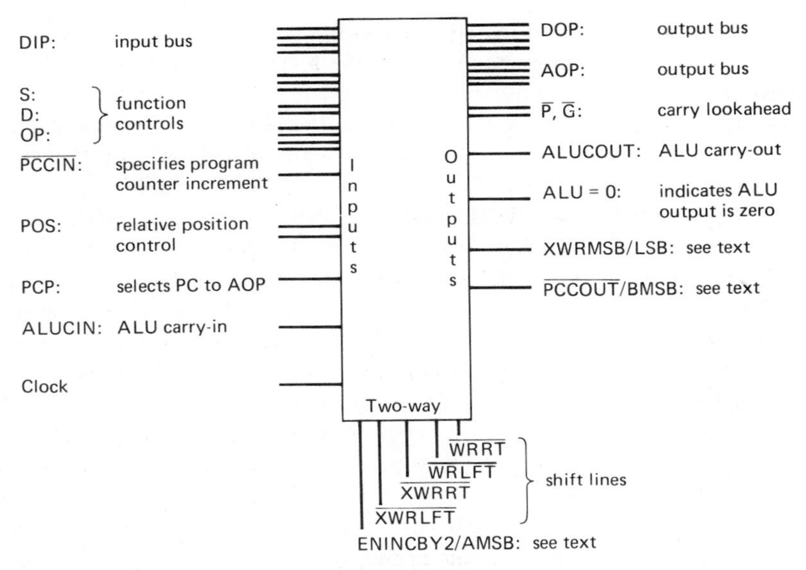

Figure 3.29. SBP0401A external connections.

the slice is the MSS. The only difference is that if POS=10, double-precision arithmetic shifts are assumed to be shifts of two single-precision signed values (i.e., two signs), and if POS=11, the shifts are assumed to be shifts of a signed double-precision number.

Although one could devise an exotic configuration where the POS inputs are controlled by the microinstruction or external logic, normally they are perma-

TABLE 3.2. SBP0401A Relative Position Control

		Multifunction Pins		
POS	Position	PCCOUT/ BMSB	ENINCBY2/ AMSB	XWRMSB/LSB
01	LSS	PCCOUT	ENINCBY2	XWRLSB
00	IS	PCCOUT	Hi-Z	Zero
10	MSS	BMSB	AMSB	XWRMSB
11	MSS	BMSB	AMSB	XWRMSB

PCCOUT - PC carry-out
ENINCBY2 - 1: PC will be incremented by 2
0: PC will be incremented by 1
AMSB - most-significant (sign) bit of ALU A-port
BMSB - most-significant (sign) bit of ALU B-port
XWRLSB - least-significant input bit of the XWR
XWRMSB - most-significant input bit of the XWR

nently wired to indicate, to each slice, its position. That is, on all intermediate slices, the two POS pins are grounded, on the lowest slice, the first POS pin is grounded and the second is connected to a logical 1 voltage level, and so on.

It is instructive to see the motivation behind the pin assignments in Table 3.2. In a system with cascaded slices, one would want to tell only the lowest slice whether to increment PC by 1 or 2; the result of this would ripple into the higher slices. Hence, on the LSS, pin ENINCBY2/AMSB is an input specifying whether the increment is by 1 or 2. Rather than having this pin go unused on all other slices, the pin is an output if the slice is specified as the MSS and contains the value of the most-significant bit (normally the sign) of the value entering the ALU's A port. Also, a carry output is needed to interconnect the PC registers on cascaded slices (only ripple carry can be used for the PC; carry lookahead outputs are provided for the ALU but not PC). On all LSS and IS slices, PCCOUT/BMSB is the PC carry-out. On the MSS, this pin contains the high-order bit (sign) of the value entering the ALU's B port. Note that this does not allow one to detect an overflow of the PC, assuming that it is an unsigned value of width equal to the width of the cascaded slices. If one wishes to do so, external logic is needed to temporarily switch the POS inputs of the MSS to 01 or 00, thus allowing this pin to output both PCCOUT and BMSB during the same clock cycle. If PC need not be as wide as the cascaded slices (e.g., there are eight cascaded slices but PC need only be 24 bits wide), one can use PCCOUT of the third slice to detect PC overflow.

The third multifunction pin allows one to examine the most-significant and least-significant bit of the value entering the XWR register, as shown in Table 3.2. This is useful, during iterative multiplication or division algorithms, for examining the bit to be shifted out during the next cycle.

The remaining status outputs in Figure 3.29 are straightforward. Notice that there is no ALU overflow indication. ALUCOUT, ALU carry-out, is not the same as overflow; overflow cannot be deduced from the carry-out value. Overflow can be deduced by external logic, but it is not trivial. It requires knowledge of the current ALU function, the signs of the ALU inputs (AMSB and BMSB on the MSS), and the sign of the ALU result (the high-order bit of DOP on the MSS). For instance, overflow occurs on an addition of two values of equal sign if the result has a different sign; overflow cannot occur on an addition of values of different signs. Overflow occurs in an A–B operation if A and B have different signs and the sign of the result is unequal to that of A; overflow cannot occur on a subtraction of values of equal signs. Because the encoding of arithmetic functions crosses all nine function-control inputs in an irregular way and because reverse subtraction (B–A) is provided, the logic needed to detect overflow can be complicated. Because overflow detection inside an ALU is simple (the carry-in to the high-order position is unequal to the carry-out), and because ALUCOUT on the MSS usually goes unused in a design, the SBP0401A could have defined ALUCOUT as a multifunction pin; on the MSS, it would indicate overflow. This lack of overflow detection in the slice can be considered a design flaw.

The remaining connections in Figure 3.29 are the four bidirectional shift lines. Their obvious purpose is to serve as the end interconnections for the two shifters, but their definition is somewhat unique. For instance, WRRT usually signals, on a right shift, the low-order bit of the ALU result. However, depending on (1) the type of shift (arithmetic, logical, circular), (2) the configuration of the shift (single or double length), and (3) the slice's position (LSS, IS, or MSS), the output of the WRRT pin can also be a 1 or the ejected bit from the XWR shifter. The other shift pins have similar characteristics.

SBP0401A Functions

The slice contains nine encoded function-control inputs: D (two bits), OP (four bits), and S (three bits). The D field specifies the manner in which OP and S are decoded. If D=00 (LL), the operation performed is

$$\text{RF ALU WR} \rightarrow \text{RF}$$

where RF is a register specified by the S inputs and ALU is an ALU operation specified by the OP inputs. The ALU functions, a rather complete set, are shown in Figure 3.30.

ALU OP-FIELD				ACTIVE-HIGH DATA	
OP3	OP2	OP1	OP0	ALUCIN = H (WITH CARRY)	ALUCIN = L (NO CARRY)
L	L	L	L	Fn = L	Fn = H
L	L	L	H	Fn = B minus A	Fn = B minus A minus 1
L	L	H	L	Fn = A minus B	Fn = A minus B minus 1
L	L	H	H	Fn = A plus B plus 1	Fn = A plus B
L	H	L	L	Fn = B plus 1	Fn = B
L	H	L	H	Fn = $\overline{B}$ plus 1	Fn = $\overline{B}$
L	H	H	L	Fn = A plus 1	Fn = A
L	H	H	H	Fn = $\overline{A}$ plus 1	Fn = $\overline{A}$
H	L	L	L	Fn = AnBn	
H	L	L	H	Fn = An $\oplus$ Bn	
H	L	H	L	Fn = $\overline{An \oplus Bn}$	
H	L	H	H	Fn = $\overline{A}$nBn	
H	H	L	L	Fn = An$\overline{B}$n	
H	H	L	H	Fn = An + $\overline{B}$n	
H	H	H	L	Fn = $\overline{A}$n + Bn	
H	H	H	H	Fn = An + Bn	

Figure 3.30. SBP0401A ALU functions (Copyright 1977 Texas Instruments Inc. Reproduced with permission.).

If D=01 (LH), the operation is

$$\text{RF ALU WR} \rightarrow \text{WR}$$

and the S and OP inputs are interpreted in the same way. If D=10, the ALU operations are restricted to addition, subtraction, and bypass, S specifies a register as before, and OP specifies, in a highly encoded way, the ALU function, sources, and destination, and a shift action. The functions are shown in Figure 3.31. RSA represents right shift arithmetic and LCIR represents left circulate. The destinations of the shift operations mean that each is a double-length shift involving both shifters.

If D=11, the register array is not used. OP and S, taken together, specify the operations as shown in Figure 3.32. When S is not LHL or HLH, the operation performed is shown in Figure 3.33.

Note that the slice has no explicit control signals for the multiplexer in the AOP output bus. In all cases but two, WR is gated to AOP. For the two cases marked with an asterisk in Figure 3.33, XWR is gated to AOP. However, if the PCP control input is set, it overrides the above and causes the PC register (register 7) to be gated to AOP.

The SBP0401A performs an extensive set of shift operations. As an illustration, Figure 3.34 represents the interconnections and signal flow for double-precision logical right shift (RSL), logical left shift (LSL), arithmetic right shift (RSA), and right circulate (RCIR) operations assuming three cascaded slices. As mentioned earlier, the shift pins differ from those on other slices in that they are internally interconnected differently depending on the type of shift and slice position. (The shift pins are driven by multiplexers, as shown in Figure 3.28.) This presents the designer with a subtle advantage. Notice that in these (and all other) cases, WRRT on the LSS is always connected to XWRLFT on the MSS. Also note how the inputs from the pins are internally routed as a function of the shift type, and how output values are generated. If this were not done (i.e., the shift pins were simply connected to the ends of each shifter as in the 2901), extra external logic would be needed to dynamically change the interconnection of the shift pins. (As a note, Advanced Micro Devices makes a chip, the Am2904, that performs these interconnections for their ALU/register slices.)

The decoding of the nine function signals is accomplished by a PLA (programmable logic array) on the chip. Texas Instruments offers to factory-program this PLA to provide the user with a tailored set of operations.

Cascading the SBP0401A

Figure 3.35 illustrates three cascaded SBP0401A slices. ALU ripple carry, rather than lookahead carry, was used to keep the diagram simple; lookahead carry would be configured identical to that of the other slices discussed. The items to notice in the diagram are the three multifunction pins and the use of the POS controls.

Function	OP3	OP2	OP1	OP0	D1	D0	S2 S1 S0
RF → DOP	L	L	L	L	H	L	RF: LLL → HHH
RF → XWR	L	L	L	H	H	L	RF: LLL → HHH
(WR minus RF minus 1 plus ALUCIN, XWR) RSA → WR, XWR	L	L	H	L	H	L	RF: LLL → HHH
RF plus WR plus ALUCIN → XWR	L	L	H	H	H	L	RF: LLL → HHH
RF plus DIP plus ALUCIN → WR	L	H	L	L	H	L	RF: LLL → HHH
RF plus DIP plus ALUCIN → XWR	L	H	L	H	H	L	RF: LLL → HHH
DIP → WR	L	H	H	L	H	L	Don't Care X X X
RF plus DIP plus ALUCIN → RF	L	H	H	H	H	L	RF: LLL → HHH
(WR minus RF minus 1 plus ALUCIN, XWR) LCIR → WR, XWR	H	L	L	L	H	L	RF: LLL → HHH
(WR plus RF plus ALUCIN, XWR) LCIR → WR, XWR	H	L	L	H	H	L	RF: LLL → HHH
(WR plus ALUCIN, XWR) RSA → WR, XWR	H	L	H	L	H	L	Don't Care X X X
(WR plus RF plus ALUCIN) RSA → WR, XWR	H	L	H	H	H	L	RF: LLL → HHH
RF plus XWR plus ALUCIN → WR	H	H	L	L	H	L	RF: LLL → HHH
RF plus XWR plus ALUCIN → XWR	H	H	L	H	H	L	RF: LLL → HHH
XWR plus ALUCIN → RF	H	H	H	L	H	L	RF: LLL → HHH
DIP → RF	H	H	H	H	H	L	RF: LLL → HHH

OP3 OP2 OP1 OP0 D1 D0 S2 S1 S0

Figure 3.31. SBP0401A functions for D = HL (Copyright 1977 Texas Instruments Inc. Reproduced with permission.).

	OP3	OP2	OP1	OP0	D1	D0	S2	S1	S0
DIP → DOP	L	L	L	L	H	H	L	H	L
DIP → XWR	L	L	L	H	H	H	L	H	L
(WR minus DIP minus 1 plus ALUCIN, XWR) RSA → WR, XWR	L	L	H	L	H	H	L	H	L
DIP plus WR plus ALUCIN → XWR	L	L	H	H	H	H	L	H	L
F_{16} plus ALUCIN → DOP	L	H	L	L	H	H	L	H	L
DIP plus WR plus ALUCIN → DOP	L	H	L	H	H	H	L	H	L
DIP → WR	L	H	H	L	H	H	L	H	L
DIP plus WR plus ALUCIN → DOP	L	H	H	H	H	H	L	H	L
(WR minus DIP minus 1 plus ALUCIN, XWR) LCIR → WR, XWR	H	L	L	L	H	H	L	H	L
(WR plus DIP plus ALUCIN, XWR) LCIR → WR, XWR	H	L	L	H	H	H	L	H	L
(WR plus ALUCIN, XWR) RSA → WR, XWR	H	L	H	L	H	H	L	H	L
(WR plus DIP plus ALUCIN, XWR) RSA → WR, XWR	H	L	H	H	H	H	L	H	L
DIP plus XWR plus ALUCIN → WR	H	H	L	L	H	H	L	H	L
DIP plus XWR plus ALUCIN → XWR	H	H	L	H	H	H	L	H	L
XWR plus ALUCIN → DOP	H	H	H	L	H	H	L	H	L
DIP → DOP	H	H	H	H	H	H	L	H	L

Function	OP3	OP2	OP1	OP0	D1	D0	S2	S1	S0
(WR plus ALUCIN) RSA → WR	L	L	L	L	H	H	H	L	H
(WR plus ALUCIN) RCIR → WR	L	L	L	H	H	H	H	L	H
(WR plus ALUCIN) LSA → WR	L	L	H	L	H	H	H	L	H
(WR plus ALUCIN) LCIR → WR	L	L	H	H	H	H	H	L	H
(WR plus ALUCIN, XWR) RSA → WR, XWR	L	H	L	L	H	H	H	L	H
(WR plus ALUCIN, XWR) RCIR → WR, XWR	L	H	L	H	H	H	H	L	H
(WR plus ALUCIN, XWR) LSA → WR, XWR	L	H	H	L	H	H	H	L	H
(WR plus ALUCIN, XWR) LCIR → WR, XWR	L	H	H	H	H	H	H	L	H
(WR plus ALUCIN)RSL → WR	H	L	L	L	H	H	H	L	H
(WR plus ALUCIN) RCIR → WR	H	L	L	H	H	H	H	L	H
(WR plus ALUCIN) LSL → WR	H	L	H	L	H	H	H	L	H
(WR plus ALUCIN) LCIR → WR	H	L	H	H	H	H	H	L	H
(WR plus ALUCIN, XWR) RSL → WR, XWR	H	H	L	L	H	H	H	L	H
(WR plus ALUCIN, XWR) RCIR → WR, XWR	H	H	L	H	H	H	H	L	H
(WR plus ALUCIN, XWR) LSL → WR, XWR	H	H	H	L	H	H	H	L	H
(WR plus ALUCIN, XWR) LCIR → WR, XWR	H	H	H	H	H	H	H	L	H

Figure 3.32. SBP0401A functions for D = HH, S = LHL or HLH (Copyright 1977 Texas Instruments Inc. Reproduced with permission.).

	OP3 → OP0	D1	D0	S2	S1	S0
RF ALU WR → RF	ALU: LLLL → HHHH	L	L	RF: LLL → HHH		
RF ALU WR → WR	ALU: LLLL → HHHH	L	H	RF: LLL → HHH		
*DIP ALU WR → DOP	ALU: LLLL → HHHH	H	H	L	L	L
*DIP ALU WR → WR	ALU: LLLL → HHHH	H	H	L	L	H
DIP ALU XWR → WR	ALU: LLLL → HHHH	H	H	L	H	H
DIP ALU WR → XWR	ALU: LLLL → HHHH	H	H	H	L	L
DIP ALU XWR → XWR	ALU: LLLL → HHHH	H	H	H	H	L
DIP ALU XWR → DOP	ALU: LLLL → HHHH	H	H	H	H	H

Figure 3.33. Remaining SBP0401A functions for D = HH (Copyright 1977 Texas Instruments Inc. Reproduced with permission.).

SBP0401A Timing

Since it is a I^2L device, the timing of the SBP0401A is not constant; rather, it operates at a constant speed-power product over a current range of three orders of magnitude. At the current-injector pin, the device looks like a diode with a voltage drop of 0.85 V. By choosing a current-limiting resistor for the supply voltage being used, the device is fed a fixed amount of injector current. If the device is fed 200 mA, it operates at its maximum speed. If fed 10 mA, it operates about eight times slower than its maximum speed. All times discussed in this section assume an injector current of 200 mA.

Representative times for the slice are shown in Figure 3.36. Note that these times are typical times at 25°C, not guaranteed times over the operating range. The typical clock period is 275 ns, meaning that this slice is considerably slower than the other slices discussed so far.

Given the times in Figure 3.36, it might be instructive at this point to calculate addition times using ripple and lookahead carries. Assume one has four cascaded slices and wishes to compute a value and bring it out on DOP. With ripple carry, it takes 180 ns for ALUCOUT to be formed on slice 1 (LSS), beginning the count at the rising clock pulse. It then takes 60 ns for the carry to be carried through slice 2, another 60 ns through slice 3, and 105 ns in slice 4 between the time that ALUCIN is present until DOP is valid, for a total of 405 ns.

For the carry-lookahead case, assume each slice is connected to a 74182 carry-lookahead generator. After 180 ns, each slice has delivered its P and G signals to the carry generator. The carry generator has a typical propagation delay of 15 ns; thus, 15 ns later, ALUCIN is sent to each slice. It then takes 105 ns before DOP is valid, resulting in an elapsed time of 300 ns.

One reason this comparison was raised was the suspicion raised earlier that, since the PC incrementing function has only a ripple-carry output (i.e., no separate set of carry-lookahead signals), this might be a barrier to the clock speed at which the slice can be operated (i.e., it might be the true bottleneck, rendering carry-lookahead logic for the ALU useless). However, one sees that this is not the case, since the propagation time for the PC ripple (PCCIN to PCCOUT) is quite fast with respect to the other times.

THE 2903 SLICE

The 2903 ALU/register slice, produced by Advanced Micro Devices [1], is the most functionally sophisticated slice currently available. In addition, its control inputs and status outputs are extremely well designed.

The organization of the 2903 is illustrated in Figure 3.37. If one compares this to Figure 3.1, the organization of the 2901, one sees a strong similarity. In fact, the 2903 is an expansion of the 2901 design. Although there are some differences between the organization of the 2903 and 2901, the principal difference, discussed shortly, is the expansion of the control logic in the 2903, enabling the 2903 to perform a set of complex functions.

Comparing the organization of the 2903 to that of the 2901, one difference is the addition of a third bus, the bidirectional DB bus. Also, the Y bus has been made bidirectional, and it is placed after the ALU shifter, rather than before the shifter as in the 2901. The 2903 also contains a sign-compare flipflop (SCFF), used in performing division. Like the 2901, the 2903 contains a two-port register array and a separate Q register and associated shifter. The ALU in the 2903 performs twice as many functions (16) as the ALU in the 2901. Figure 3.38 is the manufacturer's depiction of the 2903.

Figure 3.39 illustrates the external connections of the 2903. Logically, in comparison to the 2901, the 2903 contains 12 more connections, or 52 pins, but, by the use of relative position control and several multipurpose pins, it is packaged as a 48-pin DIP. Looking first at relative position control, $\overline{\text{LSS}}$, when zero, makes the slice the LSS (least-significant slice). When $\overline{\text{LSS}}$=1, pin $\overline{\text{WRITE}}/\overline{\text{MSS}}$ is an input pin; tying it to 0 defines the chip as the MSS, and tying it to 1 defines the chip as an IS (intermediate slice). On the LSS, $\overline{\text{WRITE}}/\overline{\text{MSS}}$ is an output; it is 0 when the current operation specifies the register array as a destination.

Recognizing that carry-lookahead outputs on the MSS, and overflow and ALU-result-sign outputs on the LSS and intermediate slices, are not normally used, the designers wisely merged these four outputs into two pins. On the MSS, pin G/N contains the high-order bit of the ALU result; on all other slices, it contains the carry-generate function. On the MSS, pin P/OVR contains the overflow signal; on all other slices, it contains the carry-propagate function. The relative position controls also affect the interpretation of the Z and shift pins; this is discussed later.

The $\overline{\text{IEN}}$ (instruction enable) input, when 0, enables the $\overline{\text{WRITE}}$ output and allows the current operation to store into the Q register and sign-compare flipflop. When 1, $\overline{\text{WRITE}}$ is forced to 1 and the Q register and SCFF are not altered. In most designs, $\overline{\text{IEN}}$ is tied to ground.

$\overline{\text{OEB}}$ (output enable B), when 0, gates the register selected as the B register to the DB bus and to the ALU S-port multiplexer. If 1, the B register is disabled as

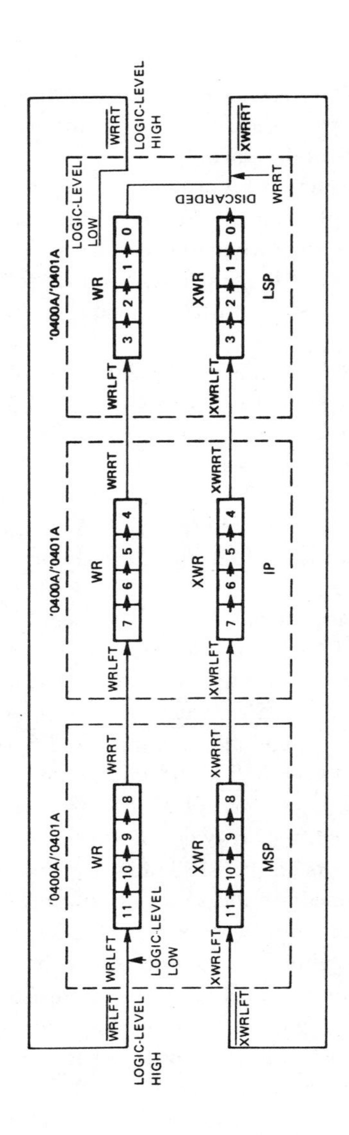

'0400A/'0401A
WR
XWR
MSP
IP
LSP
WRLFT
WRRT
XWRLFT
XWRRT
LOGIC-LEVEL HIGH
LOGIC-LEVEL LOW
DISCARDED

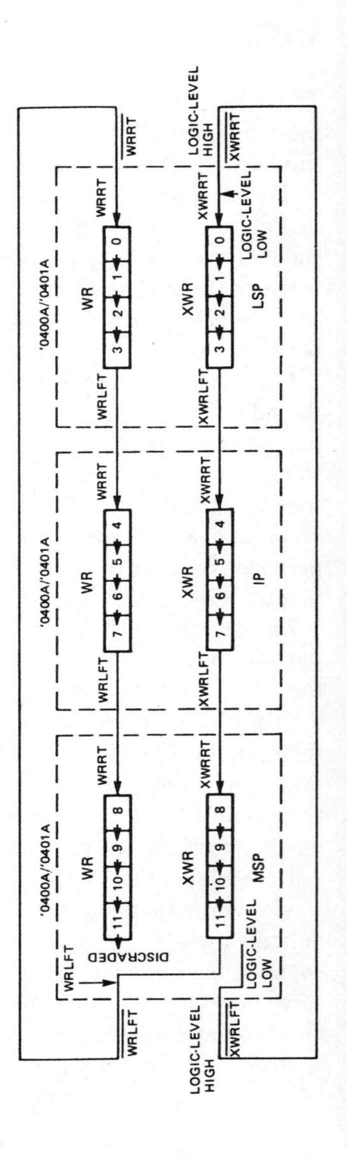

'0400A/'0401A
WR
XWR
MSP
IP
LSP
WRLFT
WRRT
XWRLFT
XWRRT
LOGIC-LEVEL HIGH
LOGIC-LEVEL LOW
DISCARDED

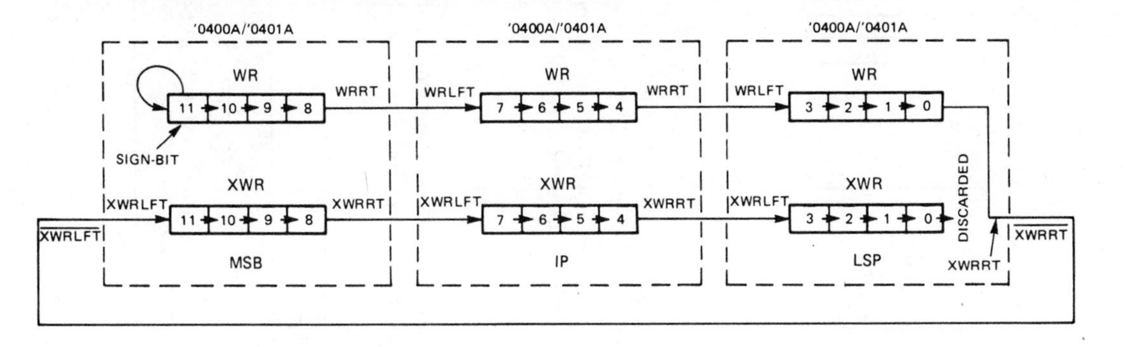

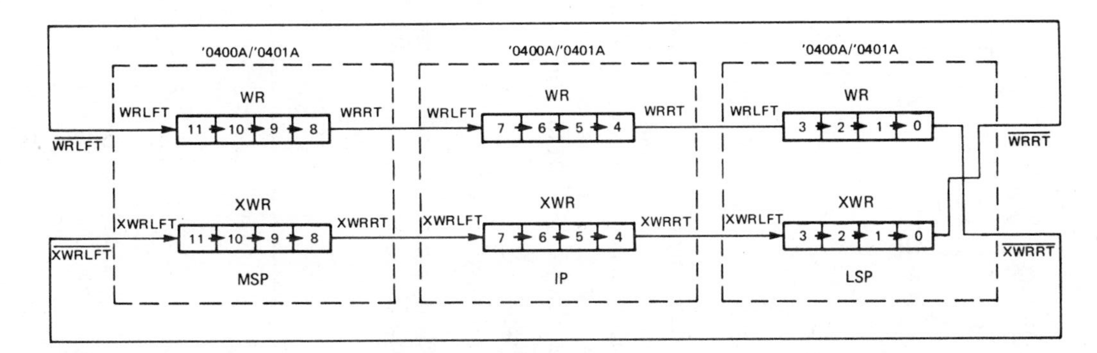

Figure 3.34. Sample internal shift connections for the SBP0401A (Copyright 1977 Texas Instruments Inc. Reproduced with permission.).

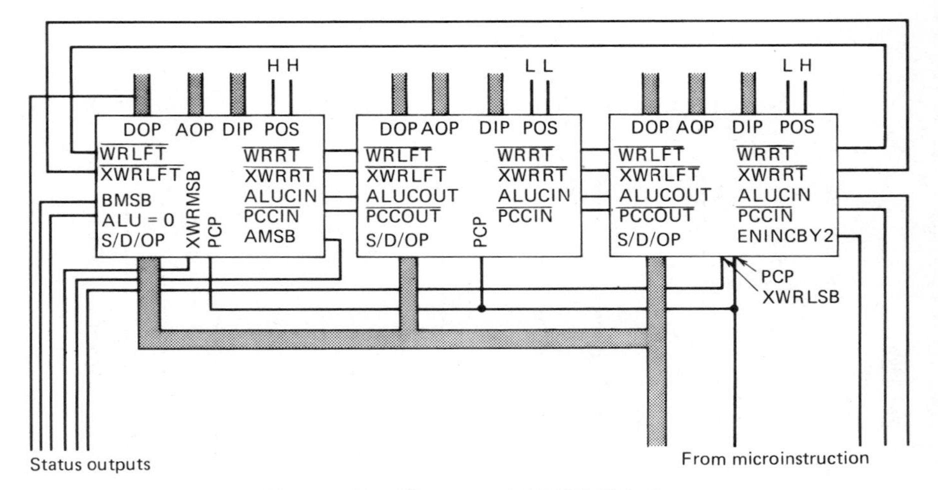

Figure 3.35. Three cascaded SBP0401A slices.

a source (allowing an ALU input to come from the DB bus). $\overline{WE}$ (write enable), if 0, causes data on the Y bus to be gated into the B register (when the clock input is low). Unless one desires to add additional registers to those in the 2903 or use the 2903 in a three-address mode (discussed later), the $\overline{WE}$ inputs on all slices should be tied to the $\overline{WRITE}$ output on the LSS. $\overline{OEY}$ enables the ALU shifter output onto the Y bus.

2903 Functions

The 2903 has a large number of control inputs, many of which were mentioned above. The A and B control lines select the registers in the array to be designated

From input	Set-up time to clock rise (ns)
DIP	180
S, D, OP	70
ALUCIN	120
$\overline{PCCIN}$	25

Note: Times are typical, not maximum, and are measured at Icc = 200 mA.

From input \ To output	Propagation times (ns) DOP	AOP	$\overline{P}$, $\overline{G}$	ALU-COUT	$\overline{PCCOUT}$ AMSB BMSB	ALU = 0
DIP	155	—	—	—	80	—
ALUCIN	105	—	—	60	—	—
S, D, OP	—	—	—	—	—	—
$\overline{PCCIN}$	—	—	—	—	35	—
PCP	—	70	—	—	—	—
CLOCK	240	105	180	180	155	215

Figure 3.36. Representative SBP0401A typical times.

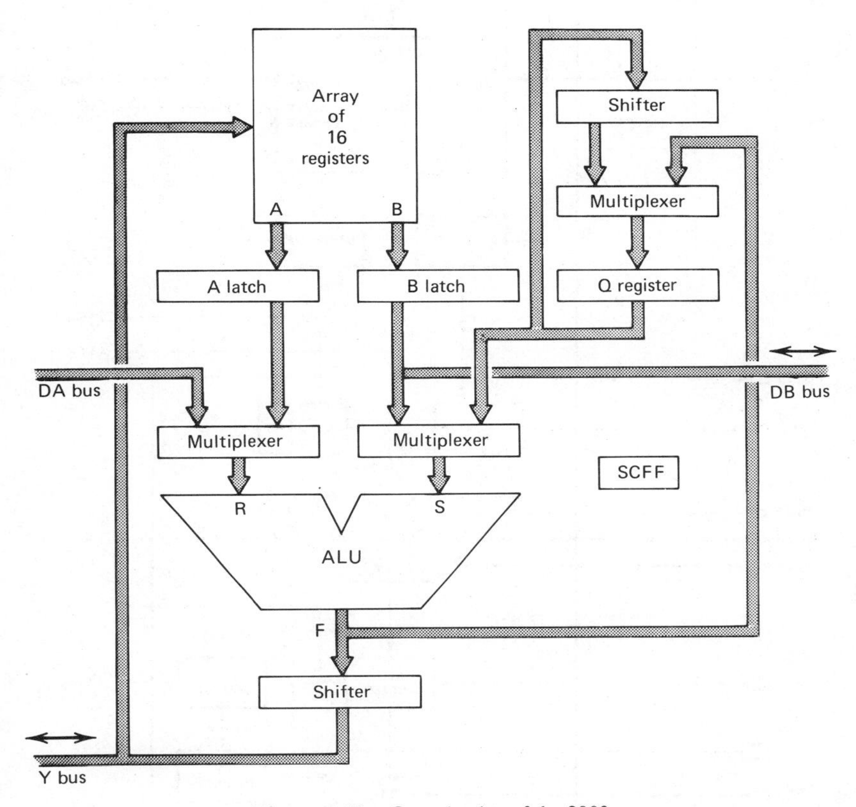

Figure 3.37. Organization of the 2903.

in the current cycle as the A and B registers. The nine I control lines select the ALU function and control the routing and shifting of data. The first I signal, with OEB, designates the input to the ALU S-port (B, Q, or the DB bus); EA specifies whether the input to the R port shall be A or the DA bus. The next four I lines specify the ALU function, as shown in Figure 3.40. If I_{0-4}=00000, one of a set of extended functions is performed, which is specified by the remaining I lines and discussed later.

Figure 3.41 specifies the routing of data (ALU and shifter outputs), providing that I_{0-4} is not 00000 (special functions). It also indicates the operation performed by both shifters, as well as the input or output state of all shift pins. Notice that the ALU shifter can perform both arithmetic and logical shifts. The difference is that, in an arithmetic shift, the high-order bit in the MSS (i.e., the sign bit) is not changed; the shift passes around this bit. Also note that for four of the settings, pin SIO_0 contains a parity output. Since, for these settings, pin SIO_3 is an input, and since these shift pins would normally be interconnected between cascaded 2903s, the SIO_0 pin of the LSS gives the parity of the entire cascaded ALU result. Although a useful feature, one can only obtain parity during a cycle in which one

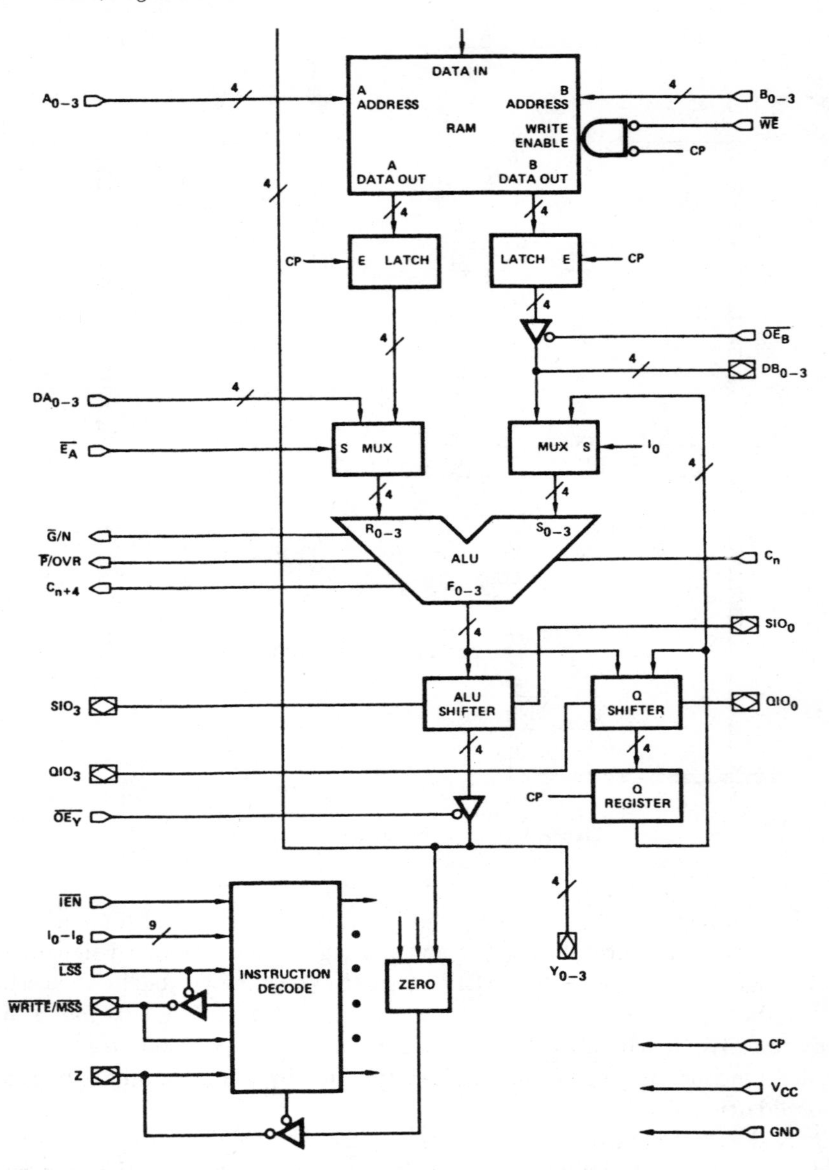

Figure 3.38. Organization of the 2903 (Copyright 1978 Advanced Micro Devices, Inc. Reproduced with permission of copyright owner.).

of these four settings is used. The MC10800 slice has dedicated parity pins, meaning that one can obtain parity after every ALU operation.

In Figure 3.41, the $\overline{\text{WRITE}}$ column is applicable to only the LSS, as this pin is an input (position indicator) on all other slices.

Notice the definition of I_{8-5} = HHHL and I_{8-5} = HHHH. These are provided for the function of *sign extension,* a necessary operation when dealing with operands of different lengths. For instance, the IBM S/370 architecture has two's-complement operands of 32 and 16 bits in length. When adding a 16-bit value to a 32-bit value, the processor must first temporarily expand the 16-bit value to 32 bits by extending or propagating its sign bit across the high-order 16 bits. To see how this is done, assume we have a system containing eight cascaded 2903 slices. If a 16-bit value is passed through the ALU, the output is

xxxx xxxx xxxx xxxx svvv vvvv vvvv vvvv

where s is the sign and v's represent the remaining value bits. What we wish to do is convert it to the form

ssss ssss ssss ssss svvv vvvv vvvv vvvv

so that it can be used as a 32-bit operand. The solution is to send I_{8-6} = HHH to all eight slices, send I_5 = L to the four high-order slices, and I_5 = H to the four low-order slices. If, for reference, the MSS is slice 7 and the LSS is slice 0, this

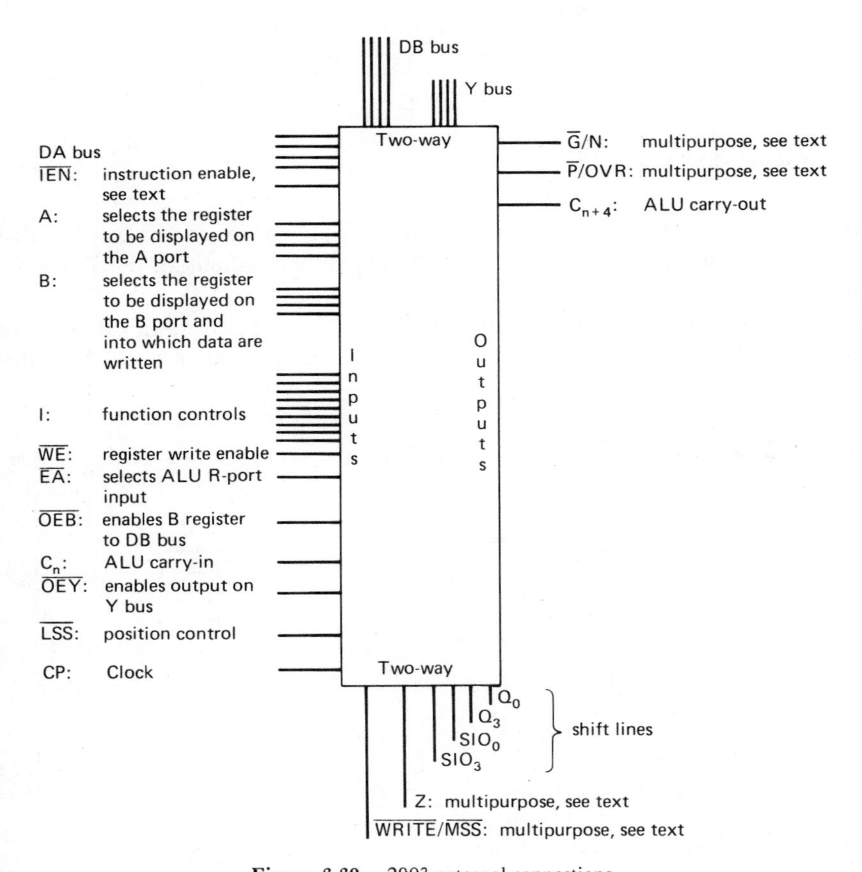

Figure 3.39. 2903 external connections.

I_4	I_3	I_2	I_1	Hex Code	ALU Functions	
L	L	L	L	0	$I_0 = L$	Special Functions
					$I_0 = H$	F_i = HIGH
L	L	L	H	1	F = S Minus R Minus 1 Plus C_n	
L	L	H	L	2	F = R Minus S Minus 1 Plus C_n	
L	L	H	H	3	F = R Plus S Plus C_n	
L	H	L	L	4	F = S Plus C_n	
L	H	L	H	5	F = $\overline{S}$ Plus C_n	
L	H	H	L	6	F = R Plus C_n	
L	H	H	H	7	F = $\overline{R}$ Plus C_n	
H	L	L	L	8	F_i = LOW	
H	L	L	H	9	$F_i = \overline{R}_i$ AND S_i	
H	L	H	L	A	$F_i = R_i$ EXCLUSIVE NOR S_i	
H	L	H	H	B	$F_i = R_i$ EXCLUSIVE OR S_i	
H	H	L	L	C	$F_i = R_i$ AND S_i	
H	H	L	H	D	$F_i = R_i$ NOR S_i	
H	H	H	L	E	$F_i = R_i$ NAND S_i	
H	H	H	H	F	$F_i = R_i$ OR S_i	

L = LOW H = HIGH i = 0 to 3

Figure 3.40. 2903 ALU functions (Copyright 1978 Advanced Micro Devices, Inc. Reproduced with permission of copyright owner.).

causes the sign bit in slice 3 to appear at its SIO_3 pin, which is connected to slice 4's SIO_0 pin. This bit, the sign, will now propagate throughout all the Y-bus bits of slices 4-7.

Figures 3.40 and 3.41 demonstrate that the 2903 can perform an extensive set of operations (and even more will be described in a subsequent section). Note that

			SIO_3		Y_3		Y_2								
I_8 I_7 I_6 I_5	Hex code	ALU shifter function	Most sig. slice	Other slices	Most sig. slice	Other slices	Most sig. slice	Other slices	Y_1	Y_0	SIO_0	$\overline{Write}$	Q reg and shifter function	QIO_3	QIO_0
L L L L	0	Arith. F/2→Y	Input	Input	F_3	SIO_3	SIO_3	F_3	F_2	F_1	F_0	L	Hold	Hi-Z	Hi-Z
L L L H	1	Log. F/2→Y	Input	Input	SO_3	SO_3	F_3	F_3	F_2	F_1	F_0	L	Hold	Hi-Z	Hi-Z
L L H L	2	Arith. F/2→Y	Input	Input	F_3	SIO_3	SIO_3	F_3	F_2	F_1	F_0	L	Log. Q/2→Q	Input	Q_0
L L H H	3	Log. F/2→Y	Input	Input	SO_3	SO_3	F_3	F_3	F_2	F_1	F_0	L	Log. Q/2→Q	Input	Q_0
L H L L	4	F→Y	Input	Input	F_3	F_3	F_2	F_2	F_1	F_0	Parity	L	Hold	Hi-Z	Hi-Z
L H L H	5	F→Y	Input	Input	F_3	F_3	F_2	F_2	F_1	F_0	Parity	H	Log. Q/2→Q	Input	Q_0
L H H L	6	F→Y	Input	Input	F_3	F_3	F_2	F_2	F_1	F_0	Parity	H	F→Q	Hi-Z	Hi-Z
L H H H	7	F→Y	Input	Input	F_3	F_3	F_2	F_2	F_1	F_0	Parity	L	F→Q	Hi-Z	Hi-Z
H L L L	8	Arith. 2F→Y	F_2	F_3	F_3	F_2	F_1	F_1	F_0	SIO_0	Input	L	Hold	Hi-Z	Hi-Z
H L L H	9	Log. 2F→Y	F_3	F_3	F_2	F_2	F_1	F_1	F_0	SIO_0	Input	L	Hold	Hi-Z	Hi-Z
H L H L	A	Arith. 2F→Y	F_2	F_3	F_3	F_2	F_1	F_1	F_0	SIO_0	Input	L	Log. 2Q→Q	Q_3	Input
H L H H	B	Log. 2F→Y	F_3	F_3	F_2	F_2	F_1	F_1	F_0	SIO_0	Input	L	Log. 2Q→Q	Q_3	Input
H H L L	C	F→Y	F_3	F_3	F_3	F_3	F_2	F_2	F_1	F_0	Hi-Z	H	Hold	Hi-Z	Hi-Z
H H L H	D	F→Y	F_3	F_3	F_3	F_3	F_2	F_2	F_1	F_0	Hi-Z	H	Log. 2Q→Q	Q_3	Input
H H H L	E	SIO_0→Y_0, Y_1, Y_2, Y_3	SIO_0	SIO_0	SIO_0	SIO_0	SIO_0	SIO_0	SIO_0	SIO_0	Input	L	Hold	Hi-Z	Hi-Z
H H H H	F	F→Y	F_3	F_3	F_3	F_3	F_2	F_2	F_1	F_0	Hi-Z	L	Hold	Hi-Z	Hi-Z

Parity = $F_3 \forall F_2 \forall F_1 \forall F_0 \forall SIO_3$ L = Low H = High Hi-Z = High impedance $\forall$ = Exclusive OR

Figure 3.41. 2903 ALU destination and shifting control (Copyright 1978 Advanced Micro Devices, Inc. Reproduced with permission of copyright owner.).

one can compare two register values (in the ALU, via subtraction) without having to store the result of the subtraction somewhere (a frequent operation that is not provided in all slices). This can be done by setting $\overline{WE}$ to 1. Notice that one can compare two register values (or a register and a value on the DA or DB bus) while simultaneously storing a value fed to the Y bus into a register. Also, in comparison to most other slices, the 2903's control signals are cleanly designed (as opposed, for instance, to the highly encoded, entangled controls in the 3002 and SBP0401A), making microinstruction design and decoding easier.

2903 Status Outputs

Although most of the output signals have been mentioned, some of them require further explanation. Figure 3.42 defines the output signals as a function of the I control signals. The last nine rows are associated with the 2903 extended functions, which are discussed in the next section.

The Z output is one of the most sophisticated features of the 2903. Notice that, in most cases, Z indicates whether the Y bus contains all zeros. In this respect, it is similar to the F=0 signal in the 2901, the ALU=0 signal in the SBP0401A, and so on. As in the other slices, the Z pins of cascaded 2903s, being open-collector outputs, are tied together to give one a status test of the cascaded Y bus being zero. Note, however, the definition of Z in the lower rows in Figure 3.42. For instance, sometimes Z of the MSS is an output containing the value of the sign-compare flipflop; in these situations, the Z pin of the other slices is an *input*. This implies that if the Z pins of all slices are wired together, in some situations the wire is used by the MSS to communicate information to the other slices. This occurs during the 2903 extended operations. Notice that, in another case, the Z line is used by the LSS to communicate the low-order bit of the Q register (Q_0) to the other slices, and in another it is used by the MSS to communicate the sign of the value entering the ALU S-port (S_3) to the other slices. Hence Z is truly a multipurpose pin.

2903 Extended Operations

As implied earlier, the 2903 also performs an extended set of operations, such operations as single-cycle iterations for signed multiplication and division, single and double-length normalization, and conversion between two's-complement and signed-magnitude representations. As will become evident shortly, many of these operations are oriented toward floating-point arithmetic, that is, numbers represented in scientific notation with a fractional mantissa and an exponent.

Figure 3.43 defines these operations, which occur when $I_{0-4} = 00000$. Note that Figure 3.43 overrides Figures 3.40 and 3.41; Figure 3.43 is a complete definition of the ALU, shifting, and data-routing operations that occur when $I_{0-4} = 00000$. Figure 3.42 (the bottom nine rows) is also needed to determine the use of the Z pin. Note that the significant difference of these operations is that, for most, the ALU function is conditional (in most cases, addition versus subtraction), based on

(Hex) $I_8I_7I_6I_5$	(Hex) $I_4I_3I_2I_1$	I_0	GI (I=0 to 3)	PI (I=0 to 3)	C_{n+4}	$\overline{P}$/OVR Most Sig. Slice	Other Slices	$\overline{G}$/N Most Sig. Slice	Other Slices	Z Most Sig. Slice	Intermediate Slice	Least Sig. Slice
X	0	H	0	1	0	0	0	F_3	$\overline{G}$	$\overline{Y_0}\,\overline{Y_1}\,\overline{Y_2}\,\overline{Y_3}$	$\overline{Y_0}\,\overline{Y_1}\,\overline{Y_2}\,\overline{Y_3}$	$\overline{Y_0}\,\overline{Y_1}\,\overline{Y_2}\,\overline{Y_3}$
X	1	X	$\overline{R_i} \wedge S_i$	$\overline{R_i} \vee S_i$	$G \vee PC_n$	$C_{n+3} \veebar C_{n+4}$	$\overline{P}$	F_3	$\overline{G}$	$\overline{Y_0}\,\overline{Y_1}\,\overline{Y_2}\,\overline{Y_3}$	$\overline{Y_0}\,\overline{Y_1}\,\overline{Y_2}\,\overline{Y_3}$	$\overline{Y_0}\,\overline{Y_1}\,\overline{Y_2}\,\overline{Y_3}$
X	2	X	$R_i \wedge \overline{S_i}$	$R_i \vee \overline{S_i}$	$G \vee PC_n$	$C_{n+3} \veebar C_{n+4}$	$\overline{P}$	F_3	$\overline{G}$	$\overline{Y_0}\,\overline{Y_1}\,\overline{Y_2}\,\overline{Y_3}$	$\overline{Y_0}\,\overline{Y_1}\,\overline{Y_2}\,\overline{Y_3}$	$\overline{Y_0}\,\overline{Y_1}\,\overline{Y_2}\,\overline{Y_3}$
X	3	X	$R_i \wedge S_i$	$R_i \vee S_i$	$G \vee PC_n$	$C_{n+3} \veebar C_{n+4}$	$\overline{P}$	F_3	$\overline{G}$	$\overline{Y_0}\,\overline{Y_1}\,\overline{Y_2}\,\overline{Y_3}$	$\overline{Y_0}\,\overline{Y_1}\,\overline{Y_2}\,\overline{Y_3}$	$\overline{Y_0}\,\overline{Y_1}\,\overline{Y_2}\,\overline{Y_3}$
X	4	X	0	S_i	$G \vee PC_n$	$C_{n+3} \veebar C_{n+4}$	$\overline{P}$	F_3	$\overline{G}$	$\overline{Y_0}\,\overline{Y_1}\,\overline{Y_2}\,\overline{Y_3}$	$\overline{Y_0}\,\overline{Y_1}\,\overline{Y_2}\,\overline{Y_3}$	$\overline{Y_0}\,\overline{Y_1}\,\overline{Y_2}\,\overline{Y_3}$
X	5	X	0	$\overline{S_i}$	$G \vee PC_n$	$C_{n+3} \veebar C_{n+4}$	$\overline{P}$	F_3	$\overline{G}$	$\overline{Y_0}\,\overline{Y_1}\,\overline{Y_2}\,\overline{Y_3}$	$\overline{Y_0}\,\overline{Y_1}\,\overline{Y_2}\,\overline{Y_3}$	$\overline{Y_0}\,\overline{Y_1}\,\overline{Y_2}\,\overline{Y_3}$
X	6	X	0	R_i	$G \vee PC_n$	$C_{n+3} \veebar C_{n+4}$	$\overline{P}$	F_3	$\overline{G}$	$\overline{Y_0}\,\overline{Y_1}\,\overline{Y_2}\,\overline{Y_3}$	$\overline{Y_0}\,\overline{Y_1}\,\overline{Y_2}\,\overline{Y_3}$	$\overline{Y_0}\,\overline{Y_1}\,\overline{Y_2}\,\overline{Y_3}$
X	7	X	0	$\overline{R_i}$	$G \vee PC_n$	$C_{n+3} \veebar C_{n+4}$	$\overline{P}$	F_3	$\overline{G}$	$\overline{Y_0}\,\overline{Y_1}\,\overline{Y_2}\,\overline{Y_3}$	$\overline{Y_0}\,\overline{Y_1}\,\overline{Y_2}\,\overline{Y_3}$	$\overline{Y_0}\,\overline{Y_1}\,\overline{Y_2}\,\overline{Y_3}$
X	8	X	0	1	0	0	0	F_3	$\overline{G}$	$\overline{Y_0}\,\overline{Y_1}\,\overline{Y_2}\,\overline{Y_3}$	$\overline{Y_0}\,\overline{Y_1}\,\overline{Y_2}\,\overline{Y_3}$	$\overline{Y_0}\,\overline{Y_1}\,\overline{Y_2}\,\overline{Y_3}$
X	9	X	$\overline{R_i} \wedge S_i$	1	0	0	0	F_3	$\overline{G}$	$\overline{Y_0}\,\overline{Y_1}\,\overline{Y_2}\,\overline{Y_3}$	$\overline{Y_0}\,\overline{Y_1}\,\overline{Y_2}\,\overline{Y_3}$	$\overline{Y_0}\,\overline{Y_1}\,\overline{Y_2}\,\overline{Y_3}$
X	A	X	$R_i \wedge S_i$	$R_i \vee S_i$	0	0	0	F_3	$\overline{G}$	$\overline{Y_0}\,\overline{Y_1}\,\overline{Y_2}\,\overline{Y_3}$	$\overline{Y_0}\,\overline{Y_1}\,\overline{Y_2}\,\overline{Y_3}$	$\overline{Y_0}\,\overline{Y_1}\,\overline{Y_2}\,\overline{Y_3}$
X	B	X	$\overline{R_i} \wedge S_i$	$\overline{R_i} \vee S_i$	0	0	0	F_3	$\overline{G}$	$\overline{Y_0}\,\overline{Y_1}\,\overline{Y_2}\,\overline{Y_3}$	$\overline{Y_0}\,\overline{Y_1}\,\overline{Y_2}\,\overline{Y_3}$	$\overline{Y_0}\,\overline{Y_1}\,\overline{Y_2}\,\overline{Y_3}$
X	C	X	$R_i \wedge S_i$	1	0	0	0	F_3	$\overline{G}$	$\overline{Y_0}\,\overline{Y_1}\,\overline{Y_2}\,\overline{Y_3}$	$\overline{Y_0}\,\overline{Y_1}\,\overline{Y_2}\,\overline{Y_3}$	$\overline{Y_0}\,\overline{Y_1}\,\overline{Y_2}\,\overline{Y_3}$
X	D	X	$\overline{R_i} \wedge \overline{S_i}$	1	0	0	0	F_3	$\overline{G}$	$\overline{Y_0}\,\overline{Y_1}\,\overline{Y_2}\,\overline{Y_3}$	$\overline{Y_0}\,\overline{Y_1}\,\overline{Y_2}\,\overline{Y_3}$	$\overline{Y_0}\,\overline{Y_1}\,\overline{Y_2}\,\overline{Y_3}$
X	E	X	$R_i \wedge S_i$	1	0	0	0	F_3	$\overline{G}$	$\overline{Y_0}\,\overline{Y_1}\,\overline{Y_2}\,\overline{Y_3}$	$\overline{Y_0}\,\overline{Y_1}\,\overline{Y_2}\,\overline{Y_3}$	$\overline{Y_0}\,\overline{Y_1}\,\overline{Y_2}\,\overline{Y_3}$
X	F	X	$\overline{R_i} \wedge \overline{S_i}$	1	0	0	0	F_3	$\overline{G}$	$\overline{Y_0}\,\overline{Y_1}\,\overline{Y_2}\,\overline{Y_3}$	$\overline{Y_0}\,\overline{Y_1}\,\overline{Y_2}\,\overline{Y_3}$	$\overline{Y_0}\,\overline{Y_1}\,\overline{Y_2}\,\overline{Y_3}$

0	0	L	0 if Z=L $R_i \wedge S_i$ if Z=H	S_i if Z=L $R_i \vee S_i$ if Z=H	$G \vee PC_n$	$C_{n+3} \veebar C_{n+4}$	$\overline{P}$	F_3	$\overline{G}$	Input	Input	Q_0
2	0	L	0 if Z=L $R_i \wedge S_i$ if Z=H	S_i if Z=L $R_i \vee S_i$ if Z=H	$G \vee PC_n$	$C_{n+3} \veebar C_{n+4}$	$\overline{P}$	F_3	$\overline{G}$	Input	Input	Q_0
4	0	L	See Note 1	See Note 2	$G \vee PC_n$	$C_{n+3} \veebar C_{n+4}$	$\overline{P}$	F_3	$\overline{G}$	$\overline{Y_0 Y_1 Y_2 Y_3}$	$\overline{Y_0 Y_1 Y_2 Y_3}$	$\overline{Y_0 Y_1 Y_2 Y_3}$
5	0	L	0	S_i if Z=L $\overline{S}_i$ if Z=H	$G \vee PC_n$	$C_{n+3} \veebar C_{n+4}$	$\overline{P}$	F_3 if Z=L $F_3 \veebar S_3$ if Z=H	$\overline{G}$	S_3	Input	Input
6	0	L	0 if Z=L $\overline{R}_i \wedge S_i$ if Z=H	S_i if Z=L $\overline{R}_i \vee S_i$ if Z=H	$G \vee PC_n$	$C_{n+3} \veebar C_{n+4}$	$\overline{P}$	F_3	$\overline{G}$	Input	Input	Q_0
8	0	L	0	S_i	See Note 3	$Q_2 \veebar Q_1$	$\overline{P}$	Q_3	$\overline{G}$	$\overline{Q}_0\overline{Q}_1\overline{Q}_2\overline{Q}_3$	$\overline{Q}_0\overline{Q}_1\overline{Q}_2\overline{Q}_3$	$\overline{Q}_0\overline{Q}_1\overline{Q}_2\overline{Q}_3$
A	0	L	0	S_i	See Note 4	$F_2 \veebar F_1$	$\overline{P}$	F_3	$\overline{G}$	See Note 5	See Note 5	See Note 5
C	0	L	$R_i \wedge S_i$ if Z=L $\overline{R}_i \wedge S_i$ if Z=H	$R_i \vee S_i$ if Z=L $\overline{R}_i \vee S_i$ if Z=H	$G \vee PC_n$	$C_{n+3} \veebar C_{n+4}$	$\overline{P}$	F_3	$\overline{G}$	Sign Compare FF Output	Input	Input
E	0	L	$R_i \wedge S_i$ if Z=L $\overline{R}_i \wedge S_i$ if Z=H	$R_i \vee S_i$ if Z=L $\overline{R}_i \vee S_i$ if Z=H	$G \vee PC_n$	$C_{n+3} \veebar C_{n+4}$	$\overline{P}$	F_3	$\overline{G}$	Sign Compare FF Output	Input	Input

L = LOW = 0
H = HIGH = 1
$\vee$ = OR
$\wedge$ = AND
$\veebar$ = EXCLUSIVE OR
$P = P_3P_2P_1P_0$
$G = G_3 \vee G_2P_3 \vee G_1P_2P_3 \vee G_0P_1P_2P_3$
$C_{n+3} = G_2 \vee G_1P_2 \vee G_0P_1P_2 \vee C_nP_pP_1P_2$

NOTES:
1. If $\overline{LSS}$ is LOW, $G_0 = S_0$ and $G_{1,2,3} = 0$
   If $\overline{LSS}$ is HIGH, $G_{0,1,2,3} = 0$
2. If $\overline{LSS}$ is LOW, $P_0 = 1$ and $P_{1,2,3} = S_{1,2,3}$
   If $\overline{LSS}$ is HIGH, $P_i = S_i$
3. At the most significant slice, $C_{n+4} = Q_3 \veebar Q_2$
   At other slices, $C_{n+4} = G \vee PC_n$
4. At the most significant slice, $C_{n+4} = F_3 \veebar F_2$
   At other slices, $C_{n+4} = G \vee PC_n$
5. $Z = \overline{Q}_0\overline{Q}_1\overline{Q}_2\overline{Q}_3\overline{F}_0\overline{F}_1\overline{F}_2\overline{F}_3$

Figure 3.42. 2903 status outputs (Copyright 1978 Advanced Micro Devices, Inc. Reproduced with permission of copyright owner.).

I_8	I_7	I_6	I_5	Hex Code	Special Function	ALU Function	ALU Shifter Function	SIO_3 Most Sig. Slice	SIO_3 Other Slices	SIO_0	Q Reg & Shifter Function	QIO_3	QIO_0	$\overline{WRITE}$
L	L	L	L	0	Unsigned Multiply	$F = S + C_n$ if $Z = L$ $F = R + S + C_n$ if $Z = H$	Log. $F/2 \rightarrow Y$ (Note 1)	Hi-Z	Input	F_0	Log. $Q/2 \rightarrow Q$	Input	Q_0	L
L	L	H	L	2	Two's Complement Multiply	$F = S + C_n$ if $Z = L$ $F = R + S + C_n$ if $Z = H$	Log. $F/2 \rightarrow Y$ (Note 2)	Hi-Z	Input	F_0	Log. $Q/2 \rightarrow Q$	Input	Q_0	L
L	H	L	L	4	Increment by One or Two	$F = S + 1 + C_n$	$F \rightarrow Y$	Input	Input	Parity	Hold	Hi-Z	Hi-Z	L
L	H	L	H	5	Sign/Magnitude-Two's Complement	$F = S + C_n$ if $Z = L$ $F = \overline{S} + C_n$ if $Z = H$	$F \rightarrow Y$ (Note 3)	Input	Input	Parity	Hold	Hi-Z	Hi-Z	L
L	H	H	L	6	Two's Complement Multiply, Last Cycle	$F = S + C_n$ if $Z = L$ $F = S - R - 1 + C_n$ if $Z = H$	Log. $F/2 \rightarrow Y$ (Note 2)	Hi-Z	Input	F_0	Log. $Q/2 \rightarrow Q$	Input	Q_0	L
H	L	L	L	8	Single Length Normalize	$F = S + C_n$	$F \rightarrow Y$	F_3	F_3	Hi-Z	Log. $2Q \rightarrow Q$	Q_3	Input	L
H	L	H	L	A	Double Length Normalize and First Divide Op.	$F = S + C_n$	Log $2F \rightarrow Y$	$R_3 \forall F_3$	F_3	Input	Log. $2Q \rightarrow Q$	Q_3	Input	L
H	H	L	L	C	Two's Complement Divide	$F = S + R + C_n$ if $Z = L$ $F = S - R - 1 + C_n$ if $Z = H$	Log. $2F \rightarrow Y$	$\overline{R_3 \forall F_3}$	F_3	Input	Log. $2Q \rightarrow Q$	Q_3	Input	L
H	H	H	L	E	Two's Complement Divide, Correction and Remainder	$F = S + R + C_n$ if $Z = L$ $F = S - R - 1 + C_n$ if $Z = H$	$F \rightarrow Y$	F_3	F_3	Hi-Z	Log. $2Q \rightarrow Q$	Q_3	Input	L

NOTES: 1. At the most significant slice only, the C_{n+4} signal is internally gated to the Y_3 output.
2. At the most significant slice only, $F_3 \forall$ OVR is internally gated to the Y_3 output.
3. At the most significant slice only, $S_3 \forall F_3$ is generated at the Y_3 output.
4. Op codes 1, 3, 7, 9, B, D, and F are reserved for future use.

L = LOW
H = HIGH
X = Don't Care
Hi-Z = High Impedance
$\forall$ = Exclusive OR
Parity = $SIO_3 \forall F_3 \forall F_2 \forall F_1 \forall F_0$

Figure 3.43. 2903 extended operations (Copyright 1978 Advanced Micro Devices, Inc. Reproduced with permission of copyright owner.).

the current value on the Z line. Also notice that, for some operations, the SIO3 pin on the MSS has an unusual output.

To give the reader an impression of these operations, the division operation will be illustrated. Since division requires the use of the sign-compare flipflop, this device must be discussed first. On each 2903, there exists a circuit in the form of Figure 3.44. Only the sign-compare flipflop on the MSS has significance. Its value defines R_3 exclusive-OR F_3 on the previous clock cycle, that is, its value is 1 if the sign of the value on the ALU's R-port was unequal to the sign of the ALU output during the previous ALU operation. Z is fed the inverse of the value in the SCFF. The SCFF is only altered during two of the extended operations (double-length normalization and two's-complement division).

The 2903 was designed for a four-quadrant nonrestoring division algorithm (meaning that it can function with negative and/or positive dividends and divisors). Although the specifications for the 2903 do not explicitly say so, the division functions were designed for floating-point number systems, that is, for fractional numbers. To divide, three extended operations are used. For a 16-bit design (four 2903s), the double-length normalization operation is performed in the first cycle (it determines the sign of the quotient), the divide operation (L_{8-5} = HHLL) is performed for the next 14 cycles, and the division-correction operation is performed for one cycle (it completes the computation of the remainder and provides bias correction by forcing the least-significant bit in the quotient to 1). The resultant quotient times the divisor, plus the remainder, will always equal the dividend.

The division algorithm requires the divisor to be placed in one register (R0 will be used here), the dividend to be placed in another register (R1 will be used), and the Q register set to zero. (If the dividend is double-length, the most-significant half is placed in R1 and the other half in Q). The algorithm also requires that the absolute value of the divisor exceed the absolute value of the dividend. If this condition is not true, the values must be scaled (shifted) before performing the algorithm. At its completion, the quotient is found in the Q register and the register originally containing the dividend (R1 in our case) contains the value of the remainder times 2^{n-1} (where n is the number of bits in the quotient). (If the reader plans to refer to the manufacturer's literature on the 2903, note that its description of division is misleading in places, incorrect in others.)

Figure 3.45 summarizes the three 2903 extended operations needed for division and shows the necessary pin connections for two slices. Notice, in the second dia-

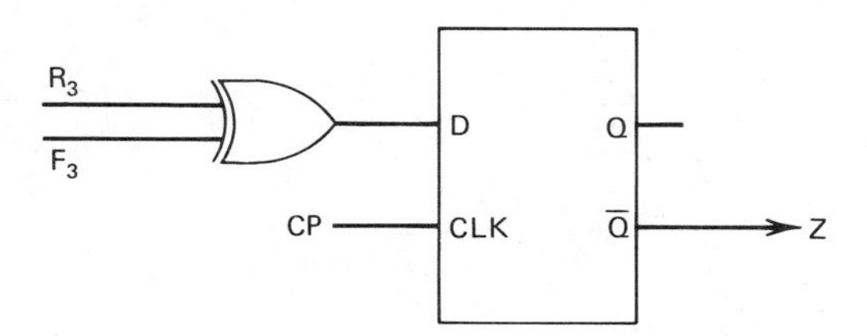

Figure 3.44. 2903 sign-compare flipflop.

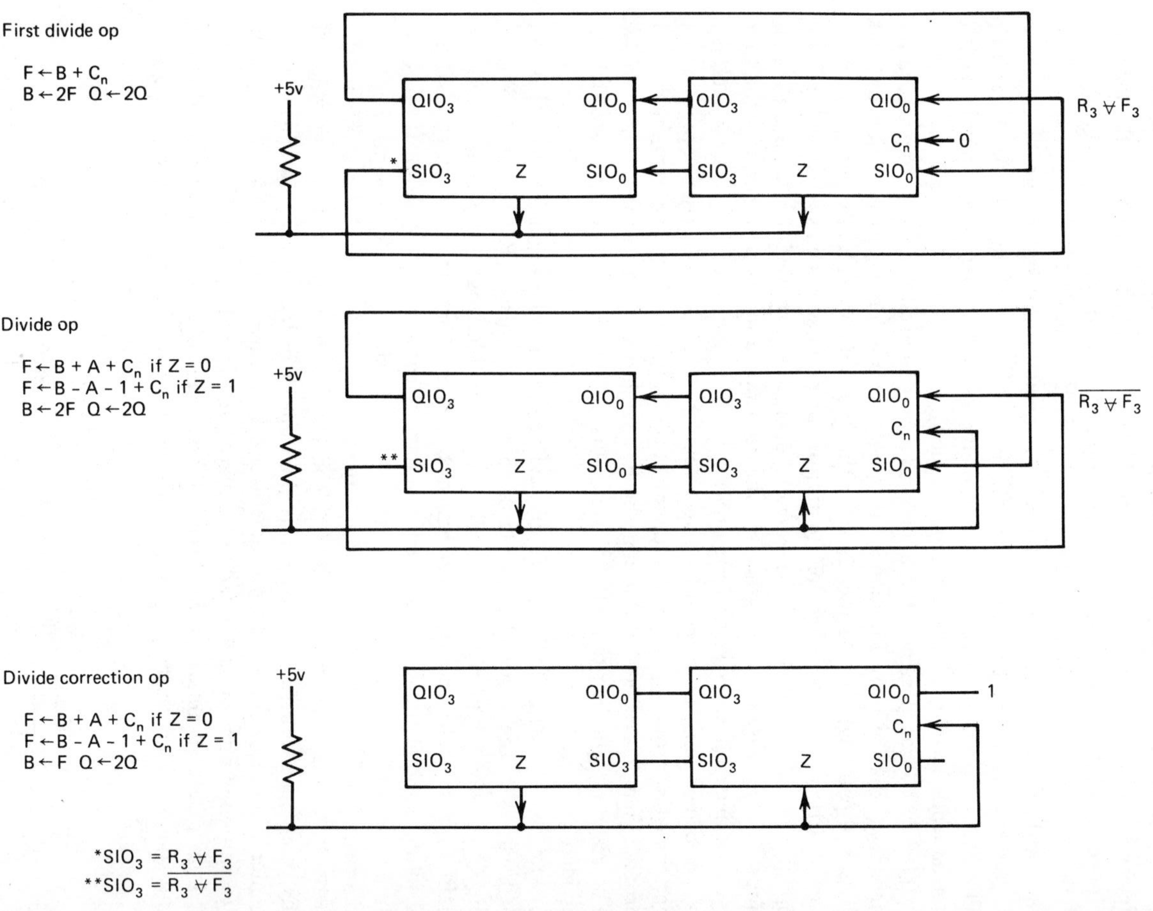

Figure 3.45. 2903 interconnections and operations for division.

gram, that Z is an output from the MSS, feeding the Z pins of the other slices. In this case, Z also feeds C_n of the LSS, meaning that if Z=0, the addition is B+A+0, and If Z=1, the subtraction is B−A−1+1.

Since division is somewhat difficult to understand, a simple example is presented. Assume an eight-bit processing section, a dividend of 00101100 (or 0.0101100, or 0.34375) in register R1, and a divisor of 10100100 (or 1.0100100, or −0.71875) in register R0. Zero should be placed in the Q register. Assuming the microinstruction sequencing logic in the processor permits a microinstruction to repeat itself (i.e., branch to itself) a specified number of times, only three microinstructions are needed

1. Double-length normalization operation,
   R0 = A register, R1 = B register
2. Divide operation,
   R0 = A register, R1 = B register
3. Divide correction operation,
   R0 = A register, R1 = B register

The microinstructions would also have to control logic (multiplexers) to make the connections shown in Figure 3.45. (Advanced Micro Devices provides a separate chip that can do this: the Am2904.) The second microinstruction would be performed six times, leading to a division time of eight cycles.

Table 3.3 is a trace of the 2903 through these cycles. At the end, Q = 1.1000011, or −0.4765625. Recalling that R1, at the end, contains the value of the remainder times 2^7, the remainder is 0.0012207. Checking the result, if one multiplies the quotient −0.4765625 by the divisor −0.71875 and adds this to the remainder 0.0012207, the result is 0.34375, the dividend.

Integer division, the more common form of division, is more difficult on the 2903. The reason is that several correction steps are need on the quotient and remainder, the steps being determined by the signs of the dividend and quotient

TABLE 3.3. 8-bit Division Operation

Operation	ALU Op	F bus	R1 (B)	Q	SCFF	Z on Next cycle
start			00101100	00000000	X	X
A (first divide)	+	00101100	01011000	00000001	1	0
C (divide)	+	11111100	11111000	00000011	0	1
C	−	01010100	10101000	00000110	1	0
C	+	01001100	10011000	00001100	1	0
C	+	00111100	01111000	00011000	1	0
C	+	00011100	00111000	00110000	1	0
C	+	11011100	10111000	01100001	0	1
E (divide corr.)	−	00010100	00010100	11000011	X	X

and the value of the remainder. If one is not interested in obtaining a valid remainder, the process is the following. Force the dividend to be positive, shift it left once, and place this value in Q. Store zero in the other dividend register (R1 in the example above). Perform a double-length normalization operation once and the divide operation n-1 times (where n is the width of the registers). At this point, if the quotient (in Q) is negative, increment it by one. Finally, if the dividend was negative at the start, take the two's complement of Q.

If one desires a valid remainder also, the process is more complicated.

The other extended operations are much more straightforward. One can multiply two n-bit two's-complement numbers in n cycles. The normalization operations are used to shift a fractional number left until the bit to the right of the sign bit is unequal to the sign bit. Examining single-length normalization, the value to be normalized is placed in Q. Looking at the status outputs in Figure 3.42 for this operation (I_{8-5}=1000), one sees that C_{n+4} of the MSS has the value Q_3 exclusive-OR Q_2 (indicating whether normalization has occurred) and Z indicates whether the entire value is zero. Figure 3.43 indicates that this operation can also increment a register (if C_n is set to 1), meaning that one can write a one-microinstruction loop to normalize a number and count the number of shifts needed to do so.

2903 Timing

The timing of the 2903 is similar to that of the 2901. When the clock input is low, the latches hold the selected values from the register array and data on the Y bus can be written into the register array. When the clock input rises, the Q register and sign-compare flipflop are loaded. Figure 3.46 contains representative set-up and propagation times. One can see that the 2903 is slower than the 2901A.

From input	Minimum set-up time to clock rise (ns)
Y	18
QIO	20

Note: All other set-up times are the propagation time to stable Y output plus the Y set-up time.

Maximum propagation times (ns)								
From input \ To output	Y	DB	C_{n+4}	$\overline{G}, \overline{P}$	N	OVR	$\overline{WRITE}$	SIO_0 SIO_3
DA, DB	62	—	59	49	63	89	—	74
C_n	40	—	34	—	33	62	—	41
A, B	99	53	95	86	99	111	—	105
I	70	—	69	51	70	84	28	81
$\overline{OEB}$	—	31	—	—	—	—	—	—
SIO_0 SIO_3	26	—	—	—	—	—	—	29
Clock	89	36	90	62	86	112	—	97

Figure 3.46. Representative Am2903 guaranteed times.

2903 Three-Register Operation

An advantage of the 2903 over many of the other ALU/register slices is its dual-port register array, allowing one to perform operations of the form R4=R4+R5. In addition, if desired, the 2903 can be expanded to permit operations on three registers, such as R4=R5+R6. In other words, rather than having two 4-bit fields in the microinstruction, A- and B-register select, the microinstruction might have three fields: A-register select, B-register source, and B-register destination.

One way to accomplish this is to time-multiplex the latter two fields into the single set of B-register-select lines in the 2903 while controlling the $\overline{\text{WE}}$ input. When the clock signal is high, the B-register-source field is gated to the B-register-select lines, causing the designated register to be used as a source for the ALU. After the clock signal drops to its low value, the value of this register is contained in the B latch. At this time, the other B field in the microinstruction is gated to the B-register-select lines on the 2903(s) and $\overline{\text{WE}}$ is set to 0, allowing the ALU output to be loaded into a third register. (The manufacturer recommends using $\overline{\text{IEN}}$ for this function instead of $\overline{\text{WE}}$, and tying all $\overline{\text{WE}}$ inputs to the $\overline{\text{WRITE}}$ output on the LSS.)

Adding Additional Registers

Another advantage of the 2903 (one not present in the 2901) is the ability to add external register arrays to allow one to address more than 16 registers, or to allow one to design a processor with multiple banks of registers (e.g., to avoid saving and restoring registers when processing interrupts in the machine-language programs). In the first case, the size of the A and B fields in the microinstruction would be expanded (e.g., to five bits if 32 registers are to be addressed). In the second case, an additional register (a 3-bit register, if there are to be eight banks of 16 registers) is used to supply the extra addressing bits.

The necessary prerequisites for adding additional registers to an ALU/register slice are (1) external busses than can be used to gate inputs and outputs in precisely the same ways that a register in the on-chip array can, and (2) control inputs that allow one to enable or disable the fetching from, and loading into, the on-chip registers. The 2903 interfaces meet these requirements. The 2903 WRITE output signal from the LSS also provides another useful feature; it tells us when an operation is being performed that loads a value into the register array. Without this, one would be faced with the task of decoding such a signal from the I lines using external logic.

Advanced Micro Devices produces a chip of 16 4-bit wide registers (Am29705) that is useful in expanding the number of registers. It is illustrated in Figure 3.47. To expand a 2903 design to 32 registers, one of these is needed per 2903 chip. Figure 3.48 shows the interconnection of such an external register array to a 2903 slice. It assumes that the microinstruction contains two 5-bit A- and B-register-select fields. Registers 00000-01111 are on the 2903; registers 10000-11111 are on

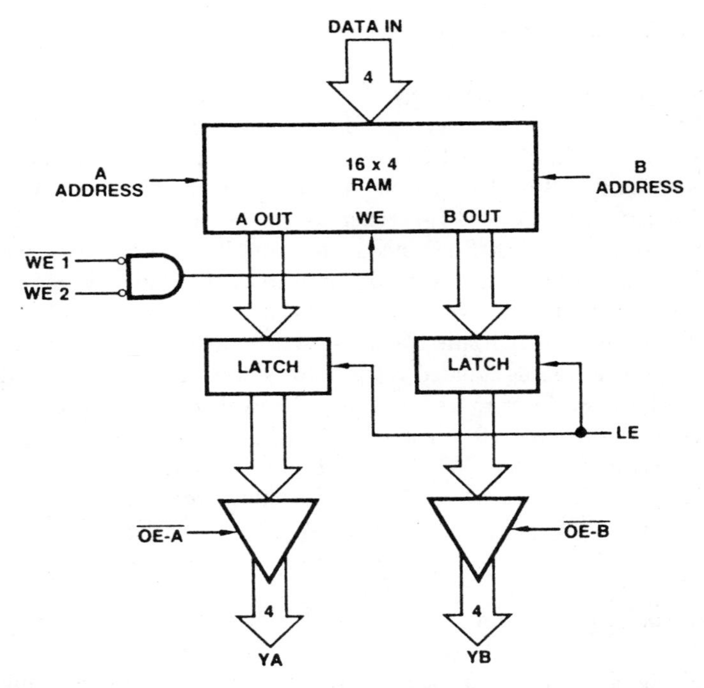

Figure 3.47. Am29705 external register array (Copyright 1978 Advanced Micro Devices, Inc. Reproduced with permission of copyright owner.).

the Am29705. When a 2903 operation is performed that stores a value into the B register, $\overline{\text{WRITE}}$ on the LSS is 0. This, in conjunction with the first bit of the B-register-select field and the clock, is used to enable a write operation into a register on either the 2903 or the external register array.

The Am2903A

At the time of the writing of this book, Advanced Micro Devices has a 2903A slice under development. The 2903A will be a faster version of the 2903. In addition, it will contain some functional changes, such as making the DA bus bidirectional and adding BCD arithmetic to the ALU.

THE 74S481/74LS481 SLICE

The 74S481 is a Schottky TTL, 4-bit wide, ALU-register slice produced by Texas Instruments [10]. The 74LS481 has an identical organization, but it is a low-power Schottky device. Both are packaged in a 48-pin quad-inline package (four rows of pins).

The organization of the 74S481 is shown in Figure 3.49. Comparing it to the previous slices discussed, the 74S481 is similar in many ways to the SBP0401 and 3002. Like the SBP0401, it has WR and XWR registers and associated shifters. Shifting operations are basically the same as in the SBP0401 (Figure 3.34). Unlike the SBP0401, the 74S481 has no register array; instead, it has an additional bidirectional bus (BI/O) that, when used in conjunction with the ALU latches and other busses, allows one to use an external register array.

Like the 3002 slice, the 74S481 has a masking device on one input to the ALU, allowing one, for instance, to AND a value on the BI/O bus with a register value before entry to the left port of the ALU. Also, the MC and PC registers are similar to the MAR and AC registers in the 3002. Like the MAR register in the 3002, the MC register cannot be internally gated through the ALU; it can only be gated to an output bus.

In addition to the shifters associated with the WR and XWR registers, the ALU output passes through a third shifter. Also, the MC and PC registers are counters, meaning that they can be independently incremented while the ALU is being used for another purpose. The manufacturer's depiction of the 74S481 is shown in Figure 3.50.

Figure 3.51 illustrates the external connections of the slice. Notice the POS input. The slice uses relative position control, but uses only one input pin to indi-

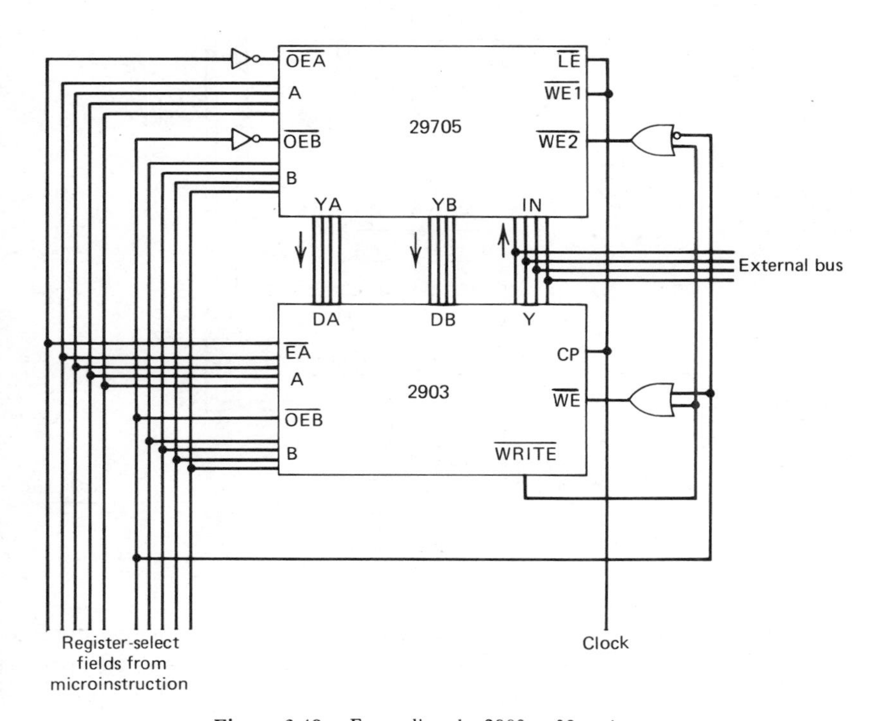

Figure 3.48. Expanding the 2903 to 32 registers.

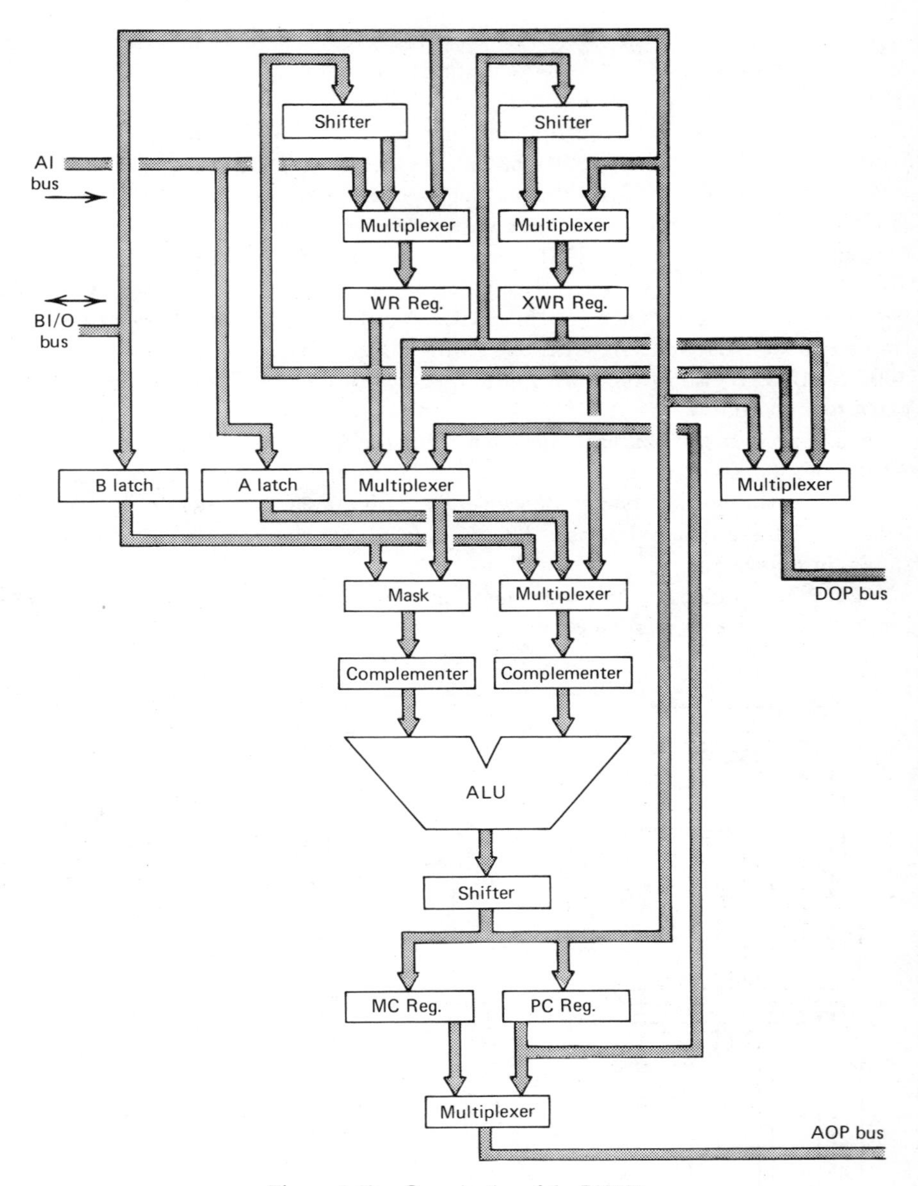

Figure 3.49. Organization of the 74S481.

cate the three possible states. A slice is defined as a LSS if the voltage at the POS pin is less than 0.8V. If the voltage is between 1.8V and 3.0V, the slice operates as an IS. If the voltage at the POS pin is greater than 3.6V, the slice operates as the MSS. The relative position of a slice influences shifting operations and interconnections, and influences the definition of some of the output pins.

The slice has 10 function-control pins, but notice that two of them, OP8 and OP9, are bidirectional. For most operations, OP8 and OP9 are input pins. However, like the 2903, the 74S481 performs a set of extended operations (e.g., multiplication and division iterations). During some of these operations, OP8 and/or OP9 are not used as inputs; rather, they become status outputs.

Examining some of the other inputs, the D lines specify which of three values should be routed to the DOP bus, or whether DOP should be held in a high-impedence state. INCMC and INCPC allow one to independently increment the MC and PC registers. On the LSS, the CCI input specifies whether the increment is one or two. On all other slices, CCI specifies an increment of 0 or 1 and is normally connected as a ripple-carry input. The BIOSEL signal controls the direction of the BI/O bus. When 0, BI/O is an output, containing the output of the ALU shifter. When 1, BI/O becomes an input bus feeding the B latch.

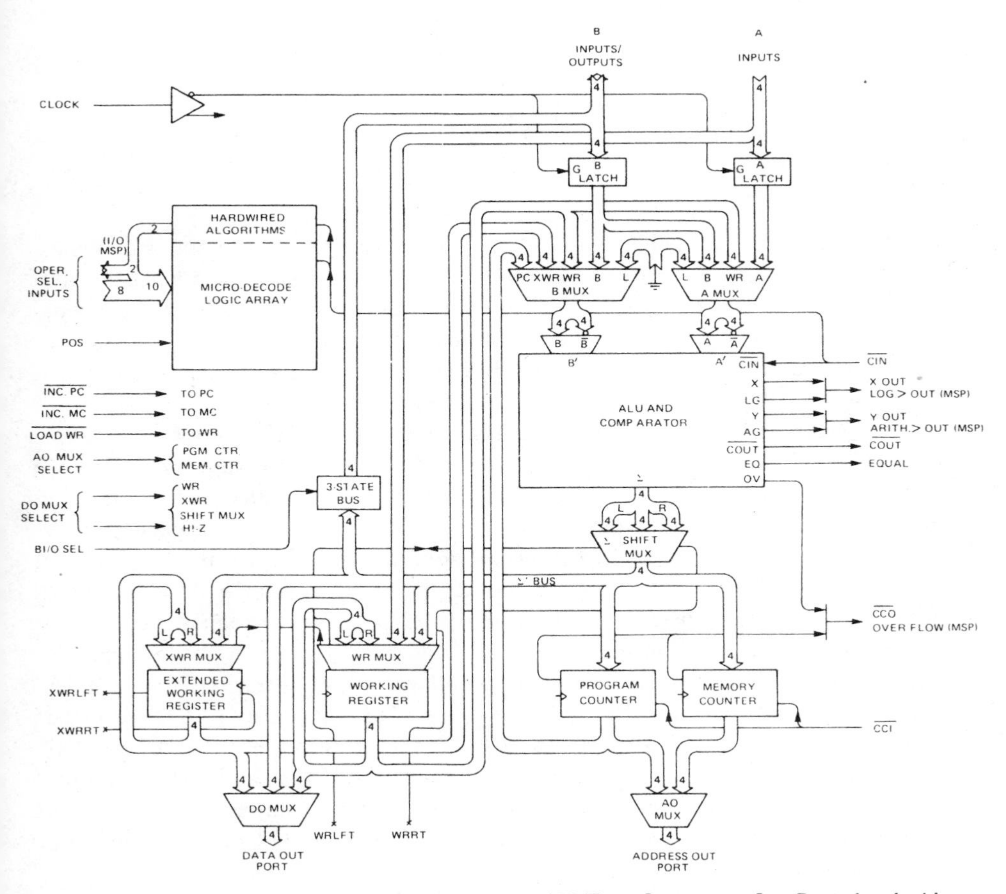

Figure 3.50. Organization of the 74S481 (Copyright 1977 Texas Instruments Inc. Reproduced with permission.).

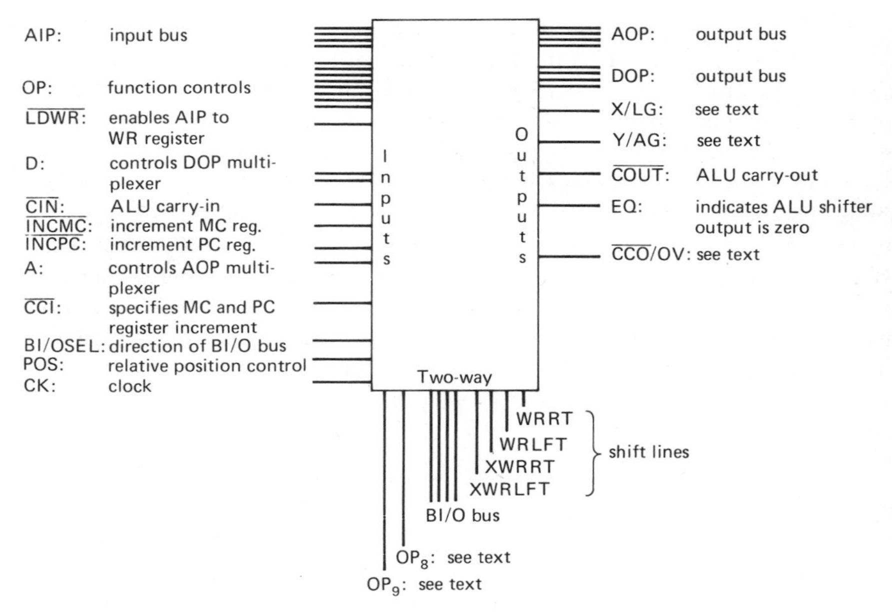

Figure 3.51. 74S481 external connections.

On all but the MSS, the X/LG and Y/AG outputs represent the carry-lookahead signals. Since the carry-lookahead signals from an MSS are not normally used, the pins have a different interpretation on the MSS. Y/AG represents "arithmetic greater than." For most ALU operations, Y/AG on the MSS is 1 if the ALU output is greater than zero. For the comparison ALU operations (internally, a subtraction with the result gated nowhere), Y/AG is 1 if one specified ALU input is arithmetically greater than the other. For all ALU operations except comparison, X/LG on the MSS is 1 if the ALU output is not zero. For the comparison operations, X/LG is 1 if one specified input is logically greater than the other. (A logical comparison means that the values are viewed as unsigned positive numbers.) The EQ pin is an open-collector output, permitting a wire-NOR across cascaded slices. For all but comparison operations. EQ is 1 if the ALU output is zero. For comparison operations, EQ is 1 if both ALU inputs are equal.

The remaining multifunction output is CCO/OV. On the MSS, it indicates whether the ALU arithmetic operation overflowed or, on an arithmetic left shift operation, whether an overflow is about to occur. On other slices, CCO/OV is the carry output of both counters (PC and MC) and is normally fed into the CCI input of the next slice to cascade the counters. The INCMC and INCPC inputs appear to allow one to increment neither, either, or both counters, but since both counters share a single carry output, only one counter should be incremented in a single clock cycle.

Note that although there are three shifters, there are only four bidirectional shift pins. The WR and ALU shifters share the WRRT and WRLFT pins, since the encoding of the function controls does not permit these two shifters to be operated simultaneously.

Also note that the definitions of some of the status outputs change when the slice's extended operations are used.

74S481 Functions

The 74S481 performs a large set of operations, which are controlled by the 10 OP inputs. (The CIN input also serves as an 11th control input in some circumstances.) The number of operations is perhaps *too* large, because they are encoded in the OP signals in a rather disorganized way, making the slice difficult to understand.

The control inputs are summarized in Figure 3.52. Operation-types IA (row 1) designate the addition or subtraction of two values. Lines OP3 and OP4 specify the input to the A port of the ALU: The AI bus, BI/O bus, the WR register, or ones. Lines OP5, OP6, and OP7 designate eight sources for the B port, three of which involve first ANDing the value on the BI/O bus to a register value. Lines OP8 and OP9 control the complementers on both ALU ports. The ALU output can be obtained on the DOP bus.

Operation-types IB are similar, except that one can also gate the ALU output to one of the four registers on the slice. Operation-types II allow one to gate the AI or BI/O busses into the A port, BI/O or WR into the B port, do an addition or subtraction, put the ALU result in WR, and do a double-length shift of WR and XWR. Operation-types III represent an addition followed by a single-length left or right, arithmetic or logical, shift into MC or XWR. Operation-types IV take the AI bus, pass it through the ALU, and perform a left or right arithmetic shift, logical shift, or circulate (circular shift). Notice that CIN is a control input here, specifying the direction of the shift.

Operation-types V allow one to shift either WR or XWR, and operation-types VI allow one to perform a double-length shift of WR and XWR. Operation-types VII are the comparison operations referred to earlier. It is difficult to see why the designers bothered to include these, since an arithmetic comparison is equivalent to a subtraction with no destination, allowing one to test the conditions equal-to-zero and greater-than-zero. However, this is provided in operation-types I. In fact, operation-types I provide more flexibility for comparisons. For instance, they allow one to compare BI/O with PC and BI/O to XWR; these combinations are not provided in operation-types VII.

Operation-types VIII represent the logical operations. Given the ability via OP8 and OP9 to complement either, neither, or both ALU inputs, an extensive set of logical operations is available.

74S481 Extended Functions

In a manner similar to that of the 2903, the 74S481 performs a set of extended operations, in particular,

OPERATION FORM		COMMAND FORMAT											TEST OUTPUTS							
												LSP	MSP		ALL		$\overline{MSP}$		MSP	$\overline{MSP}$
NO.	OPERATION	OP0 (7)[1]	OP1 (8)	OP2 (9)	OP3 (10)	OP4 (17)	OP5 (14)	OP6 (13)	OP7 (11)	OP8 (15)	OP9 (16)	$\overline{CIN}$ (18)	LG (21)	AG (20)	EQ (23)	$\overline{COUT}$ (22)	X (21)	Y (20)	OVFL (37)	$\overline{CCO}$ (37)
IA	(±A ±B + $\overline{CIN}$) → Σ' BUS ONLY[2]	H	L	L	A SOURCE		B SOURCE			A' FNCT	B' FNCT	L = CARRY H = NO CARRY	Σ'≠ZERO	Σ'>ZERO	Σ'=ZERO	$\overline{COUT}$	X[4]	Y[4]	OVFL	$\overline{CCO}$
IB	(±A ±B + $\overline{CIN}$) → REGISTER	L	REGISTER LL = Σ' → WR LH = Σ' → XWR HL = Σ' → PC HH = Σ' → MC		LL = AI → A LH = H'S → A HL = BI → A HH = WR → A		LLL = BI → B LLH = H'S → B LHL = BI · WR → B LHH = WR → B HLL = BI · XWR → B HLH = XWR → B HHL = BI · PC → B HHH = PC → B			L = A → A' H = Ā → A'	L = B → B' H = B̄ → B'									
IIA	(A + B + $\overline{CIN}$) ⇆ WR, XWR[2]	H	H	H	L	H	L	FUNCT L = A → A' H = Ā → A'	B' SRC L = BI → B' H = WR → B'	A' SRC L = AI → A' H = BI → A'	SHIFT L = LFT H = RT	L = SUB H = ADD	Σ'≠ZERO	Σ'>ZERO	Σ'=ZERO	$\overline{COUT}$	X	Y	OVFL	$\overline{CCO}$
IIB	(B − A − 1) ⇆ WR, XWR	H	H	H	L	H	L													
III	(A + B + $\overline{CIN}$) ⇆ REGISTER	H	H	L	H	A' SRC L = AI → A' H = BI → A'	REGISTER L = Σ' → MC H = Σ' → XWR	B' SOURCE LL = BI → B' LH = WR → B' HL = XWR → B' HH* = L'S → B'		SHIFT L = LOG H = *ARITH	TYPE L = LFT H = RT	L = CARRY H = NO CARRY	Σ'≠ZERO	Σ'>ZERO	Σ'=ZERO	$\overline{COUT}$	X	Y	OVFL	$\overline{CCO}$
													*DURING ARITHMETIC SHIFTS A' + C IS COMPARED TO −1							
													Σ'≠ −1	Σ'> −1	Σ' = −1	$\overline{COUT}$	X	Y	SHIFT OVFL	$\overline{CCO}$
IV	AI ⇆ Σ' BUS	H	H	H	L	H	H	REG OR AI LL = AI ⇆ Σ'		SHIFT LL = LOG LH = ARITH HL = ROTATE HH (NOT DEFINED)		TYPE L = LFT H = RT	AI≠ZERO	AI>ZERO	AI=ZERO	$\overline{CIN}$	X	Y	L (FOR LSA OVFL)	$\overline{CCO}$
VA	WR ⇆ WR	H	H	H	L	H	H	LH = WR ⇆ WR												
VB	XWR ⇆ XWR	H	H	H	L	H	H	HL = XWR ⇆ XWR												
VI	WR, XWR ⇆ WR, XWR	H	H	H	L	H	H	HH=WR,XWR→WR,XWR												
VIIA	A:B (N1:N2)	H	H	H	L	L	B' SOURCE (SAME AS FORM I ABOVE)			A' SRC L = AI → A' H = WR → A'	OPER L = A:B H = B:A	H	N1>N2	N1>N2	N1=N2	=LG	X	Y	–	$\overline{CCO}$
VIIB	B:A (N1:N2)	H	H	H	L	L														

				FUNCTION LHL = NOR LHH = OR HLL = XOR			A' SRC L = AI → A H = WR → A	REG† LL = WR LH = XWR HL = PC HH = Σ'	B SOURCE LL = BI → B LH = WR → B HL = XWR → B HH = PC → B		A' FNCT L = A → A' H = $\bar{A}$ → A'	B' FNCT L = B → B' H = $\bar{B}$ → B'	REG† (SEE UNDER OP5 COLUMN)	Σ≠ZERO	Σ'>ZERO	Σ'=ZERO	=$\overline{CIN}$	X	Y	L	$\overline{CCO}$
VIIIA	NOR/AND LOGICAL OPERATIONS		H																		
VIIIB	OR/NAND LOGICAL OPERATIONS		H																		
VIIIC	EX OR/EX NOR LOGICAL OPERATIONS		H																		
IX	NO OPERATION (ZERO → Σ' BUS)		H	H	H	H	H	H or L	H or L	H or L	H or L	H or L	H or L	L	L	H	=$\overline{CIN}$	X	Y	L	$\overline{CCO}$
X	CRC ACCUMULATION		H	H	H	H	L	L	L	L	L	O/I[5]	H	–	–	–	–	X	Y	L	$\overline{CCO}$
XI	SIGNED INTEGER DIVIDE	A. START	H	H	H	H	L	L	H	L	H	H	H	–	–	DIV=ZERO	–	–	–	–	–
		B. ITERATE (N-1 CLKS)	H	H	H	H	L	H	L	H	O/I[6]	O/I[6]	H	–	–	–	–	–	–	–	–
		C. ITERATE FINISH	H	H	H	H	L	H	H	L	O/I[6]	O/I[6]	H	–	–	–	–	–	–	–	–
		D. FIX REMAINDER	H	H	H	H	L	H	L	L	O/I[6]	O/I[6]	H	–	–	–	–	–	–	–	–
		E. ADJUST QUOTIENT	H	H	H	H	L	L	H	H	O/I[6]	O/I[6]	H	–	–	–	–	–	–	–	–
XII	UNSIGNED DIVIDE	A. START	H	H	H	H	L	L	L	L	H	H	L	–	DIV OVFL	–	–	–	–	–	–
		B. ITERATE (N-1 CLKS)	H	H	H	H	L	L	L	H	H	O/I[6]	L	–	–	–	–	–	–	–	–
		C. FINISH	H	H	H	H	L	L	L	H	L	O/I[6]	L	–	–	–	–	–	–	–	–
XIII	UNSIGNED MULTIPLY		H	H	H	H	L	L	H	L	L	O/I[6]	H	–	–	–	–	–	–	–	–
XIV	SIGNED INTEGER MULTIPLY		H	H	H	H	L	H	H	H	O/I[6]	O/I[6]	* H	–	–	–	–	–	–	–	–

AO SEL (42)
- L MC
- H PC

DO1 SEL (29)
DO2 SEL (30)
- LL Σ' BUS
- LH WR
- HL XWR
- HH HI-Z

BI/O SEL (29)
- L OUTPUT
- H INPUT

$\overline{LD WR}$ (24)
- L AI → WR
- H NO LOAD

$\overline{INC MC}$ (35)
- L INC
- H HOLD

$\overline{INC PC}$ (43)
- L INC
- H HOLD

$\overline{CCI}$ (44)
- LSP L: x 1
- LSP H: x 2
- MID OR MSP
- L CARRY
- H NO CARRY

POS (19)
- 0 V = MID
- 2.4 V = LSP
- 5 V = MSP

	DATA PORTS				
PIN NUMBER	AI	BI/O	DOP	AOP	BIT (2^n)
	(6)	(46)	(34)	(38)	0 (LSB)
	(5)	(47)	(33)	(39)	1
	(4)	(1)	(32)	(40)	2
	(3)	(2)	(31)	(41)	3 (MSB)

PIN ASSIGNMENTS

WRRT (26)	CK (45)
WRLFT (25)	V_{CC} (12)
XWRRT (28)	GND (36)
XWRLFT (27)	

NOTES:
1. NUMERALS IN PARENTHESIS ARE PIN NUMBERS
2. → DESTINED FOR ⇝ SHIFTED AND DESTINED FOR
3. H = HIGH VOLTAGE LEVEL, L = LOW VOLTAGE LEVEL
4. X AND Y ARE CARRY LOOK-AHEAD FUNCTIONS
5. O IS OUTPUT ON LSP, I IS INPUT ON $\overline{LSP}$
6. O IS OUTPUT ON MSP, I IS INPUT ON $\overline{MSP}$
7. VOLTAGE VALUES ARE NOMINAL

1. Cyclic-redundancy-character (CRC) accumulation iteration.
2. Signed integer division iteration, including two extended functions to adjust the quotient and remainder.
3. Unsigned integer division iteration.
4. Unsigned multiplication iteration.
5. Signed multiplication iteration.

Each 74S481 slice contains two flipflops, similar to the sign-compare flipflop in the 2903. Only the flipflops in the MSS are active. Examining the OP8 and OP9 pins in Figure 3.52 when an extended operation (e.g., signed division) is active, one sees that, on the MSS, they are outputs; on the other slices, they are inputs. These function much the same way as the Z pin does in the 2903; they are used by the MSS to communicate information to the other slices, information concerning the type of operation to be performed during the next cycle.

Unlike the 2903, signed division in the 74S481 is oriented toward integer division. Division proceeds as that described earlier for the 2903, but it is then followed by a one-cycle operation to correct the remainder (in WR) and another one-cycle operation to correct the quotient (in XWR). As an illustration, the last operation performs one of the following four operations, depending on the values of OP8 and OP9

$$XWR = -XWR - 1$$
$$XWR = -XWR$$
$$XWR = XWR$$
$$XWR = XWR + 1$$

This makes the 74S481 more convenient for performing integer division than the 2903, since integer division on the 2903 requires several additional cycles to adjust the remainder and quotient.

74S481 Timing

As is the case for the other slices, the clock controls the registers, counters (PC and MC), and ALU latches. When the clock input goes low, the A and B latches close. When the clock input rises, data, if specified by the control inputs, is gated into the registers (WR, XWR, PC, or MC), and PC or MC is incremented if specified by the control inputs. Figure 3.53 contains representative set-up and propagation times. The upper number in each entry is for the 74S481; the lower number is for the 74LS481.

THE 6701 SLICE

The 6701 ALU/register slice was one of the original bit slices, having been introduced by Monolithic Memories Inc. in 1974 [11]. It is a 4-bit Schottky-TTL

From input	Minimum set-up times to clock rise (ns)
AI, BI/O	65
	80
OP	90
	120
CIN	40
	50
CCI, INCMC, INCPC, LDWR	30
	40
Shift pins	30
	40

From input \ To output	Maximum propagation times (ns)								
	DOP	AOP	BI/O	COUT	X, Y	CCO/OV	EQ	AG, LG	Shift outputs
AI, BI/O	65	—	70	40	45	45	65	80	65
	80	—	100	55	55	65	85	105	85
CIN	60	—	55	40	—	30	—	—	50
	75	—	75	50	—	40	—	—	60
OP	100	—	—	80	80	90	80	105	105
	130	—	—	115	115	120	115	135	140
CCI	—	—	—	—	—	50	—	—	—
	—	—	—	—	—	70	—	—	—
BI/OSEL, D	30	—	30	—	—	—	—	—	—
	40	—	30	—	—	—	—	—	—
A	—	30	—	—	—	—	—	—	—
	—	40	—	—	—	—	—	—	—
CLOCK	65	45	65	65	—	70	—	—	65
	85	55	85	85	—	100	—	—	85

Figure 3.53. Representative 74S481/74LS481 guaranteed times.

slice in a 40-pin DIP. Since the 6701 and 2901 are almost identical, the 6701 is not discussed in detail; only its differences from the 2901 are discussed.

The organization of the 6701 is almost identical to that of the 2901. Referring to Figure 3.1, the only differences are

1. The ALU S-port multiplexer has two, rather than three, inputs in the 6701. The A latch cannot be routed to the ALU S port.
2. The output bus multiplexer has three, rather than two, inputs in the 6701. The B latch can be routed directly to the output bus.

The external connections are also almost identical to those of the 2901. The only differences (refer to Figure 3.4) are

1. The 6701 has eight, rather than nine, I (function control) inputs.
2. The 6701 has no F_3 status output.
3. The 6701 has an additional open-collector status output indicating whether the value on the F bus is 1111.

This leaves the 6701 one pin short, which is an unused pin.

Since the 6701 has one fewer control-input pins, one might expect that its major differences lie in the area of on-chip functions, which is the case. Basically, its functions are a subset of those of the 2901. Recall that the nine I controls in the 2901 are subdivided into three groups of three. The first group controls the ALU input multiplexers, the second specifies the ALU function, and the third controls the ALU destination and shifters. In the 6701, the last group is essentially identical, but the first two groups (six pins) in the 2901 are replaced with a single group of five pins in the 6701. This might imply that the 6701 performs 32 fewer ALU functions and input-source combinations, which is true, but several of these combinations in the 2901 are not useful (e.g., 0 AND Q, 0 OR Q, 0 AND B, D AND 0, etc.). Hence, the 6701 does perform fewer operations, but the number of useful eliminated operations is less than 32. For instance, the 6701 has no exclusive-NOR operation. The only AND operations are A AND B and D AND B (i.e., one cannot specify A AND Q or D AND A). Two subtraction combinations permitted in the 2901 (Q–D and D–A) are not provided in the encoding of control signals in the 6701.

6701 Timing

The 6701 has a guaranteed minimum clock-cycle time of 175 ns; representative set-up and propagation times are illustrated in Figure 3.54. Since it is approximately 75% slower than the 2901A, consumes slightly more power, and performs fewer functions than the 2901, the 6701 appears to have little utility today.

From input	Minimum set-up time to clock rise (ns)
A, B	185
D	70
C_n	70
I	120
Shift inputs	45

	Maximum propagation times (ns)						
To output → / From input ↓						Shift outputs	
	Y	C_{n+4}	$\overline{G}, \overline{P}$	F=0	OVR	RAM	Q
A, B	140	110	100	140	125	135	—
D	75	45	40	75	55	70	—
C_n	40	25	—	40	35	35	—
I	100	80	75	105	95	95	40
Clock	140	125	115	140	135	135	40

Figure 3.54. Representative 6701 guaranteed times.

THE 9405 SLICE

Another family of bit-slice devices is Fairchild's 9400 Macrologic series [12]. The less-dense components (less than 300 gates/chip) are low-power Schottky devices; the more-dense components (up to 1500 gates/chip) use I^3L (isoplanar integrated injection) logic, a variant of I^2L. The devices are also available in CMOS technology as the 4700 Macrologic series.

The heart of the 9400 family is the 9405 (or faster 9405A) ALU/register slice, illustrated in Figure 3.55. The 4-bit-wide slice consists primarily of an ALU with limited capabilities and a single-port array of eight registers. Note the absence of multiplexers and shifters; data routing and shifting functions must be supplied by external logic, for which the 9400 family has some additional components.

9405 External Connections

The 9405 has only 24 pins, and considerable use of multifunction pins was employed to achieve this. The external connections are shown in Figure 3.56. Notice that the 9405 uses relative position control; if input MSS is high, the slice is defined as the most-significant slice.

A glaring omission from Figure 3.56 is a carry-input pin. The 9405 has a carry input, but it is a multifunction pin within the I, or function-control, group. The functions specified by the I pins are listed in Figure 3.57. The only arithmetic function is addition. In the case of addition, I_0 is, in effect, the carry input. Hence one cascades the 9405 is the following way. Assuming a 3-bit control field exists in the microinstruction, it is connected to the I inputs of the least-significant 9405 and to only the I_1 and I_2 inputs of the other 9405 slices. The I_0 pin of each slice is connected to the W (carry) output pin of the preceding slice, or to the output of a carry-lookahead generator.

This is sufficient for the addition operation, but not for the other operations.

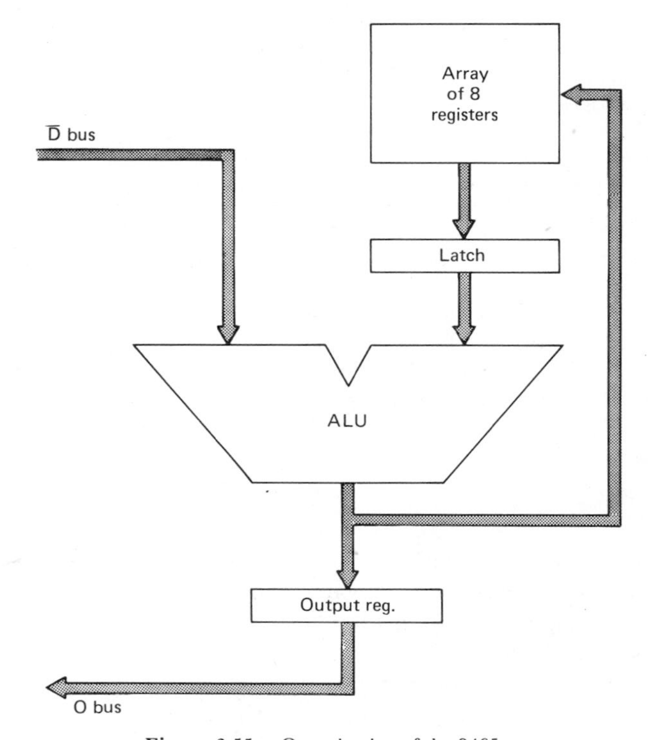

Figure 3.55. Organization of the 9405.

Since no carries are produced for the other (nonarithmetic) operations, the 9405 generates the appropriate signals on the W, X, and Y outputs to give the I_0 pin on the next slice the value of the third bit of the microorder. For these operations, the $\overline{W}$ output is equal to the slice's I_0 input, $\overline{X}$ is forced to 0, and $\overline{Y}$ is forced to 1. If ripple carry is being used, the third microorder bit is propagated throughout the

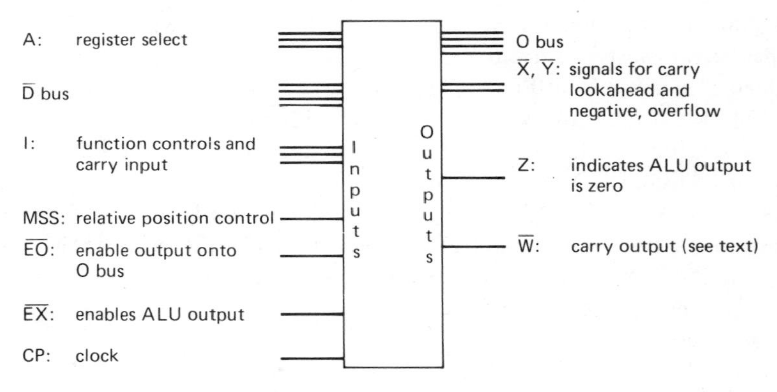

Figure 3.56. 9405 external connections.

I_2 I_1 I_0	INTERNAL OPERATION	
L L L	R_X plus $\overline{\text{D-Bus}}$ plus 1 → R_X	Accumulate
L L H	R_X plus $\overline{\text{D-Bus}}$ → R_X	Accumulate
L H L	$R_X \cdot \overline{\text{D-Bus}}$ → R_X	Logical AND
L H H	$\overline{\text{D-Bus}}$ → R_X	Load
H L L	R_X → Output Register	Output
H L H	$R_X + \overline{\text{D-Bus}}$ → R_X	Logical OR
H H L	$R_X \oplus \overline{\text{D-Bus}}$ → R_X	Exclusive OR
H H H	D-Bus → R_X	Load Complement

Figure 3.57. 9405 function controls (Reproduced with permission of Fairchild Camera and Instrument Corp.).

cascaded slices as a carry would be. If the I_0 pin is fed from a carry-lookahead generator, the forced values of $\overline{X}$ and $\overline{Y}$ cause the same to happen.

Like the 2903 and 74S481 slices, the 9405 takes advantage of the fact that the carry-lookahead outputs on the most-significant slice are not needed. On the most-significant slice, X indicates whether the ALU output is negative and Y indicates whether an overflow has occurred. Z is an open-collector output indicating a zero ALU result.

The $\overline{EX}$ input, when low, enables the ALU result into the output register and the register array. One can use this to select different registers as the source and destination registers in the following manner. When the clock input is high, the register value selected by the A inputs is gated through the latch into the ALU. One would hold $\overline{EX}$ high during this time. After the clock goes low, one would gate a different value to the A inputs (to select the destination register) and then hold the $\overline{EX}$ input low.

9405 Functions

As shown in Figure 3.57, the 9405 provides a very limited number of functions. To overcome this, external logic is needed. Another member of the family, the 9404 data-path switch, is provided for this purpose. The 9404 is a 4-bit combinatorial-logic device, combining the functions of a two-input multiplexer, a left/right shifter, a complementer, and a constant generator (+1, 0, −1, −2). The functions of the 9404 are illustrated in Figure 3.58.

The 9404 is usually placed on the 9405's input bus as shown in Figure 3.59. The 9404 provides an input shifter, and its complement operations allow one to subtract in the 9405.

9405 Timing

Figure 3.60 illustrates representative guaranteed times of the Fairchild 9405A slice. In analyzing the timing situations of a 9405-based design, one must be aware that part of the function-control signals (I_0) must propagate through the slices. Also, since the 9405 is usually used in conjunction with the 9404, one must

Inputs	Outputs	Function	Inputs	Outputs	Function
$I_4\ I_3\ I_2\ I_1\ I_0$	$O_3\ O_2\ O_1\ O_0$		$I_4\ I_3\ I_2\ I_1\ I_0$	$LO\ O_3\ O_2\ O_1\ O_0\ RO$	
L L L L L	L L L L	Byte mask	H L L L L	RI RI RI RI RI	K-bus sign extend
L L L L H	H H H H	Byte mask	H L L L H	$K_3\ K_3\ K_2\ K_1\ K_0$	K-bus sign extend
L L L H L	L L L H	Minus "2" in 2s comp(1)	H L L H L	RI RI RI RI RI	D-bus sign extend
L L L H H	L L L L	Minus "1" in 2s comp(1)	H L L H H	$D_3\ D_3\ D_2\ D_1\ D_0$	D-bus sign extend
L L H L L	$D_3\ D_2\ D_1\ D_0$	Byte mask, D-bus	H L H L L	$D_3\ D_2\ D_1\ D_0$ RI	D-bus shift left
L L H L H	H H H H	Byte mask, D-bus	H L H L H	$K_3\ K_2\ K_1\ K_0$ RI	K-bus shift left
L L H H L	$D_3\ D_2\ D_1\ D_0$	Byte mask, D-bus	H L H H L	LI $D_3\ D_2\ D_1\ D_0$	D-bus shift right
L L H H H	L L L L	Byte mask, D-bus	H L H H H	$D_3\ D_3\ D_2\ D_1\ D_0$	D-bus shift right arith(2)
L H L L L	L H H H	Negative byte sign mask	H H L L L	LI $K_3\ K_2\ K_1\ K_0$	K-bus shift right
L H L L H	H H H H	Positive byte sign mask	H H L L H	$K_3\ K_3\ K_2\ K_1\ K_0$	K-bus shift right arith(2)
L H L H L	$K_3\ K_2\ K_1\ K_0$	Byte mask, K-bus	H H L H L	$K_3\ K_2\ K_1\ K_0$	Byte mask, K-bus
L H L H H	L L L L	Byte mask, K-bus	H H L H H	H H H H	Byte mask, K-bus
L H H L L	$D_3\ D_2\ D_1\ D_0$	Load byte	H H H L L	$D_3\ D_2\ D_1\ D_0$	Complement D-bus
L H H L H	$K_3\ K_2\ K_1\ K_0$	Load byte	H H H L H	$K_3\ K_2\ K_1\ K_0$	Complement K-bus
L H H H L	H H H L	Plus "1"	H H H H L		Undefined (reserved)
L H H H H	H H H H	Zero	H H H H H		Undefined (reserved)

H = High Level L = Low Level (1) Comp = Complement (2) Arith = Arithmetic

Figure 3.58. 9404 function controls (Reproduced with permission of Fairchild Camera and Instrument Corp.).

consider the propagation delay added by the 9404. The longest path through the 9404 (I inputs to data and shift outputs) is 27 ns (maximum).

THE 4705 SLICE

As mentioned earlier, the Macrologic family also exists in CMOS technology, where the devices are labelled 47xx. The 4705 ALU/register slice is the CMOS equivalent of the 9405.

Since Figures 3.55 to 3.59 also describe the 4705 slice (and the 4704, the CMOS equivalent of the 9404), no further discussion of the 4705 is needed. However, because this is the first nonbipolar device discussed in the book, and because the reader is probably less familiar with CMOS logic than with TTL and ECL logic, a short discussion of CMOS is warranted.

CMOS devices, when compared to TTL devices, exhibit considerably lower power consumption, higher noise immunity, wider supply-voltage ranges, and slower speeds. Because of these characteristics, CMOS was originally developed for aerospace applications.

CMOS devices can operate over a supply-voltage range of approximately 3 to 15 V. Over this range, the system designer has a choice of speeds and power consumption. For instance, running a 4700-series CMOS device at 5 V yields a typical gate propagation delay of 40 ns and a quiescent power consumption of 10 nW per gate. At a supply voltage of 10 V, the per-gate characteristics are 20 ns and 20 nW. At a supply voltage of 15 V, the corresponding characteristics are approximately 15 ns and 100 nW.

One of the disadvantages of CMOS is its high output impedence, which makes the switching speeds of CMOS gates considerably more sensitive to load capaci-

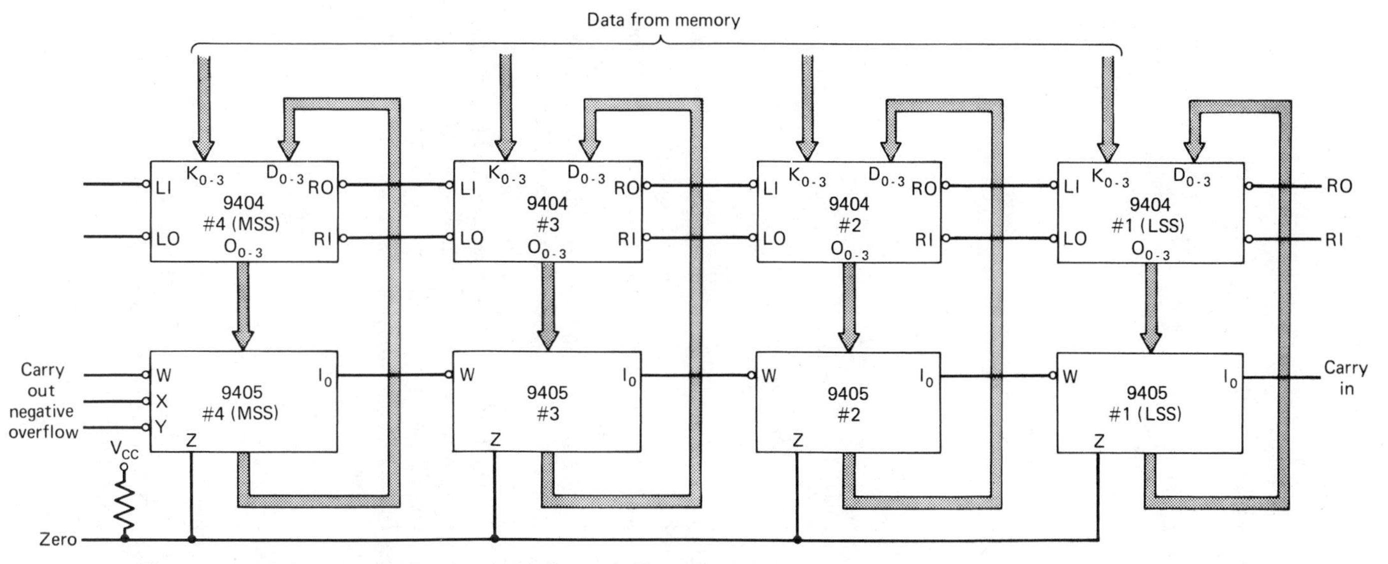

Figure 3.59. 9404 and 9405 slices in a 16-bit data path (Reproduced with permission of Fairchild Camera and Instrument Corp.).

From input	Minimum set-up times to clock rise (ns)
A	75
$\overline{EX}$	15
D	45
I_0	40^3 / 18^4
$I_{1,2}$	40

Notes: 1. MSS high
2. MSS low
3. Least-significant slice
4. Not least-significant slice

Max. propagation times (ns)				
From input \ To output	O	$\overline{W}$	$\overline{X}, \overline{Y}$	Z
D		40	54^1 / 40^2	60
A		60	75^1 / 58^2	75
I_0		22	40^1	40
$I_{1,2}$		18	55^1 / 30^2	60
EO	16			
CP	25			

Figure 3.60. Representative 9405 guaranteed times.

tances than TTL gates. The times in the above paragraph are measured at a load capacitance of 15 pF. At 150 pF, 5 V CMOS is about four times slower and 10 V CMOS is about three times slower, where TTL speeds would slow down by only a factor of two.

The high output impedence also makes CMOS gates considerably more sensitive to capacitively coupled noise. On the other hand, CMOS is much less sensitive to supply-voltage noise and to temperature variations (commercial operating-range temperatures for the CMOS 4700-series devices are –40°C to +85°C).

The 4705 is identical to the 9405 in all respects except electrical characteristics. Figure 3.61 lists a few representative characteristics. As shown, the 4705 running with a 10 V supply is about five to six times slower than the 9405A, but with a significantly lower power consumption.

THE F100220 SLICE

At the time of writing this book, Fairchild Camera and Instrument Corp. was in the process of introducing a new family of bit-slice devices, the F100220 family. The devices are ECL LSI devices, contain a substantial amount of error-handling circuitry, and most are *8-bit* slices.

The F100220 address and data interface unit is the ALU/register slice. All elements and busses are 9-bits wide: eight data bits and a parity bit. Its major elements are a binary/decimal ALU, shifter, result register, auxiliary latch, and parity generation and checking logic. The ALU can perform operations on both packed and unpacked binary-coded-decimal values.

The F100220 contains three bidirectional 9-bit data busses. Registers can be added by use of the F100222 dual-access stack, an array of 32 9-bit registers with two independent ports, parity logic, and a maximum access time of 10 ns.

Given the width of the devices in this new family, one would expect that conventional packages (i.e., DIPs) are not used. The devices are supplied in a 1-inch square 68-pin ceramic package.

COMPARISONS

A useful way to summarize the discussion of ALU/register slices is to compare their design and physical characteristics, arithmetic and logical capabilities, speeds, and relative advantages and disadvantages. Table 3.4 compares the physical and organizational characteristics of the slices discussed in this chapter. Most of the characteristics in the table are self-explanatory; the ones that require further explanation are listed below.

Package —DIP = dual inline, QIP = quad inline. Most of the slices are also available in a flat package.

Power —the maximum power dissipation over the slice's operating range.

Expandable registers —indicates whether the size of the register array can be easily extended with additional chips.

Data paths —the number of paths between devices on the slice, excluding input and output busses.

Control inputs —the number of input pins controlling the slice (e.g., function control, register selection, bus control), excluding carry, clock, shift, and relative-position-control inputs.

Function encoding —the degree to which the function-control inputs are encoded.

	9405A	4705 5V	4705 10V	4705 15V
Max. power consumption (mW)	800	1	5	15
Max. prop. delay A to $\overline{X}$, $\overline{Y}$ (ns)	58	1014	410	288
Max. prop. delay I_0 to Z (ns)	40	516	234	164
Typ. prop. delay I_0 to Z (ns)	30	258	117	82
Max. set-up time D (ns)	45	500	198	140

Figure 3.61. Representative times of the 9405A versus the 4705.

TABLE 3.4. Comparison of ALU/Register Slice Characteristics

	2901	3002	MC 10800	SBP 0400	2903	74S 481	6701	9405	F100-220
Width (bits)	4	2	4	4	4	4	4	4	8
Technology	LSTTL[1]	STTL	ECL	I^2L	LSTTL	STTL[4]	STTL	LSTTL[5]	ECL
Pins	40	28[2]	48	40	48	48	40	24	68
Package	DIP	DIP	QIP	DIP	DIP	QIP	DIP	DIP	
Maximum power (mW)	1325[1]	950[2]	1650	1000[3]	1750	2125[4]	1400	800[5]	4000
Registers	16+1	10+3	0+1	8+2	16+1	0+4	16+1	8	0+2
Input ports	1	3	1	1	1	1	1	1	0
Output ports	1	2	0	2	0	2	1	1	0
Bidirectional ports	0	0	2	0	2	1	0	0	3
Register-array ports	2	1	N/A	1	2	N/A	2	1	N/A
Expandable registers	no	no	yes	no	yes	yes	no	no	yes
Relative position control	no	no	no	yes	yes	yes	no	yes	no
Data paths	9	10	14	14	7	18	9	3	
Control inputs	18	9	17	11	22	18	17	8	11
Function encoding	S	M	SM	M	SM	H	S	S	
Shifters	2	1	1	2	2	3	2	0	1
Independent shifters	2	0	1	2	2	2	2	0	0

Notes: [1] - A version is also available in a combination of LSTTL and ECL.
[2] - In comparing these numbers, remember that one needs twice as many packages.
[3] - Power consumed by chip and injector resister, assuming 5V supply, 200mA injector current.
[4] - An LSTTL version is also available, consuming 1625 mW.
[5] - A CMOS version is also available, consuming under 15 mW.

H highly encoded.
M the controls are partitioned into distinct groups, but there is considerable interaction among the groups.
SM the controls are partitioned into distinct groups with a small amount of interaction among the groups.
S the controls are partitioned into distinct groups, each controlling an independent set of functions.

Independent shifters —the number of shifters on the device that can be used independently of the ALU and each other.

Table 3.5 compares the arithmetic and logical capabilities of the slices. The list of attributes is not necessarily complete (i.e., some of the slices perform operations not in the list); it was developed to compare the slices across a major set of arithmetic and logical functions. Note that to fully understand these operations, one must also examine the function-control encoding on the particular slice. For instance, because of its encoding, a slice might be able to perform a given operation across only a subset of its on-chip registers.

Most of the attributes should be self-explanatory. *Reverse subtraction* is the ability for either ALU input port to serve as the subtrahend (i.e., A-B and B-A). *Comparison* is the ability to compare two values (e.g., via subtraction) without having to modify a register (i.e., the ability to route the output of the ALU nowhere). *Independent incrementer* indicates whether the slice has one or more registers that can be incremented in parallel with the use of the ALU for another operation.

The number of operations performed is totalled in Table 3.5. Of course, the total is not very meaningful since it is not weighted by the importance of each operation. One must consider the significance of each included and excluded operation with respect to a particular application to select the most-appropriate slice.

Table 3.6 summarizes the status outputs provided by each of the slices.

One remaining consideration is speed. Although representative set-up and propagation times were listed earlier for each slice discussed, it is difficult to deduce relative slice speeds via a direct comparison of these times. Hence, the devices will be compared with respect to the following attributes:

1. Minimum clock-period time.
2. In a 16-bit design, the time to add the values of two registers and produce the resultant status (e.g., overflow).
3. In a 16-bit design, the time to add the values of two registers and shift the result right by one bit before storing it into a register.

TABLE 3.5. Comparison of Arithmetic and Logical Capabilities

Function	2901	3002	MC 10800	SBP 0400	2903	74S 481	6701	9405[1]	F100-220
Addition	X	X	X	X	X	X	X	X	X
Subtraction	X		X	X	X	X	X		X
Reverse subtraction	X		X	X	X	X	X		X
−A	X	X	X	X	X	X	X		X
A+1	X	X	X	X	X	X	X		X
A+2			X		X				
A−1	X	X	X			X	X		X
A−2			X						
Add and shift	X		X	X	X	X	X		
Arithmetic shifts			X	X	X	X			
Comparison	X		X		X	X			
Independent incrementer				X		X			
BCD add/subtract			X						X
Multiply iteration					X	X			
Divide iteration					X	X			
Normalization iteration					X				
Signed-magnitude conversion					X				
CRC accumulation						X			
Sign propagation					X				
Inversion	X	X	X	X	X	X	X	X	X
AND	X	X	X	X	X	X	X	X	X
OR	X	X	X	X	X	X	X	X	X
NAND			X		X	X			X
NOR			X		X	X			X
Exclusive OR	X		X	X	X	X	X	X	
Exclusive NOR	X	X	X	X	X	X			
Mask before ALU		X	X			X			
$\bar{A}$ AND B	X		X	X	X	X			
$\bar{A}$ OR B			X	X		X			
Total	14	9	22	15	22	23	11	6	12

Notes: [1] The capabilities of the 9405 can be significantly extended with the 9404 data-path switch.

All calculations will be for (1) guaranteed times over the device's operating range and (2) typical times at 25°C, giving a total of six times for each device.

A carry-lookahead configuration will be used, assuming the 74S182 as the carry-lookahead generator. In the case of the 3002, eight slices are needed because of its 2-bit width; hence, two cascaded carry-lookahead generators are assumed in this case.

TABLE 3.6. Comparison of Status Outputs

Status	2901	3002	MC 10800	SBP 0400	2903	74S 481	6701	9405	F100-220
ALU carry-out	X	X	X	X	X	X	X	X	X
ALU overflow	X		X		X	X	X	X	X
ALU output = 0	X		X	X	X	X	X	X	X
ALU output sign	X		X		X			X	
Carry lookahead	X	X	X	X	X	X	X	X	X
Upper shift lines	X		X	X	X	X	X		X
Lower shift lines	X	X	X	X	X	X	X		X
Parity			X		X[1]				X
ALU input signs				X					
Greater/less than						X			

[1] limited circumstances only

The MC10800 and 74S481 have no register arrays. For the MC10800, the source and destination register will be assumed to reside in a separate set of MC10145 register arrays, as shown in Figure 3.62a; the other source is the accumulator in the MC10800. For the 74S481 (and 74LS481), a source and the destination will be assumed to reside in a separate register array, a set of 74170s (Figure 3.62b). The other source will be the WR register. For the 9405/4705, the addition operation measured is the addition of the values in the output register and an array register, storing the result in the array register. The data-path switch (9404/4704) is assumed in the design.

In analyzing the times of the two situations (addition, and addition/shift), the stopwatch will start when the input signals (e.g., function controls, register selections, carry input to the LSS) arrive at the slices (or the slices and the register

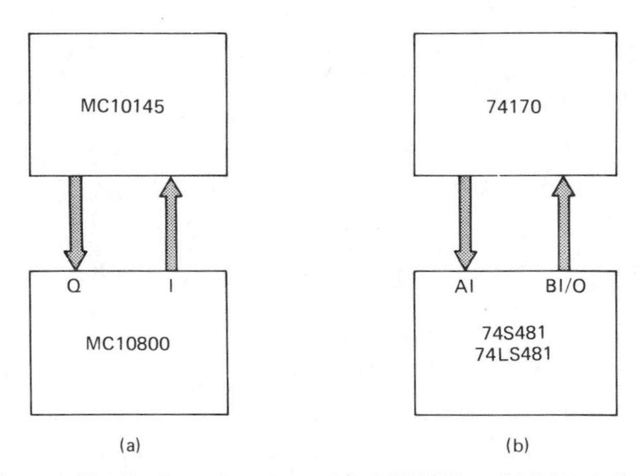

Figure 3.62. Configurations assumed for MC10800 and 74S481 timings.

arrays in the cases of the MC10800 and 74S481). For the addition case, the stopwatch will stop when the last status output is available. In the case of add/shift, the stopwatch will stop when enough time has elapsed to permit the trailing clock edge (the one which causes the destination register to be loaded) to occur.

It is instructive to illustrate how these times are calculated. Consider the case of addition in the Am2901A, using the guaranteed times in Figures 3.9 and 3.10. One sees that, beginning with the arrival of the A, B, and I inputs, it takes each slice 65 ns to generate its carry generate and propagate outputs. Neglecting interchip signal delays, the 74S182 takes 7 ns to generate the carry inputs to the slices. Examining Figure 3.10, the latest status output (F=0) is generated 47 ns after the carry input (C_n) is stable. Hence, the time in this case is 119 ns.

Considering the case of add/shift, the configuration of Figure 3.63 will be assumed. The sign of the ALU result in the MSS (F_3 in the case of the 2901) will be tied to the upper shift input, thus causing the sign to be retained (i.e., creating an arithmetic shift). (In slices that perform arithmetic shifts internally, this connection is not needed.) The times for the add/shift path in the Am2901 are

2901A generation of carry-lookahead outputs	—65 ns
74S182 generation of carry inputs	— 7 ns
2901A generation of the sign, (F_3 from C_n or generation of the shift output (RAM_0) from C_n, whichever is greater	—50 ns
2901A set-up time required on shift input (RAM_3)	—25 ns

In other words, the time between the receipt of the control inputs and the trailing clock edge (loading the shifted ALU result into the register array) is 147 ns. Table 3.7 lists these calculated times for the slices discussed in this chapter. A missing

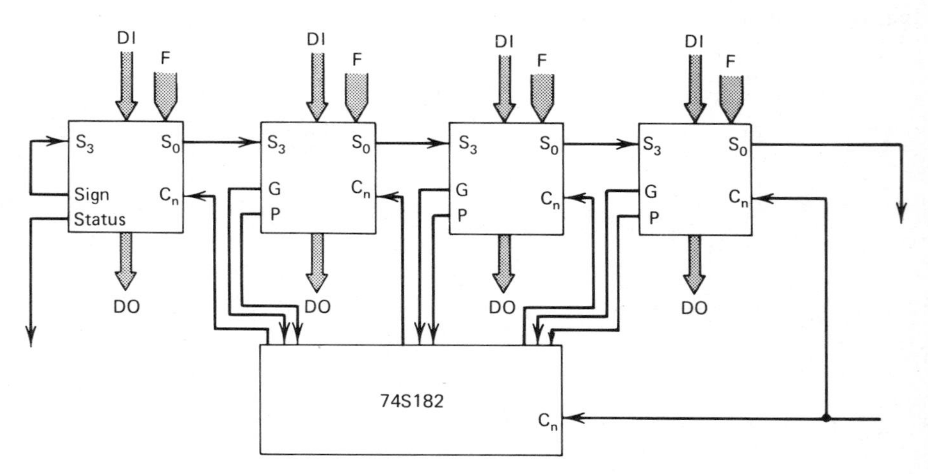

Figure 3.63. Configuration assumed for add/shift times.

TABLE 3.7. Comparative ALU/Register Slice Speeds

	Typical			Maximum		
Slice	Minimum clock period	Addition	Add/shift	Minimum clock period	Addition	Add/shift
Am2901A	75	80	90	100	119	147
Am2901B	60	60	73	77	94	110
IDM2901A					107	122
IDM2901A-1					80	
I3002	70	63	[a]	100	95	[a]
N3002	45	53	[a]	70	79	[a]
MC10800[b]		52	70		76	92
SBP0401A[c]	275	317				
Am2903		99	116		155	185
74S481		87	137	100	122	182
74LS481		105	150	125	142	212
6701	110	120	170	175	147	222
9405A/9404	50	98	[e]	75	137	[e]
4705/4704[d]	263	336	[e]	526	669	[e]

Notes: Blank entries indicate that the data needed to calculate the time is not available.

[a] - The 3002 has no independent shifter; therefore the shift must be performed in a separate cycle.

[b] - MC10179 ECL carry-lookahead generator assumed.

[c] - When operated at an injector current of 200 mA.

[d] - When operated at 10 V.

[e] - Because the 9404 places the shifter at the ALU input, add/shifts cannot be easily done.

value means that all or part of the timing information needed to calculate the value was not available from the manufacturer at the time of the writing of this book.

As one last form of comparison, the following is a list of the major advantages and disadvantages of each slice when comparing each slice to the collection of other slices.

Advantages	*2901*	*Disadvantages*
Two-port register array		Limited number of external busses
Clean definition of control signals		
Excellent documentation		
Large family of support devices		

Advantages	*3002*	*Disadvantages*
Large number of external busses		Two-bit width
Bit masking on ALU input		One-port register array
		No independent shifter
		No overflow detection
		Limited ALU functions
		Nondestructive comparisons are difficult

Advantages	*MC10800*	*Disadvantages*
Fast		Difficult to interface with TTL
Parity outputs		No register array
BCD and binary arithmetic		
Bit masking on ALU input		
Extensive functions		
Clean definition of control signals		

Advantages	*SBP0401A*	*Disadvantages*
Shift connection logic		Slow
Independent incrementer		External logic needed for overflow detection
Relative position control		One-port register array

Advantages	*2903*	*Disadvantages*
Two-port register array		High power consumption
Extended operations		Slower than most
Floating-point orientation		
Clean definition of control signals		
Relative position control		
Extensive functions		
Large family of support devices		

Advantages	*74S481*	*Disadvantages*
Extended operations		No register array
Independent incrementers		High power consumption
Large number of external busses		Slower than most
Relative position control		No output for sign of ALU result
Extensive functions		Extensive control-signal encoding
Bit masking on ALU input		

Advantages	*6701*	*Disadvantages*
		Generally inferior to the 2901

Advantages	*9405*	*Disadvantages*
Relative position control		Usually requires a separate chip (9404) per 9405 One-port register array

Advantages	*4705*	*Disadvantages*
Extremely low power consumption Relative position control		Usually requires a separate chip (4704) per 4705 One-port register array Very slow

REFERENCES

1. *The Am2900 Family Data Book.* Sunnyvale, Cal.: Advanced Micro Devices, 1978.
2. J. R. Mick, "Am2900 Bipolar Microprocessor Family," *Proceedings of the Eighth Annual Workshop on Microprogramming.* New York: IEEE, 1975, pp. 56–63.
3. *Series 3000 Reference Manual.* Santa Clara, Cal.: Intel, 1976.
4. *Microprogramming the Series 3000.* Santa Clara, Cal.: Intel, 1976.
5. *Introducing the Series 3000 Bipolar Microprocessor.* Sunnyvale, Cal.: Signetics, 1978.
6. S. Y. Lau, "Design High-Performance Processors with Bipolar Bit Slices," *Electronic Design,* 25(7), 1977, pp. 86–95.
7. A. Colin, "Intel 3000 and Am2900 Microprocessors—A Comparison," *Microprocessors,* 1(5), 1977, pp. 287–292.
8. *M10800 High Performance MECL LSI Processor Family.* Phoenix: Motorola, 1976.
9. *MC10800 Data Sheet.* Phoenix: Motorola, 1977.
10. *The Bipolar Microcomputer Components Data Book for Design Engineers.* Dallas: Texas Instruments, 1977.
11. *5701/6701 4-Bit Expandable Bipolar Microcontroller.* Sunnyvale, Cal.: Monolithic Memories, 1975.
12. *Macrologic Bipolar Microprocessor Databook.* Mountain View, Cal.: Fairchild Camera and Instrument Corp., 1978.

4

Microprogram Sequencing Devices

As discussed in Chapter 2, a processor usually consists of two major sections—the processing section and the control section. Given a microprogrammed processor, the control section consists of three major subsections: the control storage, the microinstruction decoding logic, and the microinstruction sequencing logic. The purpose of the latter is to change the state of the control section by determining the address of the active microinstruction for the next machine cycle.

Thus, the function of a microprogram sequencer is to receive one or more control-storage addresses and one or more status conditions, and, from these, send to control storage the address of the next microinstruction. In most designs, one control-storage address comes directly from the current microinstruction and another is the current microinstruction address plus one, and the sequencer selects between them. In other words, the microprogram sequencer performs the task of microinstruction branching or jumping.

The desirable forms of branching in control-section designs, all of which have parallels in the control constructs in high-level programming languages, are listed below. Assume that CA represents the control-storage address of the current microinstruction, NA represents the address generated by the sequencer (i.e., the address of the microinstruction for the next cycle), MIF represents an address value supplied by the current microinstruction, and C represents a condition to be tested (e.g., ALU overflow, ALU result = 0). Also, STK represents a pushdown stack (last-in, first-out memory). Gating a value to STK places the value on the top of the stack and pushes all other values in the stack down one position. Gating a value from STK removes the top value from the stack and moves all other values in the stack up one position.

Sequence	NA = CA + 1
Branch	NA = MIF
Conditional branch	If C is true, NA = MIF else, NA = CA + 1

Repeat	If a counter is not yet zero, counter = counter − 1 NA = MIF else, NA = CA + 1
Subroutine call	NA = MIF STK = CA + 1
Subroutine return	NA = STK
Multiway branch	NA = CA or CA + 1 or CA + 2 • • • depending on some input value

The need for most of these forms should be obvious. The repeat operation is needed for iterative functions, such as multiplication and division algorithms and searches. The subroutine mechanism allows one to branch to a sequence of microinstructions and eventually transfer control back to the point of the branch, allowing one to invoke this sequence of microinstructions from multiple places within the microprogram. The use of a stack allows subroutines to call other subroutines. The multiway branch has many uses, such as decoding operation codes in machine instructions and decoding interrupt codes from an interrupt controller.

Note that these branching forms imply a set of desired facilities in a microinstruction-sequencing device:

1. A register to hold CA.
2. An output bus to deliver NA to the control storage, preferably a 3-state bus (to allow one to override the device, for instance, during system debugging).
3. An input bus to receive MIF from the current microinstruction.
4. An incrementer to increment CA.
5. A counter.
6. An input pin supplying C.
7. A pushdown stack.
8. Function inputs to indicate the desired branching form.
9. To implement the multiway branch, either a full adder or an OR network.

In this chapter, three types of sequencing devices are discussed. One consists of microprogram-sequencing slices that can be cascaded together to form a sequencer of any width. The second consists of support devices that can be coupled to sequencing slices to add additional branching capabilities. The third set consists of wider sequencers (e.g., 12 bits, to address a control-storage of 4096 words) that, not being designed as slices, cannot be cascaded.

It should be noted that although the primary emphasis here is the use of these sequencers in the control section of a processor, they have other applications. For instance, if one is designing a high-speed controller that needs no arithmetic capabilities and a microprocessor-based solution proves to be too slow, a microprogram sequencer and a small control storage is a possible alternative. That is, microprogram sequencers need not be used in conjunction with ALU/register slices. Furthermore, although semiconductor manufacturers produce sequencers and ALU/register slices as part of particular bit-slice families, one should realize that the two are quite independent, and, when it makes sense, a sequencer from one family can be used with an ALU/register slice of another family.

THE 2909 SEQUENCER SLICE

The 2909 microprogram sequencer is a 4-bit, low-power Schottky TTL slice in a 28-pin package. It was originally produced by Advanced Micro Devices [1–2] as part of their Am2900 bit-slice family. It is now second-sourced by other manufacturers and is a component (the IDM2909A) in National Semiconductor's IDM2900 family [3].

The organization of the 2909 is shown in Figure 4.1. It contains three input

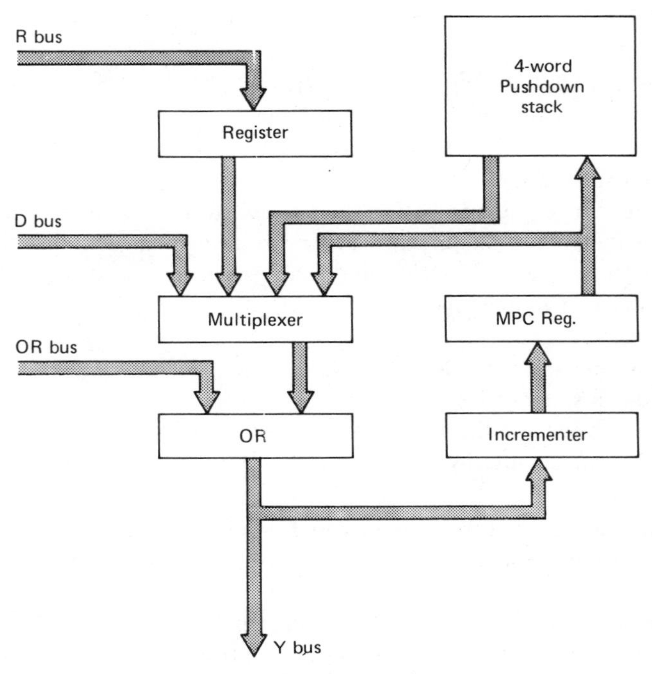

Figure 4.1. Organization of the 2909.

busses, which are usually fed, directly or indirectly, from the branching micro-orders in the current microinstruction, and one output bus, which normally feeds the address inputs of control storage. The 2909 contains two registers, one of which is fed from the R input bus and the other of which (MPC, or microprogram counter) is always fed from the output (Y) bus. Before Y is gated into MPC, it passes through an incrementer, which performs an addition of 0 or 1.

The 2909 also contains a 4 × 4 register array used as a pushdown stack. Given that MPC contains the address of the current microinstruction plus one, the stack is used to remember the return address when branching to a microprogram subroutine. That is, when a subroutine call is performed, the value of MPC is pushed onto the stack. When a subroutine returns, the return address is obtained from the top of the stack.

The central part of the 2909 is the multiplexer. As shown in Figure 4.1, the multiplexer selects the address of the next microinstruction (the value placed on the Y bus) from one of four sources: the D bus (e.g., for branching), the register (REG) (e.g., for branching to the beginning of a loop), the stack (e.g., for returning to the point of a subroutine call), and MPC (for continuing to the next location in control storage).

The remaining component in the 2909 is four OR gates, which OR each bit from the multiplexer with each corresponding bit on the OR bus. As will be shown later in the chapter, this is useful in implementing multiway branches.

Figure 4.2 is the manufacturer's block diagram of the 2909. The diagram contains some references to another similar slice, the 2911; these notes can be ignored for now. Note that this diagram shows an additional register: the stack pointer. Rather than actually moving data from word to word within the stack whenever a value is pushed or popped, the stack pointer is operated as an up/down counter to keep track of the word in the array that is logically the top value. The stack pointer was eliminated from Figure 4.1 for simplicity, since it is not externally visible.

The pins (except for the supply voltage and ground) of the 2909 are illustrated in Figure 4.3. The S inputs select one of the four multiplexer inputs. PUP is active when $\overline{\text{FE}}$ is zero or low. When PUP is 1, the value in MPC is pushed onto the stack. When PUP is 0, the top value on the stack is popped (discarded). $\overline{\text{RE}}$, when 0, causes the register (REG) to be loaded from the R bus. $\overline{\text{ZERO}}$, when 0, forces zeros onto the Y bus. This might be used during system initialization, where one might want to begin at the microinstruction in control-storage location 0.

The OE input allows one to place the Y bus in a high-impedence state. Although not normally used in a system design, it is useful in system debugging because it allows one to force a branch to a particular control-storage address by feeding the control storage address inputs from an external source. The C_n input is the carry input to the incrementer, allowing one to load Y or Y+1 into MPC.

The 2909 has only five output pins. Four of these, the Y bus, are normally gated to the address inputs of control storage. The other, the carry-out of the incrementer, is used to interconnect cascaded 2909s.

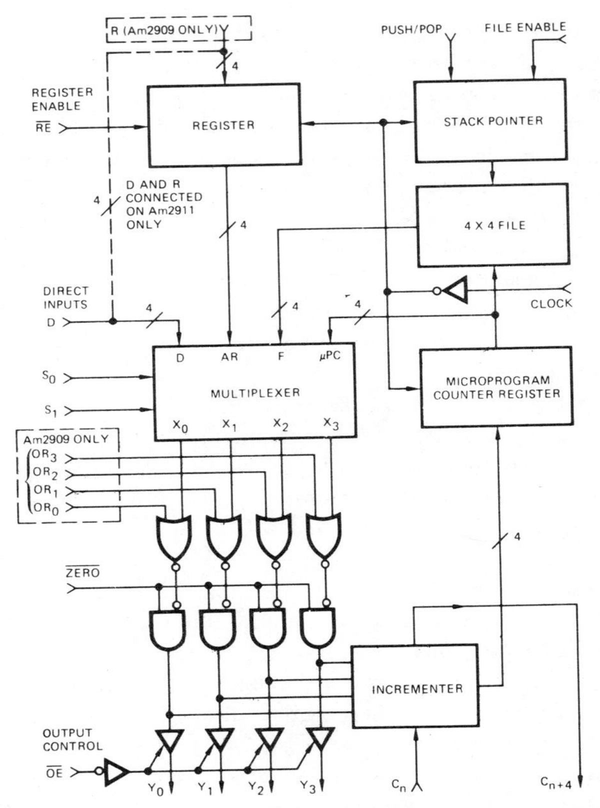

Figure 4.2. Organization of the 2909 (Copyright 1978 Advanced Micro Devices, Inc. Reproduced with permission of copyright owner.).

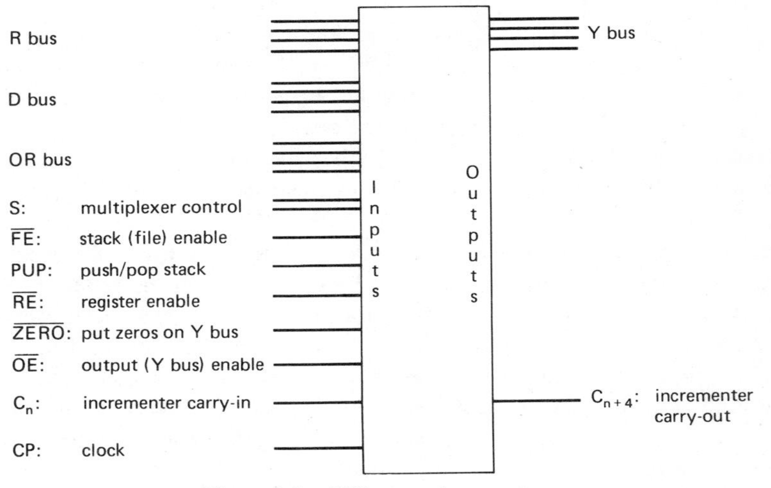

Figure 4.3. 2909 external connections.

2909 Functions

The only function input that requires further explanation is S. The S input selects the source for the Y bus as follows

00—MPC
01—REG
10—Stack
11—D bus

The S, FE, and PUP inputs, when taken together, provide the branching controls to the 2909. Table 4.1 illustrates these controls by showing the value produced on the Y bus, the value placed in MPC, and the changes to the stack. In the table, C_n is assumed to be 1, TOP represents the top of the stack, and A(X) represents the address in X.

Documentation on the 2909 does not clearly define the results of stack underflow (popping an empty stack) and overflow (pushing a fifth value onto the stack). However, because of the manipulation of the stack pointer, there appears to be a wrap-around effect. That is, after a pop, the popped item actually becomes the bottom item on the stack, meaning that if one performed four successive pops, the stack would have the value it had immediately preceding the first pop. On a push to a full stack, the bottom item logically falls off the end of the stack. Both conditions would normally be considered to be microprogram errors (bugs). Detection of these conditions, which is accomplished via status outputs in other sequencers, is an advantage in system debugging.

TABLE 4.1. 2909 Branching Functions

S, $\overline{FE}$, PUP	New Y	New MPC	New Stack	Function
0000	MPC	MPC+1	TOP discarded	Pop stack and continue
0001	MPC	MPC+1	MPC pushed	Push MPC and continue
001x	MPC	MPC+1	no change	Continue
0100	REG	REG+1	TOP discarded	Pop stack and branch to A (REG)
0101	REG	REG+1	MPC pushed	Call A (REG)
011x	REG	REG+1	no change	Branch to A (REG)
1000	TOP	TOP+1	TOP discarded	Return from subroutine
1001	TOP	TOP+1	MPC pushed	Branch to A (TOP), push MPC
101x	TOP	TOP+1	no change	Loop (branch to A (TOP))
1100	D	D+1	TOP discarded	Pop stack and branch to A (D)
1101	D	D+1	MPC pushed	Call A (D)
111x	D	D+1	no change	Branch to A (D)

2909 Timing

The timing within the 2909 is straightforward; when the clock input (CP) rises from low to high, values (if so indicated by the input signals, namely RE and FE) are stored in REG, MPC, and the stack. Representative set-up and propagation times over the operating range are given in Figure 4.4. The times are for Advanced Micro Devices' Am2909. Since the 2909 incrementer has no carry-lookahead outputs, ripple carry must be used between cascaded 2909s. This is rarely a problem, since the incrementer is unlikely to be on a critical speed path; the incrementing is overlapped with the access delay of control storage (i.e., the incrementer and MPC register are not on the path from the address source to the Y bus).

Figure 4.4 lists two sets of propagation delays from the rising clock edge. The larger values apply when the operation is a push or pop.

Using the 2909

As an initial example of the use of a sequencer slice, the 2909 will be used in a simple design. Assume that one is designing a processor that needs 256 words of control storage. Two cascaded 2909s provide sufficient addressing range (their combined Y bus is 8-bits wide). Assume a microinstruction design in which the first microorder (three bits) specifies the type of branch (i.e., the mechanism used to generate the address of the next microinstruction), the second microorder (eight bits) is used as a branch address, and the remaining microorders are sent to the processing section (if we are designing a processor) or to control points (if we are

From input	Minimum set-up time to clock rise (ns)
R bus	10
D, OR busses	30
S, $\overline{ZERO}$	45
$\overline{RE}$	22
PUP, $\overline{FE}$	26
C_n	28

From input \ To output	Maximum propagation times (ns) Y	C_{n+4}
D, OR	17	30
C_n	—	14
S, $\overline{ZERO}$	30	48
$\overline{OE}$	25	—
CP, S=LX	43	55
CP, S=HL	80	95

Figure 4.4. Representative Am2909 guaranteed times.

designing a simple high-speed controller). One should keep in mind that the design of the control section, for example, the number and types of sequencer slices, is largely independent from the design of the remainder of the system. The control-section design is largely a function of just the control-storage size and the definition of the branching microorders.

Keeping things simple, the branch microorder specifies the following eight types of branches. Assume that MIF represents the value of the second microorder in the current microinstruction, NA the generated address of the next microinstruction, STK the stack, and CA the address of the current microinstruction. An operation on STK pushes a value onto the stack or obtains and removes the top value from the stack.

GO	NA = CA + 1
RETURN	NA = STK
CALL	NA = MIF, STK = CA + 1
BRANCH	NA = MIF
BRANCH ON OVR	If OVR is true, NA = MIF else, NA = CA + 1
BRANCH ON NOT OVR	If OVR is false, NA = MIF else, NA = CA + 1
BRANCH ON ALU=0	If ALU=0 is true, NA = MIF else, NA = CA + 1
BRANCH ON NOT ALU=0	If ALU=0 is false, NA = MIF else, NA = CA + 1

The conditions OVR and ALU=0 are assumed to come from outside the control section.

Figure 4.5 is a diagram of the design of the control section. Note that the 2909 register (REG), OR, and ZERO facilities are not being used; therefore they have been disabled. The Y busses from the two 2909 slices are connected to the 8-bit address port of control storage. The second microorder (the branch address) of the current microinstruction is connected to the D busses of the 2909s.

In this design, the sequencing is performed by controlling the S, FE, and PUP inputs of the 2909s. These inputs receive the output of the block of decoding logic, whose inputs are the branch-type microorder and the two conditions.

Table 4.2 lists each branch type, its bit representation in the microoder (chosen arbitrarily), and the resultant signals that should be sent to the 2909s from the decoding logic. Hence, Table 4.2, when taken with the two conditions, forms a truth table for the decoding logic.

The decoding logic is customized for this particular design. A likely form in this design is a high-speed 32 × 4 PROM, where the OVR and ALU=0 conditions and the first microorder form a 5-bit address, and the addressed word feeds the 2909s. For instance, the GO setting, since it functions independently of the OVR

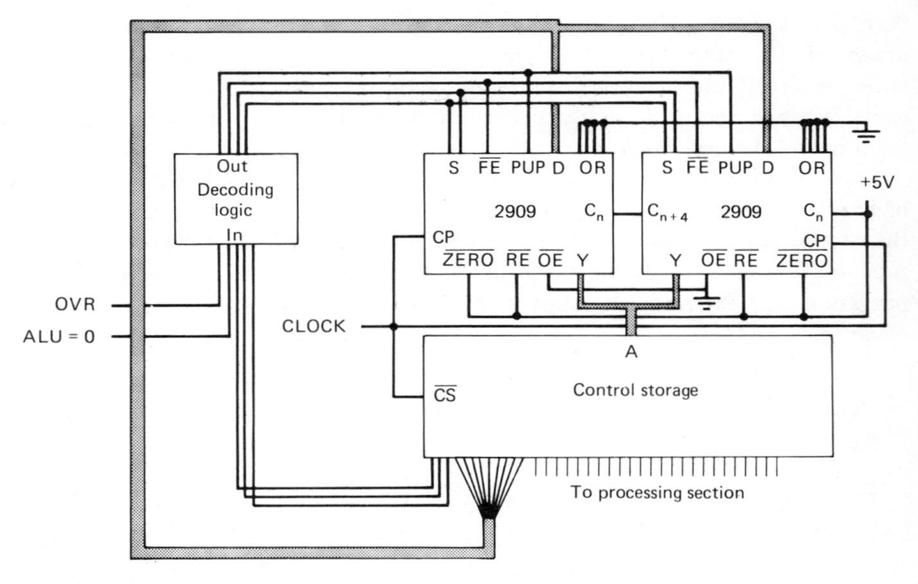

Figure 4.5. Simple control-section design.

and ALU=0 conditions, would result in words xx000 (0, 8, 16, and 24) being addressed. Hence the value 001x would be placed in these four words. The BRANCH-ON-OVR value of the microorder would result in words 1x100 and 0x100 being addressed. Hence, the two words at locations 20 and 28 in the decoding PROM would contain 111x (causing a branch to occur) and the words at locations 4 and 12 would contain 001x (causing the microinstruction beyond the current microinstruction to be selected).

To quickly analyze the timing in Figure 4.5, assume that the decoding PROM has a maximum propagation delay of 40 ns from the address input to output, and the control-storage block has a propagation delay of 60 ns from the address input. Starting at the time when the two conditions and the branch-type microorder are present at the address inputs of the decoding PROM, the time before control storage can deliver the next microinstruction is

40 ns—propagation delay through the decoding PROM
30 ns—propagation delay from S to Y through the 2909s
60 ns—propagation delay through control storage

Hence the smallest possible machine-cycle time is 130 ns + X, where X is the time needed, from the beginning of the availability of a new microinstruction, to the generation of OVR and ALU=0. If OVR and ALU=0 are conditions generated by the processing section in the current machine cycle, X is likely to be large (e.g., 100 ns). If a technique known as *microinstruction pipelining* is used (discussed in Chapter 5), OVR and ALU=0 are conditions left in a register from the prior machine cycle, and X is likely to be negligible. (However, in this case one would also have to consider the 80 ns propagation delay from the prior clock edge to the Y output of the 2909.)

Examining the ripple carry through the 2909 incrementers, the carry-out of the lowest 2909 is available 48 ns after the receipt of the S inputs from the decoding PROM, and a minimum set-up time of 28 ns for the carry-input to the upper 2909 must pass before the clock is allowed to rise. Hence, this is not a critical-speed path because the clock would not rise until after the propagation time through control storage (i.e., 48+28 ns is smaller than 30+60 ns).

THE 2911 SEQUENCER SLICE

The 2911 microprogram sequencer, originally produced by Advanced Micro Devices [1], is a 4-bit slice in a 20-pin package. The 2911 is identical to the 2909 except for two omissions:

1. The 2911 has no R bus. Instead, the D bus feeds both the multiplexer and the register.
2. The 2911 has no OR bus, and the OR gates in the output (Y) bus are not present.

These two busses account for the 8-pin difference between the 2909 and 2911. Figure 4.6 illustrates the organization of the 2911.

TABLE 4.2. Branch Microorder Encoding and Decoding

		Signals to be sent to S, $\overline{FE}$, PUP	
Microorder	Setting	Condition true	Condition false
Go	000	001x	001x
Return	001	1000	1000
Call	010	1101	1101
Branch	011	111x	111x
Branch on OVR	100	111x	001x
Branch on not OVR	101	001x	111x
Branch on ALU=0	110	111x	001x
Branch on not ALU=0	111	001x	111x

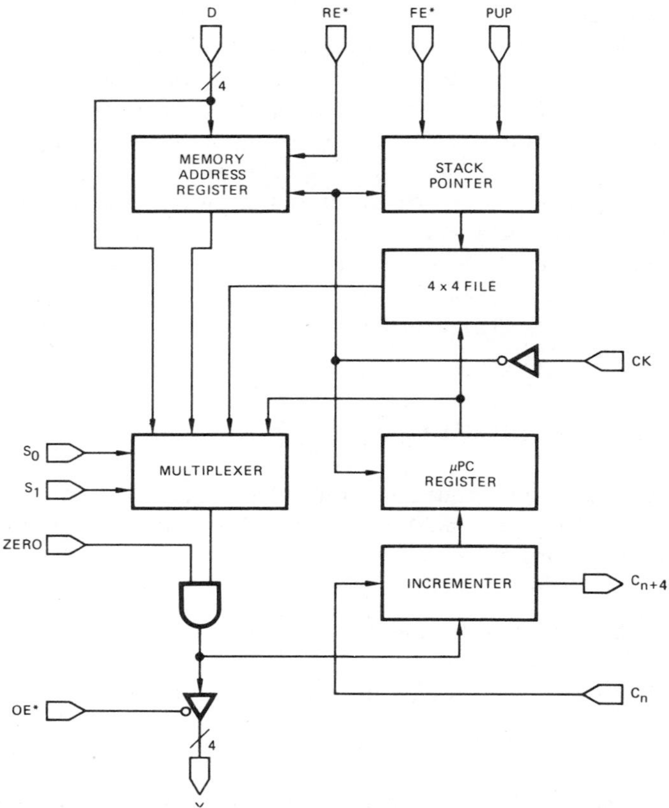

Figure 4.6. Organization of the 2911 (Copyright 1978 Advanced Micro Devices, Inc. Reproduced with permission of copyright owner.).

With respect to the definition of the remaining pins, the functions, and timing, the 2911 is identical to the 2909. The 2911 is physically smaller and somewhat less expensive.

In a design where a separate bus is not needed to the on-chip register and where the OR gates on the Y bus are not needed, the 2911 is obviously a better choice than the 2909. Such is the case in the example, Figure 4.5. In a case where the OR facility is needed, the 2909 and 2911 can be used together. The OR gates on the output bus are useful in implementing a multiway branch. Consider a design needing a control storage of 4096 words and the ability to perform a 16-way branch. A 2909 can be used as the least-significant slice, cascaded to two 2911s. To perform a 16-way branch, one generates a 12-bit address in the sequencer whose low-order four bits are zero, and gates a 4-bit value (e.g., from an external source, such as a register) into the OR bus of the 2909. Hence, the next microinstruction is taken from the generated address or one of the next 15 locations, depending on the 4-bit value. Incidentally, this is the first situation we have examined where different types of slices are cascaded together.

THE 29811 NEXT-ADDRESS CONTROL UNIT

Unlike the 2909 and 2911, the 29811 is not a slice and contains no sequential logic. Rather, it is a combinatorial-logic device designed to enhance the branching capabilities of the 2909 and 2911 sequencer slices. For instance, the 29811 could serve as the decoding logic shown in Figure 4.5. Designed by Advanced Micro Devices as another member of their Am2900 family of components, the 29811 is a low-power Schottky device in a 16-pin package [1,2].

The organization of the 29811 is shown in Figure 4.7. It has six input pins and eight output pins. Although the designer is free to use the device in any way that seems appropriate, the four I inputs are intended to be fed from a 4-bit micro-order, giving the microinstruction 16 distinct branching forms. The TEST input is fed the condition on which the branch is dependent. Four of the outputs feed the function-control inputs on one or more 2909 or 2911 sequencer slices; the other four outputs can be used to control a counter and a PROM to give the design a "repeat" branch (iterative loop) and a multidirection branch.

The heart of the 29811 is obviously the combinatorial decoding matrix. Before defining this, however, the 16 intended branch types are described. After the decoding matrix is defined, the use of the 29811 in a typical design is illustrated.

In describing the branching forms that can be implemented using the 29811, the notation at the beginning of the chapter will be used. In particular,

CA —address of current microinstruction.
NA —generated address of the next microinstruction.
MIF —address field supplied by the current microinstruction.
STK —a pushdown stack. An operation on STK pushes or pops a value.
TOP —the value on the top of the stack.
CNTR—a counter
REG —the register in the 2909/2911

Table 4.3 specifies the branching forms normally implemented by the 29811, assuming that the 29811 is connected to 2909/2911 slices in a manner described later. Note that each is a function of the four I inputs and, in most cases, the TEST input. Each form specifies how NA is formed and, in many cases, how the states of the pushdown stack and counter are altered.

As a step toward seeing how these branching operations are accomplished, Figure 4.8 illustrates the mapping performed by the 29811 from its five inputs to its eight outputs. ("JSB," or "jump to subroutine," is the call operation.)

To illustrate the use of the 29811, the microinstruction in Figure 4.9 will be used. The first microorder specifies the type of branch (i.e., from Table 4.3) to be performed. The third microorder allows one of eight conditions to be specified (e.g., overflow, zero result from ALU, ALU result is negative). The second mi-croorder specifies whether the specified condition is to be tested for a true or false

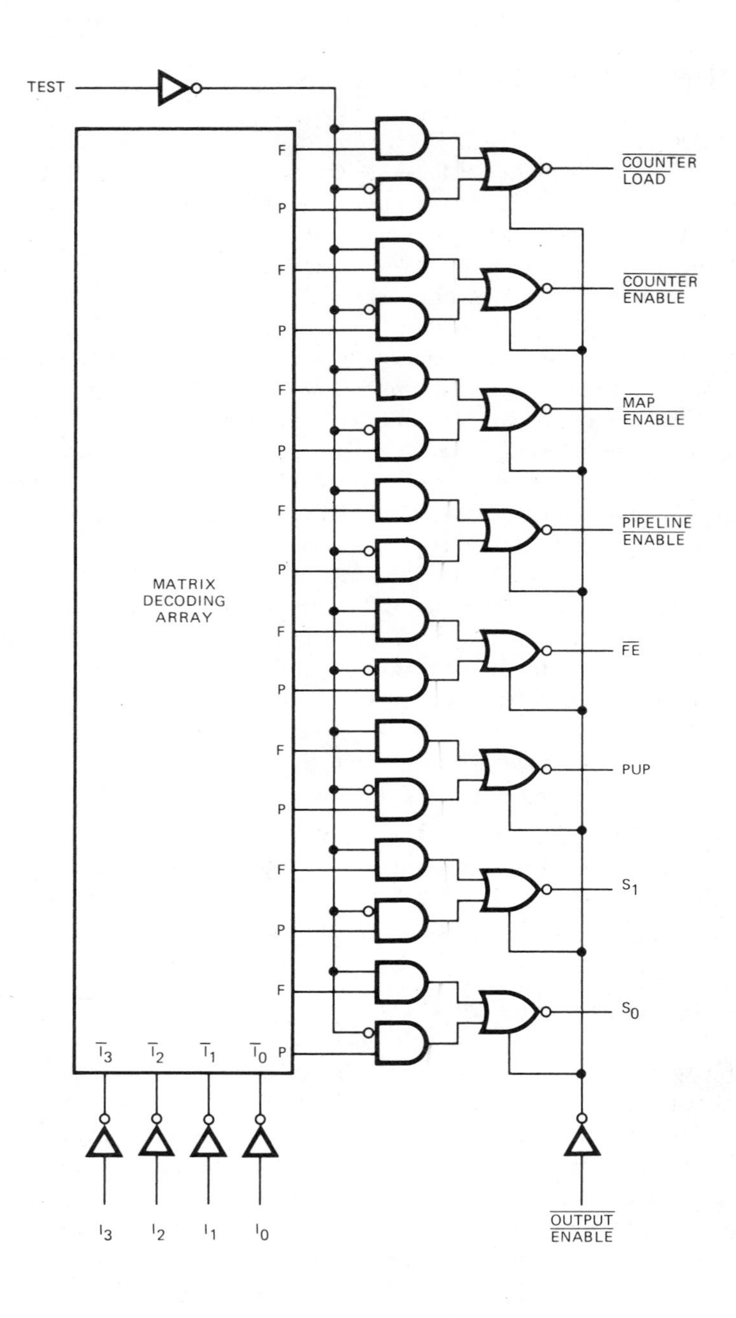

P = Pass

F = Fail

Figure 4.7. Organization of the 29811 (Copyright 1978 Advanced Micro Devices, Inc. Reproduced with permission of copyright owner.).

TABLE 4.3. 29811 Branching Functions

I	Abbreviation	Branch type	Function: TEST=0	Function: TEST=1
0000	JZ	Jump to zero	NA = 000...	same
0001	CJS	Conditional call	NA = CA+1	NA = MIF STK = CA+1
0010	JMAP	Jump to map location	NA = external source	same
0011	CJP	Conditional jump	NA = CA+1	NA = MIF
0100	PUSH	Push, conditional load counter	NA = CA+1 STK = CA+1	NA = CA+1 STK = CA+1 CNTR = MIF
0101	JSRP	Conditional call using REG or MIF	NA = REG STK = CA+1	NA = MIF STK = CA+1
0110	CJV	Conditional jump vector	NA = CA+1	NA = external source
0111	JRP	Conditional jump using REG or MIF	NA = REG	NA = MIF
1000	RFCT	Repeat using stack	NA = TOP CNTR = CNTR−1	NA = CA+1 POP (STK)
1001	RPCT	Repeat using MIF	NA = MIF CNTR = CNTR−1	NA = CA+1
1010	CRTN	Conditional return	NA = CA+1	NA = STK
1011	CJPP	Conditional jump and pop	NA = CA+1	NA = MIF POP (STK)
1100	LDCT	Load counter and continue	NA = CA+1 CNTR = MIF	same
1101	LOOP	Conditional loop	NA = TOP	NA = CA+1 POP (STK)
1110	CONT	Continue	NA = CA+1	same
1111	JP	Jump	NA = MIF	same

value. For instance, if the overflow condition in the third microorder is encoded as 001, a microinstruction beginning as

0011 1 001 00000111

is interpreted as "jump to location 7 if overflow is present." A microinstruction beginning as

0011 0 001 00000100

is interpreted as "jump to location 4 if overflow is absent."

As implied above, the fourth microorder contains a branch address. (Here we are assuming a design with a maximum control-storage size of 256 words.) However, the field has two purposes in the design about to be described; it is also used as a source for an initial value to be loaded into a counter. Defining fields in

MNEMONIC	FUNCTION	INPUTS					OUTPUTS							
							NEXT ADDR SOURCE		FILE		COUNTER			
		I_3	I_2	I_1	I_0	TEST	S_1	S_0	$\overline{FE}$	PUP	$\overline{LOAD}$	$\overline{EN}$	$\overline{MAP\ E}$	$\overline{PL\ E}$
	PIN NO.	14	13	12	11	10	4	5	3	2	6	7	1	9
JZ	JUMP ZERO	L	L	L	L	L	H	H	H	H	L	L	H	L
		L	L	L	L	H	H	H	H	H	L	L	H	L
CJS	COND JSB PL	L	L	L	H	L	L	L	H	H	H	H	H	L
		L	L	L	H	H	H	H	L	H	H	H	H	L
JMAP	JUMP MAP	L	L	H	L	L	H	H	H	H	H	H	L	H
		L	L	H	L	H	H	H	H	H	H	H	L	H
CJP	COND JUMP PL	L	L	H	H	L	L	L	H	H	H	H	H	L
		L	L	H	H	H	H	H	H	H	H	H	H	L
PUSH	PUSH/COND LD CNTR	L	H	L	L	L	L	L	L	H	H	H	H	L
		L	H	L	L	H	L	L	L	H	L	H	H	L
JSRP	COND JSB R/PL	L	H	L	H	L	L	H	L	H	H	H	H	L
		L	H	L	H	H	H	H	L	H	H	H	H	L
CJV	COND JUMP VECTOR	L	H	H	L	L	L	L	H	H	H	H	H	H
		L	H	H	L	H	H	H	H	H	H	H	H	H
JRP	COND JUMP R/PL	L	H	H	H	L	L	H	H	H	H	H	H	L
		L	H	H	H	H	H	H	H	H	H	H	H	L
RFCT	REPEAT LOOP, CTR ≠ 0	H	L	L	L	L	H	L	H	L	H	L	H	L
		H	L	L	L	H	L	L	L	L	H	H	H	L
RPCT	REPEAT PL, CTR ≠ 0	H	L	L	H	L	H	H	H	H	H	L	H	L
		H	L	L	H	H	L	L	H	H	H	H	H	L
CRTN	COND RTN	H	L	H	L	L	L	L	H	L	H	H	H	L
		H	L	H	L	H	H	L	L	L	H	H	H	L
CJPP	COND JUMP PL & POP	H	L	H	H	L	L	L	H	L	H	H	H	L
		H	L	H	H	H	H	H	L	L	H	H	H	L
LDCT	LD CNTR & CONTINUE	H	H	L	L	L	L	L	H	H	L	H	H	L
		H	H	L	L	H	L	L	H	H	L	H	H	L
LOOP	TEST END LOOP	H	H	L	H	L	H	L	H	L	H	H	H	L
		H	H	L	H	H	L	L	L	L	H	H	H	L
CONT	CONTINUE	H	H	H	L	L	L	L	H	H	H	H	H	L
		H	H	H	L	H	L	L	H	H	H	H	H	L
JP	JUMP PL	H	H	H	H	L	H	H	H	H	H	H	H	L
		H	H	H	H	H	H	H	H	H	H	H	H	L

L = LOW
H = HIGH

Figure 4.8. Truth table for the 29811 (Copyright 1978 Advanced Micro Devices, Inc. Reproduced with permission of copyright owner.).

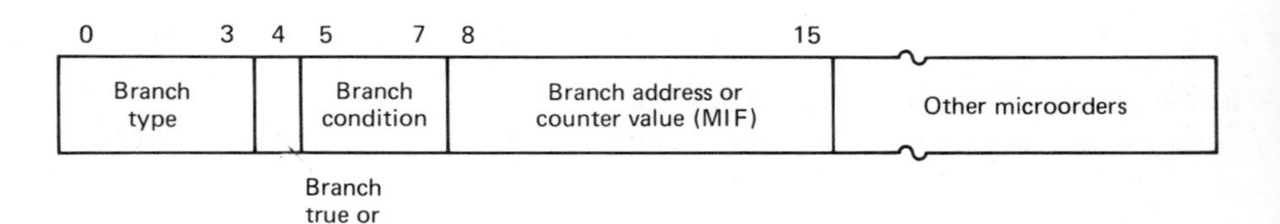

Figure 4.9. Format of the example microinstruction.

the microinstruction as being multipurpose is common practice, the motivation being to reduce the width of the microinstruction (and control storage).

Figure 4.10 depicts the logic of a control section for the microinstruction described above. However, for simplicity, not all the branch types in Table 4.3 have been incorporated into the design. The "jump to map location" and "conditional jump vector" branch types normally involve deriving the address of the next microinstruction from external sources (e.g., a mapping PROM and an interrupt vector). They are not supported in the design in Figure 4.10, although these branch types are illustrated in a later section discussing the 2910 sequencer. Also, the "jump to zero" branch type is not provided in the design. Last, the two branch types using REG (the register in the 2911) are not incorporated into the design since, on the 2911s in Figure 4.10, the loading of REG is permanently disabled. (Actually these two branch types could still be used. The system start-up

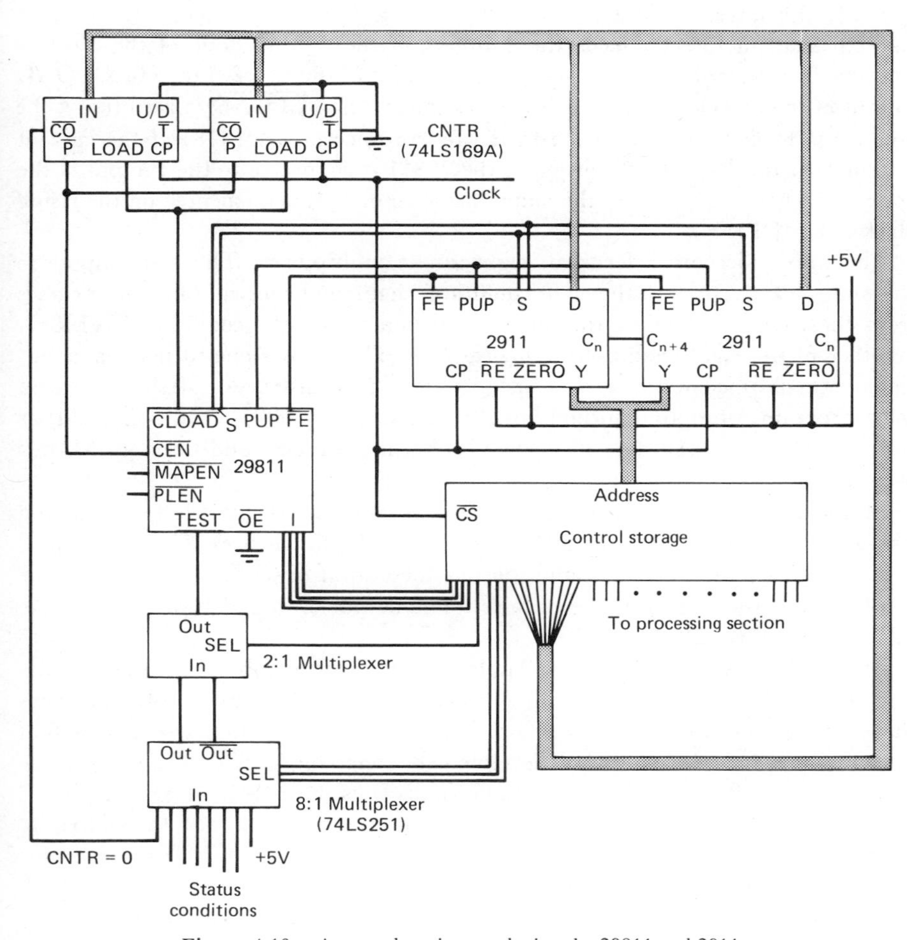

Figure 4.10. A control section employing the 29811 and 2911.

logic could force a predetermined value in REG by temporarily lowering $\overline{RE}$; a flipflop would then be used to raise $\overline{RE}$. This might be useful if there were some fixed control-storage location to which one branched often, such as the beginning of the machine-instruction-fetch routine in a processor. Hence, the branch-type 0111 would allow one to branch to the instruction-fetch microcode or to the microinstruction specified by MIF, depending on the condition being tested.)

Figure 4.10 uses two 2911 sequencer slices. The 29811 is used to extend the design to the branching types of Table 4.3 (minus the exclusions mentioned above). The two lower boxes are multiplexers used to decode the second and third microorder. The two boxes at the top are two 4-bit downward counters (e.g., two 74LS169A devices). They are interconnected to form an 8-bit counter, used to implement iterative loops in the microprogram. (Although both the counter and the two 2911s are 8-bits wide, this need not be the case. If one desired a counting capability of only 0–15, one 4-bit-wide counter could have been used.)

Study the interconnections of the 29811. The first microorder of the current microinstruction directly feeds the I inputs of the 29811. Four of the 29811's outputs directly feed the corresponding input pins of the two 2911s. The $\overline{CLOAD}$ output of the 29811 is connected to the counter; when $\overline{CLOAD}$ is zero (low), the counter is loaded with the value in the fourth microorder (MIF) of the current microinstruction. The $\overline{CEN}$ output of the 29811 is connected to the P input of the counter. When $\overline{CEN}$ is low, the value in the counter is decremented on the rising clock pulse (CP).

The third microorder feeds an eight-to-one multiplexer. The eight inputs to the multiplexer represent the eight conditions that may be tested. One input comes from the counter, allowing the microinstruction to test the condition CNTR=0. Another of the eight conditions is logical *true*. This allows one to make a conditional branch unconditional. For instance, if a microinstruction desires to call a microprogram subroutine unconditionally, it can specify 0001 (conditional call) as the branch type and specify "true" as the branch condition (third microorder).

In this design, the eight-to-one multiplexer produces both the selected input and its inverse as outputs (e.g., the multiplexer is a 74LS251 device). The two outputs are connected to a two-to-one multiplexer, the output of which is selected by the second microorder (branch true or false). The output of this multiplexer feeds the TEST input of the 29811.

At this point we can go back to Table 4.3 and determine how the branch types function. Consider conditional call (I=0001). If TEST=1, Figure 4.8 indicates that S=11, $\overline{FE}$=0, PUP=1, $\overline{CLOAD}$=1, and $\overline{CEN}$=1. The latter two indicate that the counter is neither loaded nor decremented. Referring back to Table 4.1 in the section on the 2909 (which has the same controls as the 2911), we see that MPC (which would have the value CA+1) is pushed onto the stack, and the address of the next microinstruction is taken from the D bus (the value of MIF). Hence, we have performed a subroutine call by branching to the location specified by MIF and saving the return point on the stack. If TEST were zero, Figure 4.8 indicates that the S outputs of the 29811 would be 00 and $\overline{FE}$ would be 1. Referring again to Table 4.1, this would cause the call to be bypassed; control would pass to the next microinstruction (at address CA+1).

To examine another case, assume that the branch type in the current microinstruction is 1001 (repeat using MIF). Such a microinstruction might be the last in an iterative loop of microinstructions, and it is likely that the condition being tested is "CNTR=0 is true." If CNTR is not zero, TEST would be 0. Referring to Figure 4.8 given I=1001 and TEST=0, we see that the outputs of the 29811 are S=11, $\overline{FE}$=1, PUP=1, $\overline{CLOAD}$=1, and $\overline{CEN}$=0. The latter enables the counter to decrement by one. Referring to Table 4.1, the signals sent to the 2911s would cause the next microinstruction to be selected from the address specified by MIF (which is probably the address of the microinstruction at the beginning of the loop). Hence, we have branched back to the beginning of the loop and decremented the iteration counter. For the case where the iteration counter reaches zero (TEST=1), the reader should perform the same analysis and confirm that the action performed is to fetch the next microinstruction from CA+1.

Microprogramming with the 29811 Branching Types

Given an understanding of how the branching operations in Table 4.3 are achieved, it is now worthwhile to leap again into the world of microprogramming to see the usefulness of these operations.

The 29811 presents one with three types of branches associated with the use of subroutines. Generally, subroutines in the microprogram have two motivations. First, if there is a sequence of microinstructions that is needed in multiple places in the microprogram, defining the sequence as a subroutine and calling it from all the places it is needed saves control-storage space, since the sequence appears only once in control storage. Second, subroutines are useful in partitioning the microprogram into independent functional units, making the microprogram easier to understand and modify. The *conditional call* and *conditional call using REG or MIF* branching operations are used to call (branch to) subroutines. If an address is previously stored in REG, the latter form allows one to perform a two-way call. Depending on the condition being tested, the subroutine called is the one whose address is in REG or MIF (the address in the current microinstruction). *Conditional return* is normally used in the last microinstruction in a subroutine; it branches to the microinstruction beyond the one that performed the last call operation. The presence of the pushdown stack in the 2909 and 2911 means that subroutines can be nested, that is, one subroutine can call another subroutine.

Several of the branching types are associated with simple branching operations. *Continue* is used when no branch is desired; the next microinstruction is taken from the next location in control storage. *Conditional jump* allows one to branch to a particular microinstruction if a specified condition is present. *Jump* specifies an unconditional branch to a particular microinstruction. However, because making one of the branch conditions (as was done in Figure 4.10) a permanently true or false condition is a necessity in almost all designs (it is needed, for instance, to perform unconditional calls and returns), the *jump* operation is redundant, because a microinstruction specifying *conditional jump* could specify the permanent condition *true*. The *conditional jump using REG or MIF* is a "forking" operation. From the current microinstruction, one proceeds to one of two locations.

Most of the remaining branching operations are associated with loops. These can be subdivided into three types of loops: (1) using the stack and the counter, (2) using the counter, and (3) using the stack.

The first loop form uses the stack to store the address of the first microinstruction in the loop and uses the counter to specify the number of iterations through the loop. The *push, conditional load counter* operation is used in the microinstruction that is just prior to the first microinstruction in the loop. The address of the next microinstruction (the top of the loop) is pushed onto the stack. The branch condition in the microinstruction is likely to be set to *true*, causing the value of the MIF field to be gated into the counter. MIF should specify the number of times the loop should be repeated (number of iterations minus one). The last microinstruction in the loop should contain the operation *repeat using stack*, probably specifying the branch condition CNTR=0. If the counter is zero, the loop is terminated (control transfers to the subsequent microinstruction and the stack is popped). If the counter is not zero, the counter is decremented and a branch is taken to the top of the loop (the address specified by the top value on the stack).

In this loop form, one might also have a requirement to perform a test in the middle of the loop and exit if a condition is present. *Conditional jump and pop* is provided for this purpose. If the condition is present, a branch is taken to the address specified in MIF (normally a location outside of the loop) and the loop address is popped from the stack.

The second loop form is similar, except that the stack is not used. Some time prior to the loop, the *load counter and continue* operation is performed to load the repetition count from MIF into the counter. In the last microinstruction in the loop, the *repeat using MIF* operation is used, probably specifying CNTR=0 as the condition. If the counter is not zero, it is decremented and a branch is taken to the location specified by the MIF field (normally the address of the first microinstruction in the loop).

These two loop forms are essentially equivalent. The first form does not use the MIF field in the last microinstruction in the loop, meaning that if MIF is defined as a multipurpose field (e.g., it can also serve as an "emit" field for the processing section), it can be used for another purpose in this microinstruction. On the other hand, the second form does not use the stack, meaning that the microprogrammer is likely to use this form in a loop within several nested levels of subroutine calls (i.e., it does not run the danger of causing a stack overflow), or if preceding the loop with a microinstruction specifying the *push, conditional load counter* is undesirable.

The third loop form does not use the counter. It is used for loops that do not iterate a fixed number of times (e.g., loops that normalize a value, search through storage, wait for an event to occur). The microinstruction immediately prior to the first microinstruction in the loop specifies the *push, conditional load counter* operation. However, since the counter is not used in this type of loop, the condition specified in this microinstruction should be false, causing only the address of the next microinstruction to be pushed onto the stack. The last microinstruction specifies *conditional loop* and a condition to be tested (e.g., ALU overflow). If the condition is present, the stack is popped and control is transferred to the microinstruction beyond this one. Otherwise, control is transferred to the top of the loop.

At this point, it is worthwhile to consider what has been achieved. The use of the 29811 gives one an extremely flexible and powerful set of microprogram sequencing operations, but doing so in the design in Figure 4.10 involved using only three LSI packages and four MSI packages (excluding control storage itself). If we needed to address up to 4096 words of control storage, we would need only eight packages for the control section (one more 2911 slice).

29811 Timing

The guaranteed propagation times for Advanced Micro Devices' Am29811 are shown in Figure 4.11. The Am29811 is actually a preprogrammed Am29751A 32 × 8 PROM.

THE 29803 16-WAY BRANCH CONTROL UNIT

Like the 29811, the 29803 is not a sequencer slice. Rather, it is a combinatorial device intended to enhance the branching capabilities of the 2909 sequencer slice by providing a multitude of 2-, 4-, 8-, and 16-way branch operations. Like the 29811, the 29803 is a low-power Schottky device in a 16-pin package and was originally produced by Advanced Micro Devices as part of their Am2900 family [1,2],

The organization of the 29803 is shown in Figure 4.12. The device has 10 input pins and 4 output pins. The intended use for the device is that the four I inputs are supplied by the current microinstruction, the four T inputs are connected to a register or bus, and the OR outputs are connected to the OR inputs of

Max. prop. time	
From input \ To output	All
I	40
TEST	40
$\overline{OE}$	25

Figure 4.11. Am29811 guaranteed times.

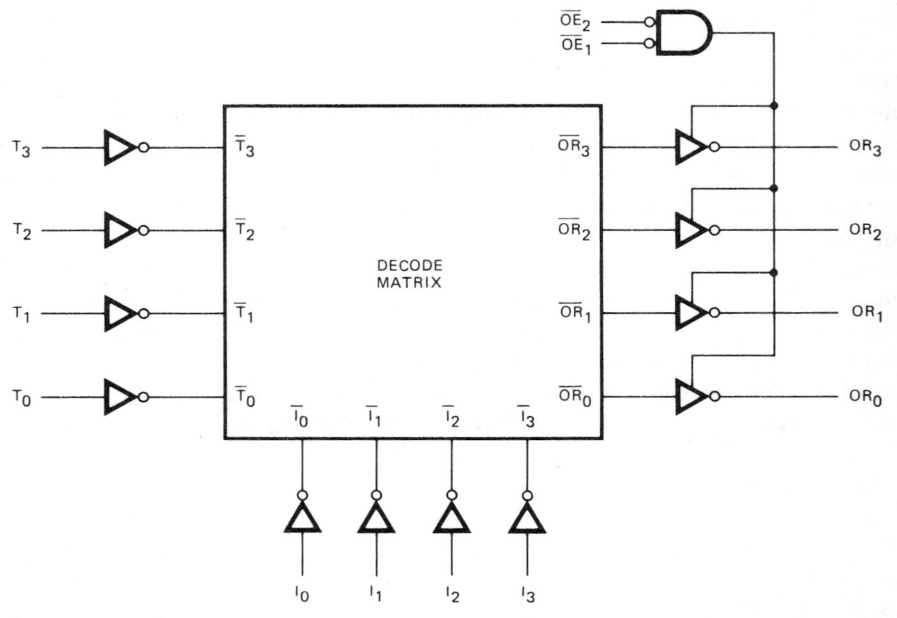

Figure 4.12. Organization of the 29803 (Copyright 1978 Advanced Micro Devices, Inc. Reproduced with permission of copyright owner.).

the least-significant 2909 slice. The I inputs cause the T inputs to be gated to the OR outputs in a variety of combinations. This allows one to take a 2, 4, 8, or 16-way branch in control storage.

The heart of the 29803 is the decoding matrix, which is specified by the truth table in Figure 4.13.

The connection of a 29803 to the least-significant 2909 slice in a sequencer is shown in Figure 4.14. The 29803 is used in the following fashion. Suppose, for instance, that one wanted to branch to 16 different microinstructions, based on the value in a 4-bit register, four status conditions, or some other 4-bit value. This value would be gated to the T inputs of the 29803, and 1111 (HHHH) would be gated to the I inputs (e.g., from a 4-bit microorder in the current microinstruction). In this situation, Figure 4.13 shows that the T inputs are passed directly to the OR outputs. The 2909/2911s would most likely be directed to jump to a particular location, and this location's address would be pre-assigned by the microprogrammer such that its four low-order bits are zero (i.e., it would be aligned on a 16-word boundary in control storage). Referring to Figure 4.1 or 4.2, the effect is to OR the T inputs into the address before it reaches control storage, causing the jump to be taken to the location, location+1, . . . , location+15, depending on the value of the T inputs.

This case does not illustrate the utility of the 29803, since the 29803 did nothing but pass the T inputs to the OR outputs. However, assume that one wishes, in another part of the microprogram, to perform a 4-way branch based on the first and third bits of the value gated to T. The microinstruction would set I to 0101 (Test TO & T2), and direct the 2909 to jump to a location in control storage. This location's address would be aligned when the microprogram was written such that its two low-order bits are zero. Referring to Figure 4.13, one can see that this jump becomes a jump to the location, the location + 1, the location + 2, or the location + 3. Hence, the 29803 allows one to perform a 2-, 4-, 8-, or 16-way branch based on any combination of the T inputs. One I (instruction) input (0000) disables the 29803 by causing the OR outputs to be 0000.

One should note, however, that the branching flexibility provided by the 29803 is not needed in most designs. If used, its primary drawback is that it requires four precious bits in the microinstruction (for the I inputs). The ability to perform a 16-way branch (but not all the other combinations), however, is often useful. This is more easily implemented with a quadruple two-to-one multiplexer, such as a 74LS157, as shown in Figure 4.15. This requires only a 1-bit microorder (e.g., 0 = no address modification, 1 = 16-way address modification). If 0, 0000 is sent to the 2909 OR inputs. If 1, a 4-bit value is sent to the OR inputs.

The guaranteed propagation times for Advanced Micro Devices' Am29803 are shown in Figure 4.16. Like the 29811, the Am29803 is actually only a PROM programmed by the manufacturer. Based on its characteristics, it appears to be a preprogrammed Am29761A 256×4 PROM.

THE 2910 SEQUENCER

The 2910 sequencer is another device that was produced by Advanced Micro Devices as part of their Am2900 bit-slice family [1,2,4]. Unlike the 2909 and 2911, the 2910 is not a bit-slice device; rather, it is a self-contained 12-bit-wide sequencer in a 40-pin package.

The 2910 is most easily explained by referring back to Figure 4.10. Essentially, the 2910 is a one-chip replacement for much of the logic in Figure 4.10. The 2910 is roughly equivalent to three 2911 bit-slice sequencers, a 29811 control device, and three cascaded counter registers (e.g., 74LS169A). In addition to containing all this function and more on a single chip, the 2910, as will be shown later, also has advantages in terms of speed and power.

The organization of the 2910 is illustrated in Figure 4.17. The organization is similar to that of the 2911, except that all devices and data paths are 12 bits wide. Also the pushdown stack in the 2910 is five words deep, as opposed to the four-word stack in the 2909 and 2911. The counter register in the 2910 functions in much the same way as the external counter added to the 2911s in the design in Figure 4.10. The counter is used to keep track of loop iterations; it may also be loaded with a control-storage address that is to be used as a branch address in some subsequent machine cycle.

The manufacturer's diagram of the 2910 is shown in Figure 4.18. Notice that the 2910 contains considerably more control logic in the chip than in the 2909 and 2911. In particular, note the instruction PLA (programmed logic array) in the

Function	I_3	I_2	I_1	I_0	T_3	T_2	T_1	T_0	OR_3	OR_2	OR_1	OR_0
No Test	L	L	L	L	X	X	X	X	L	L	L	L
Test T_0	L	L	L	H	X	X	X	L	L	L	L	L
					X	X	X	H	L	L	L	H
Test T_1	L	L	H	L	X	X	L	X	L	L	L	L
					X	X	H	X	L	L	L	H
Test T_0 & T_1	L	L	H	H	X	X	L	L	L	L	L	L
					X	X	L	H	L	L	L	H
					X	X	H	L	L	L	H	L
					X	X	H	H	L	L	H	H
Test T_2	L	H	L	L	X	L	X	X	L	L	L	L
					X	H	X	X	L	L	L	H
Test T_0 & T_2	L	H	L	H	X	L	X	L	L	L	L	L
					X	L	X	H	L	L	L	H
					X	H	X	L	L	L	H	L
					X	H	X	H	L	L	H	H
Test T_1 & T_2	L	H	H	L	X	L	L	X	L	L	L	L
					X	L	H	X	L	L	L	H
					X	H	L	X	L	L	H	L
					X	H	H	X	L	L	H	H
Test T_0, T_1 & T_2	L	H	H	H	X	L	L	L	L	L	L	L
					X	L	L	H	L	L	L	H
					X	L	H	L	L	L	H	L
					X	L	H	H	L	L	H	H
					X	H	L	L	L	H	L	L
					X	H	L	H	L	H	L	H
					X	H	H	L	L	H	H	L
					X	H	H	H	L	H	H	H
Test T_3	H	L	L	L	L	X	X	X	L	L	L	L
					H	X	X	X	L	L	L	H
Test T_0 & T_3	H	L	L	H	L	X	X	L	L	L	L	L
					L	X	X	H	L	L	L	H
					H	X	X	L	L	L	H	L
					H	X	X	H	L	L	H	H
Test T_1 & T_3	H	L	H	L	L	X	L	X	L	L	L	L
					L	X	H	X	L	L	L	H
					H	X	L	X	L	L	H	L
					H	X	H	X	L	L	H	H
					L	X	L	L	L	L	L	L
					L	X	L	H	L	L	L	H

Test T_0, T_1 & T_3	H	L	H	H	L	X	H	L	L	L	H	L
					L	X	H	H	L	L	H	H
					H	X	L	L	L	H	L	L
					H	X	L	H	L	H	L	H
					H	X	H	L	L	H	H	L
					H	X	H	H	L	H	H	H
Test T_2 & T_3	H	H	L	L	L	L	X	X	L	L	L	L
					L	H	X	X	L	L	L	H
					H	L	X	X	L	L	H	L
					H	H	X	X	L	L	H	H
Test T_0, T_2 & T_3	H	H	L	H	L	L	X	L	L	L	L	L
					L	L	X	H	L	L	L	H
					L	H	X	L	L	L	H	L
					L	H	X	H	L	L	H	H
					H	L	X	L	L	H	L	L
					H	L	X	H	L	H	L	H
					H	H	X	L	L	H	H	L
					H	H	X	H	L	H	H	H
Test T_1, T_2 & T_3	H	H	H	L	L	L	L	X	L	L	L	L
					L	L	H	X	L	L	L	H
					L	H	L	X	L	L	H	L
					L	H	H	X	L	L	H	H
					H	L	L	X	L	H	L	L
					H	L	H	X	L	H	L	H
					H	H	L	X	L	H	H	L
					H	H	H	X	L	H	H	H
Test T_0, T_1, T_2 & T_3	H	H	H	H	L	L	L	L	L	L	L	L
					L	L	L	H	L	L	L	H
					L	L	H	L	L	L	H	L
					L	L	H	H	L	L	H	H
					L	H	L	L	L	H	L	L
					L	H	L	H	L	H	L	H
					L	H	H	L	L	H	H	L
					L	H	H	H	L	H	H	H
					H	L	L	L	H	L	L	L
					H	L	L	H	H	L	L	H
					H	L	H	L	H	L	H	L
					H	L	H	H	H	L	H	H
					H	H	L	L	H	H	L	L
					H	H	L	H	H	H	L	H
					H	H	H	L	H	H	H	L
					H	H	H	H	H	H	H	H

L = LOW, H = HIGH, X = Don't care

Figure 4.13. Truth table for the 29803 (Copyright 1978 Advanced Micro Devices, Inc. Reproduced with permission of copyright owner.).

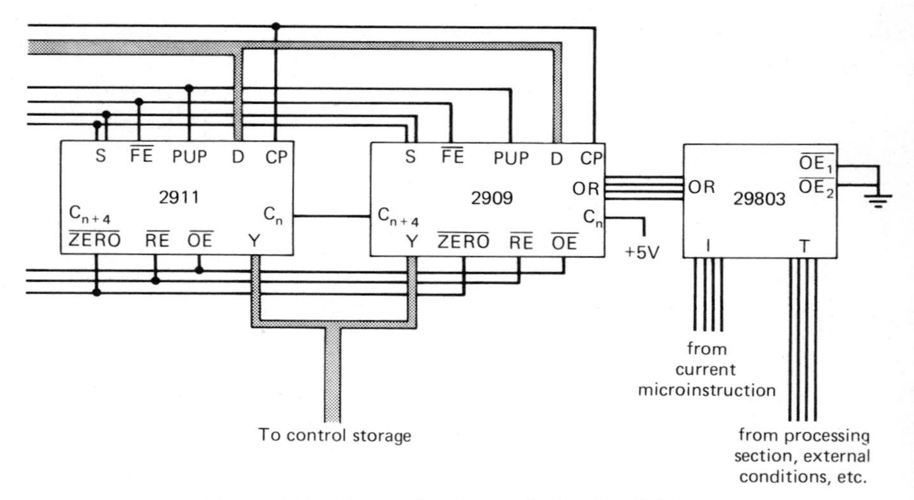

Figure 4.14. A control section employing the 29803.

diagram; this performs the same function as the separate 29811 chip did in Figure 4.10. In addition, the function-control logic in the 2910 provides an additional extremely useful branching operation that is not provided by the 29811.

The external connections of the 2910 sequencer are shown in Figure 4.19. The 12-bit D bus is the bus on which branch addresses are introduced from an external source (e.g., the current microinstruction) and the bus from which the counter is loaded. The 12-bit Y bus is the output bus feeding the address inputs of control storage. The I inputs specify the function (e.g., branching operation) to be performed by the 2910; with a few exceptions, they are identical to the I inputs of the 29811. The $\overline{\text{CC}}$ input feeds, to the 2910, the branch condition to be tested during the current cycle; it is equivalent to the TEST input of the 29811. The $\overline{\text{CCEN}}$ input, if 1, causes the 2910 to ignore the $\overline{\text{CC}}$ input and operate as though $\overline{\text{CC}}$ were 0. Normally, either $\overline{\text{CCEN}}$ is permanently grounded, or it can be fed from the current microinstruction to turn any conditional function (e.g., a conditional branch) into an unconditional function. That is, it can replace the need for the "permanently true" status condition that was used in the design in Figure 4.10.

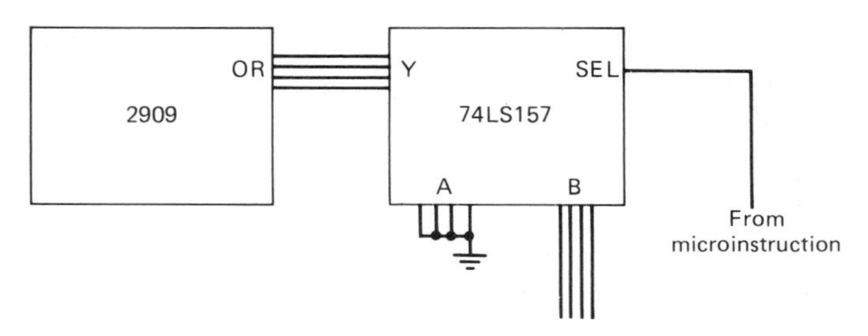

Figure 4.15. Using a multiplexer instead of the 29803.

Max. prop. time	
From input \ To output	OR
I	45
T	45
$\overline{OE}$	30

Figure 4.16. Am29803 guaranteed times.

Some of the functions encoded in the I inputs specify manipulations of the counter. However, the $\overline{RLD}$ (counter/register load) input allows one to control the counter independently of the current function input. If $\overline{RLD}$ is 0, the counter is loaded from the D bus. This is useful when the microinstruction design provides one with the ability to load the counter from an external source such as the processing section (e.g., the output of the ALU in the processing section). (However, since the 2910 has a single input bus, one must be careful to avoid trying to use the bus for two purposes simultaneously.)

Being a self-contained sequencer, the 2910 provides few output signals. The $\overline{FULL}$ output signal goes to 0 when five items are in the stack. The remaining

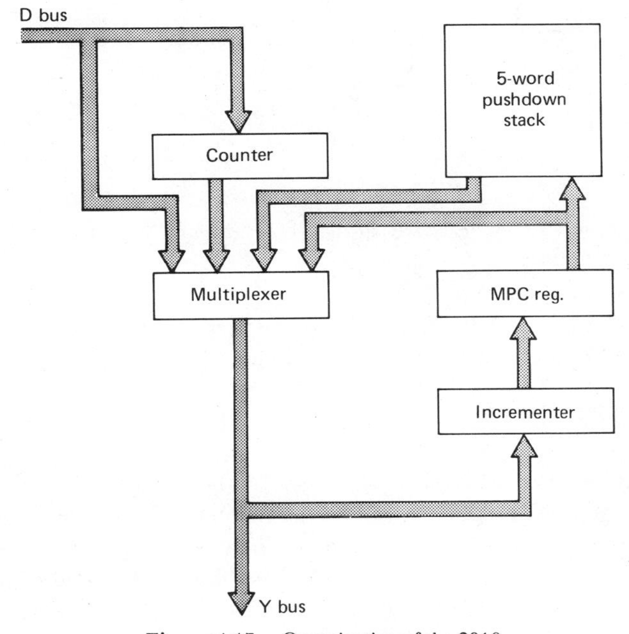

Figure 4.17. Organization of the 2910.

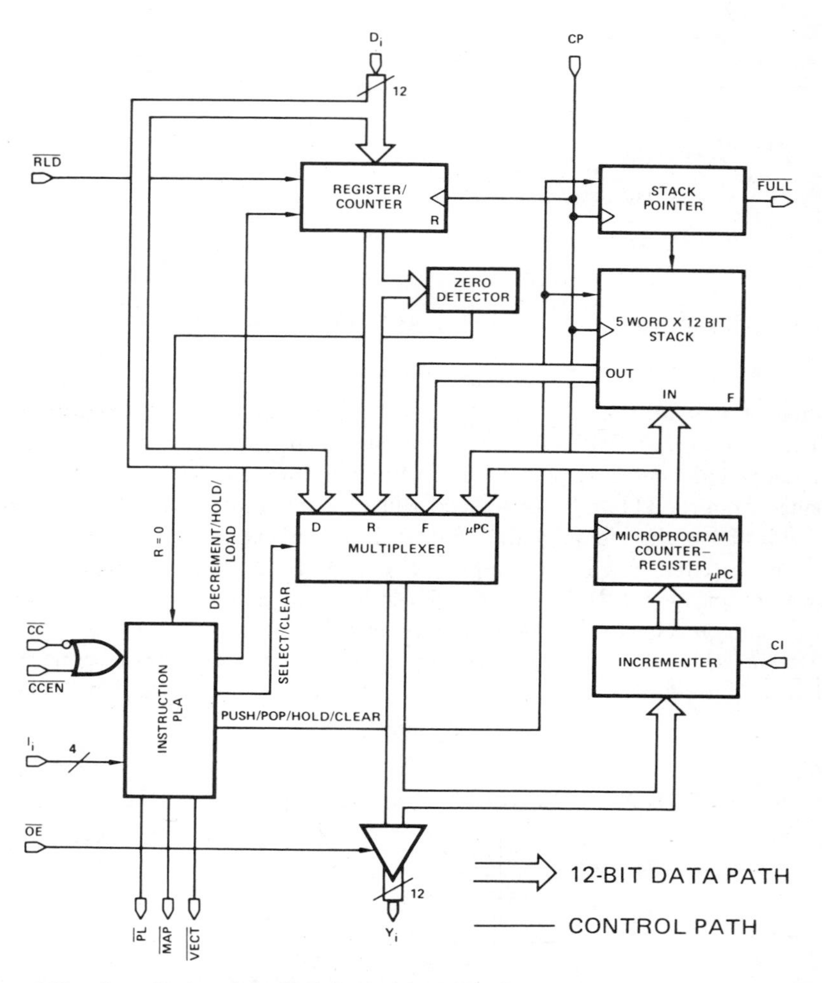

Figure 4.18. Organization of the 2910 (Copyright 1978 Advanced Micro Devices, Inc. Reproduced with permission of copyright owner.).

outputs (PL, MAP, and VECT) contain signals generated as a result of particular function-control inputs. These outputs are discussed later.

2910 Functions

Figure 4.20 specifies the functions performed by the 2910 sequencer as a result of the I, CCEN, and CC inputs, as well as the current value of the counter. With a few small exceptions to be discussed momentarily, the functions are the same as those of a 29811 coupled to 2909 or 2911 slices. For all but two of the I inputs, the $\overline{PL}$ output is signalled (dropped to 0). Normally this output is connected to gate the branch-address field in the current microinstruction (MIF) to the D inputs of the 2910. For the other two I functions, either MAP or VECT is signalled. This allows one to enable a branch-address input from elsewhere.

Table 4.4 is another illustration of the functions specified by the I inputs. The symbols NA, STK, CA, and so on are the same as those used earlier in the chapter. In this table, it is assumed that the CI (incrementer carry) input to the 2910 is permanently connected to the logic-level 1, causing the MPC register to always be filled with the current output address on the Y bus plus 1. Also, it is assumed that MIF (the branch address in the current microinstruction) is permanently connected to the D bus input; alternatively, this connection can be enabled by the $\overline{\text{PL}}$ output.

The functions in Table 4.4 are the same as the functions described in the section on the 29811 with four exceptions. First, the *jump to zero* branch operation, in addition to forcing zeros on the Y bus, also clears (empties) the pushdown stack. The two repeat functions are slightly different from those in the 29811. In the 2910, the counter is tested, if one of these two functions is specified, for zero value. In the design incorporating the 29811 in Figure 4.10, the zero output signal of the counter had to be connected to the status-condition multiplexer and explicitly tested in the microinstruction with the branch-condition field.

Finally, one new, extremely powerful, branch operation was added to the 2910. This is the *three-way branch*, denoted by the I input 1111. In the 29811, I=1111 designated an unconditional jump. As noted earlier, a separate unconditional-jump operation is redundant, since one can easily cause the conditional-jump function to perform an unconditional jump. Hence, in the 2910 the unconditional jump was replaced by the three-way branch. The three-way branch causes one to three control-storage addresses to be selected based on the value of the counter and the condition-test input. The three branch addresses come from (1) the D input bus (usually connected to MIF in the current microinstruction), (2) the address

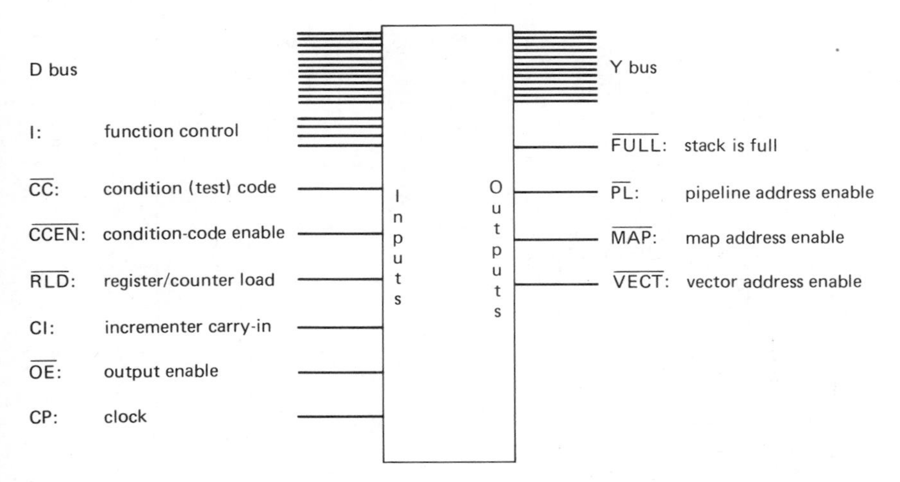

Figure 4.19. 2910 external connections.

HEX I_3-I_0	MNEMONIC	NAME	REG/ CNTR CON-TENTS	FAIL $\overline{CCEN}$ = LOW and $\overline{CC}$ = HIGH		PASS $\overline{CCEN}$ = HIGH or $\overline{CC}$ = LOW		REG/ CNTR	ENABLE
				Y	STACK	Y	STACK		
0	JZ	JUMP ZERO	X	0	CLEAR	0	CLEAR	HOLD	PL
1	CJS	COND JSB PL	X	PC	HOLD	D	PUSH	HOLD	PL
2	JMAP	JUMP MAP	X	D	HOLD	D	HOLD	HOLD	MAP
3	CJP	COND JUMP PL	X	PC	HOLD	D	HOLD	HOLD	PL
4	PUSH	PUSH/COND LD CNTR	X	PC	PUSH	PC	PUSH	Note 1	PL
5	JSRP	COND JSB R/PL	X	R	PUSH	D	PUSH	HOLD	PL
6	CJV	COND JUMP VECTOR	X	PC	HOLD	D	HOLD	HOLD	VECT
7	JRP	COND JUMP R/PL	X	R	HOLD	D	HOLD	HOLD	PL
8	RFCT	REPEAT LOOP, CNTR ≠ 0	≠0	F	HOLD	F	HOLD	DEC	PL
			=0	PC	POP	PC	POP	HOLD	PL
9	RPCT	REPEAT PL, CNTR ≠ 0	≠0	D	HOLD	D	HOLD	DEC	PL
			=0	PC	HOLD	PC	HOLD	HOLD	PL
A	CRTN	COND RTN	X	PC	HOLD	F	POP	HOLD	PL
B	CJPP	COND JUMP PL & POP	X	PC	HOLD	D	POP	HOLD	PL
C	LDCT	LD CNTR & CONTINUE	X	PC	HOLD	PC	HOLD	LOAD	PL
D	LOOP	TEST END LOOP	X	F	HOLD	PC	POP	HOLD	PL
E	CONT	CONTINUE	X	PC	HOLD	PC	HOLD	HOLD	PL
F	TWB	THREE-WAY BRANCH	≠0	F	HOLD	PC	POP	DEC	PL
			=0	D	POP	PC	POP	HOLD	PL

Note 1: If $\overline{CCEN}$ = LOW and $\overline{CC}$ = HIGH, hold; else load. X = Don't Care

Figure 4.20. 2910 I functions (Copyright 1978 Advanced Micro Devices, Inc. Reproduced with permission of copyright owner.).

TABLE 4.4. 2910 Branching Functions

I	Abbre- viation	Branch type	Function	
			$\overline{CC}$=1	$\overline{CC}$=0
0000	JZ	Jump to zero	NA = 000... STK cleared	same
0001	CJS	Conditional call	NA = CA + 1	NA = MIF STK = CA + 1
0010	JMAP	Jump to map location	NA = D	same
0011	CJP	Conditional jump	NA = CA + 1	NA = MIF
0100	PUSH	Push, conditional load counter	NA = CA + 1 STK = CA + 1	NA = CA + 1 STK = CA + 1 CNTR = MIF
0101	JSRP	Conditional call using CNTR or MIF	NA = CNTR STK = CA + 1	NA = MIF STK = CA + 1
0110	CJV	Conditional jump vector	NA = CA + 1	NA = D
0111	JRP	Conditional jump using CNTR or MIF	NA = CNTR	NA = MIF
1000	RFCT	Repeat using stack	If CNTR = 0, NA = CA + 1 POP(STK) Else, NA = TOP CNTR = CNTR−1	same
1001	RPCT	Repeat using MIF	If CNTR = 0, NA = CA + 1 Else, NA = MIF CNTR = CNTR−1	same
1010	CRTN	Conditional return	NA = CA + 1	NA = STK
1011	CJPP	Conditional jump and pop	NA = CA + 1	NA = MIF POP (STK)
1100	LDCT	Load counter and continue	NA = CA + 1 CNTR = MIF	same
1101	LOOP	Conditional loop	NA = TOP	NA = CA + 1 POP (STK)
1110	CONT	Continue	NA = CA + 1	same
1111	TWB	Three-way branch	If CNTR = 0, NA = MIF POP (STK) Else, NA = TOP CNTR = CNTR−1	If CNTR = 0, NA = CA + 1 POP (STK) Else, NA = CA + 1 POP (STK) CNTR = CNTR−1

specified by the top value on the stack, and (3) the address of the current microinstruction plus one. The three-way branch can be used at the end of an iterative loop. It allows one to test both a condition and the value of the counter in one cycle. For instance, suppose the microprogram contains a loop in which, during each iteration, one wishes to test a bit shifted out of the processing section. In the last microinstruction in the loop, one might specify the three-way branch and the branch condition "shift output is 1." If the condition is false ($\overline{CC}$=1) and the counter is not zero, a branch is taken back to the beginning of the loop (assuming this is the address placed on the stack) and the counter is decremented. If the condition is false and the counter is zero (meaning that the loop has ended without encountering a shift output of 1), a branch is taken to the value specified in the MIF field of the current microinstruction, and the top value on the stack is removed. If, at the end of an iteration, the condition is true, the loop terminates and the next microinstruction is the one immediately beyond the end of the loop.

2910 Timing

The storage devices within the 2910 are controlled by the rising clock edge. When the clock input rises, depending on the value of I, an incremented value is stored in the MPC register, a value is stored in the counter, the counter is decremented, and/or a value is stored in the stack. Worst-case set-up and propagation times over the operating range of the 2910 are illustrated in Figure 4.21. There are two propagation times specified from the rising clock edge to the Y output; the larger time (100 ns) is applicable when the current branch operation is one that involves decrementing the counter.

Earlier it was mentioned that the single-chip 2910 is a faster alternative to a 29811 and cascaded 2911 slices. To see this, let us take one branch operation, con-

Max. propagation times (ns)			
From input \ To output	Y	$\overline{PL}$ $\overline{MAP}$ $\overline{VECT}$	$\overline{FULL}$
D	20	—	—
I	70	51	—
$\overline{CC}$	43	—	—
$\overline{CCEN}$	45	—	—
$\overline{OE}$	35	—	—
CP, I=8,9,15	100	—	—
CP, all other I	55	—	60

From input	Minimum set-up time to clock rise (ns)
D	58
I	104
$\overline{CC}$	80
$\overline{CCEN}$	80
$\overline{RLD}$	36
CI	46

Figure 4.21. Representative Am2910 guaranteed times.

ditional jump, and calculate the time through the sequencing logic. Assume we start the computation when the condition being tested by the current microinstruction is present at the TEST input of the 29811 or the CC input of the 2910 (i.e., we will omit the time through the branch-condition multiplexers). In calculating times, one must find all possible logic paths through the design, compute the times for each, and take the largest time as the answer. Rather than doing this here, we will just try a few paths, calculating the elapsed time that must pass before the rising clock edge can occur.

To calculate the 29811/2911 times, assume the design in Figure 4.10, but assume a 12-bit-wide sequencer (i.e., three 2911 slices). Also assume that control storage has an access time of 45 ns. Since the rising clock edge in this design will disable the control-storage outputs and also enable the incremented value in the 2911 slices into the MPC register, we must analyze two logic paths: (1) the time from the presence of the TEST input on the 29811 to the time that the control-storage outputs are available and (2) the time from the TEST input to the time that the MPC register can be loaded from the incrementer. For case 1, there is a propagation time of 40 ns in the 29811 from TEST input to S output (Figure 4.11). Neglecting interchip propagation delays, there is a 30 ns propagation time in the 2911 from the S input to the Y output (Figure 4.4). Finally, there is a 45 ns delay in control storage, leading to a delay along this path of 115 ns. Looking at case 2, again there is a 40 ns delay in the 29811. Then, in the low-order 2911, there is a propagation delay of 48 ns from S to C_{n+4}, a propagation delay of 14 ns in the middle 2911 from C_n to C_{n+4}, and a set-up time in the most-significant 2911 of 28 ns for C_n before the clock edge can appear, leading to a total time along this path of 130 ns. Therefore, neglecting the time needed at the beginning to select the condition being tested, the 29811/2911 design has a path length of 130 ns for the conditional-jump operation.

For the 2910, there is a 43 ns delay between the CC input and the Y output, and a 45 ns access time in the control storage, leading to a path length of 88 ns. Viewing the minimum set-up times in the 2910, the largest is 80 ns, which is less than 88 ns; hence the conditional-jump path through the 2910-based sequencer is 88 ns. (Notice, in Figure 4.21, that there is a 104 ns minimum set-up time from I inputs to the rising clock edge. This was not used in the calculation above, because it was assumed that the I inputs arrive at the 2910 earlier than the CC input; that is, the propagation time through the external branch-condition logic would probably cause the I set-up time to not be on the critical path.)

Comparisons of other situations, such as the repeat operation using the counter, also show the 2910 to be faster than the design in Figure 4.10.

Another advantage of the 2910 is reduced power consumption. The typical power consumed by a single 2910 is one watt. The power consumed by a 12-bit sequencing section consisting of three 2911s, one 29811, and three 74LS169A counters is over two watts. The other advantages of the 2910, of course, are a reduction in physical component space, as well as a reduction in design time. Hence, for most applications, the 2910 seems to obsolete the 29811, 2909, and 2911 components. Exceptions would be situations where more than 12 bits of control-storage addressibility are needed (although Chapter 5 describes techniques to sidestep this) or cases where one wanted to use 2909 or 2911 slices and design different branch-operation microorders using discrete logic.

Using the 2910

Figure 4.22 illustrates a straightforward, but powerful, control-section design employing the 2910. For the sake of illustration, a microinstruction format similar to that of Figure 4.9 will be assumed.

In this example, an Am2922 multiplexer (Advanced Micro Devices) is used to replace the two multiplexers in Figure 4.10. The reason is that the Am2922 encompasses both multiplexers, as it has a POL (polarity) input which can be used to invert the selected status condition.

This control-section design is assumed to be part of a central processor. It is also assumed that the processor design has provided for, in one of the processing-

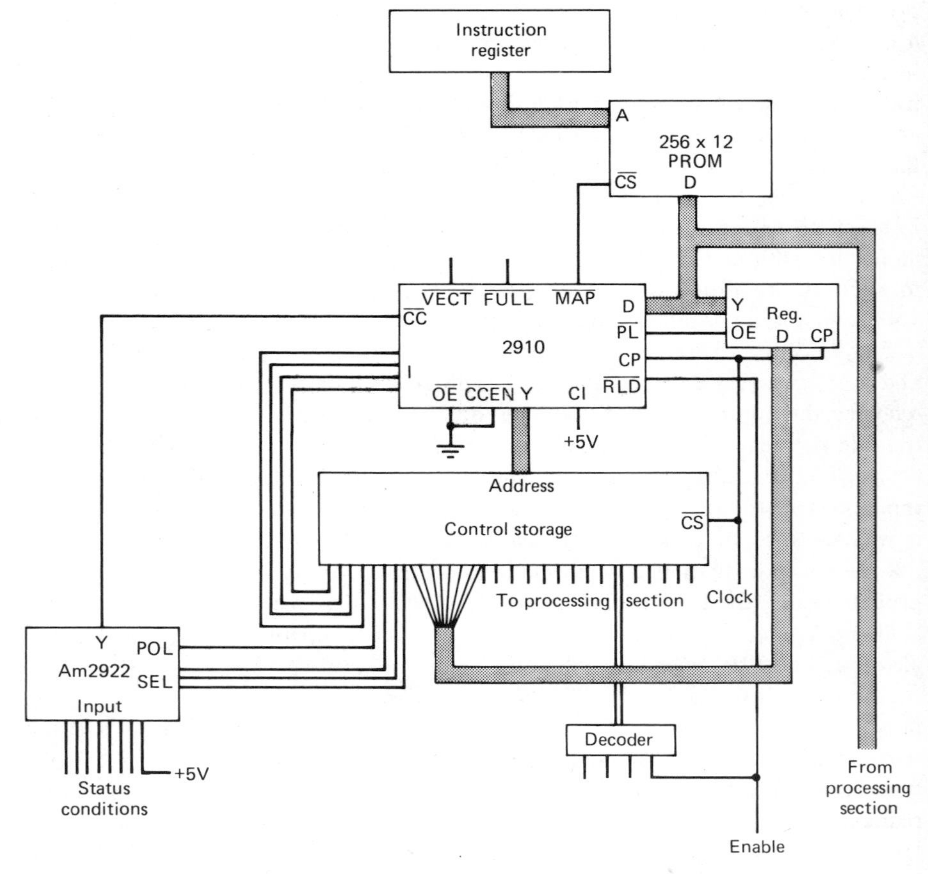

Figure 4.22. A control-section design using the 2910.

section microorders, the ability to gate data from the processing section (e.g., from the Y outputs of a set of 2901 ALU/register slices) into the counter register of the 2910. This is useful in situations where the counter is to be loaded at the beginning of an iterative loop and the number of iterations is not constant (i.e., cannot be "wired" into the MIF field in the microinstruction). Rather, the iteration count is data dependent and comes from the processing section. A processing-section microorder is shown gated into a decoder; one of the outputs of the decoder indicates that the counter in the 2910 is to be loaded from a bus from the processing section. In this case, if $\overline{RLD}$ becomes 0, the counter in the 2910, regardless of the I inputs to the 2910, is loaded directly from the D bus.

Note, with such a facility, that the microprogrammer must be cautious. For instance, in a microinstruction that is loading the counter from the processing section, the microinstruction cannot also specify a branch operation that requires another value to enter the 2910 on its D inputs. For instance, the counter could be loaded from the processing section by a microinstruction that specifies the *continue* or *conditional return* branch operation, but not if the microinstruction specifies *conditional jump* (i.e., one would be enabling more than one output onto the bus feeding the 2910's D input, leading to unpredictable results).

In this example, a use for the *jump-to-map-location* operation is illustrated. Assume that the processor contains a register into which the current machine instruction being decoded and executed is placed. Also assume that, in the system, the first eight bits of a machine instruction represent its operation code. In Figure 4.22 the first eight bits of the instruction register are gated into the address lines of a 12-bit-wide PROM. Encoded in each word of the PROM would be the address, in control storage, of the beginning of the microcode for each type of machine instruction. After the microprogram has fetched the next machine instruction into the instruction register, it might send control to a microinstruction that specifies the *jump-to-map-location* branch operation. In this situation, the 2910 enables its MAP output and obtains the next microinstruction from the address specified on the D bus. The MAP output of the 2910 has been connected to the select input of the PROM. When the *jump-to-map-location* operation is specified in the current microinstruction, the PROM will be enabled and the word in the PROM currently being selected specifies the address of the microinstruction to which control will be sent. Hence, the PROM and the *jump-to-map-location* branch operation are used here to decode the operation code of machine instructions.

Because the 2910's D inputs are being fed from several sources, the MIF field in the current microinstruction cannot be directly connected to the D input of the 2910; rather, this microorder is staged in a register. When a branch operation is specified that delivers the MIF value to the 2910, the PL output of the 2910 is enabled. PL is connected to the output-enable input of this register.

The VECT output of the 2910 has not been used in this design; the *conditional-jump-vector* operation is not supported. This facility is similar to the *jump-to-map-location* operation and MAP output. *Conditional-jump-vector* can be used to obtain the address of the next microinstruction from yet another external source, such as an external mapping PROM. This PROM could be addressed by the set of interrupt conditions in the processor, thus allowing the processor, in one cycle, to test for the presence of various interrupt conditions and branch to various places within the microprogram. The *conditional-jump-vector* branch is different from the *jump-to-map-location* branch, because it is a conditional branch. This additional external source of branch locations would only be activated if the condition being tested by the current microinstruction is true.

THE 3001 SEQUENCER

The 3001 microprogram control unit is a Schottky TTL, 9-bit sequencer in a 40-pin package [5–6]. Like the 2910, the 3001 is not a bit-slice device, although it was originally produced by Intel as part of their 3000 bit-slice family.

Figure 4.23 is an overview of the organization of the 3001 sequencer. Unlike the sequencers and sequencer slices discussed this far, the 3001 is largely a combinatorial device. It has only a single register called the MAR, or microprogram address register. Notably missing from the diagram are a pushdown stack and an incrementer. Rather, the 3001 is oriented toward a different form of microprogram sequencing. The form discussed so far in this chapter and also in Chapter 2 involves some type of incrementing logic in the control section, such that, in

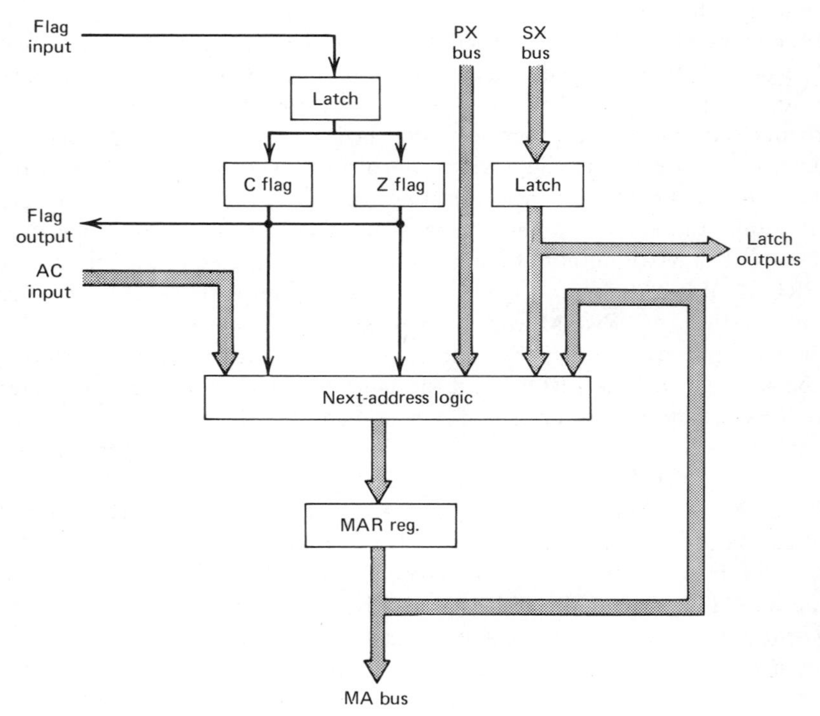

Figure 4.23. Organization of the 3001.

general, each microinstruction would implicitly transfer control to the microinstruction in the next control storage location, or explicitly jump to another microinstruction. The second form, the one assumed by the design of 3001, involves no correlation between the order of microinstructions in control storage and the order in which the microinstructions become active. Rather, this sequencing form requires each microinstruction to explicitly specify the address or addresses of the microinstruction that is to be activated in the next machine cycle.

The heart of the 3001 is the block of combinatorial logic termed the "next address logic." The primary input to this logic is the AC input, a set of seven control lines that specify, in a rather unwieldy way, the manner in which the next address should be formed and also several bits that should appear directly in the next address. In relation to the 2910, the AC inputs serve roughly the same purpose as the I function-control inputs and the D bus in the 2910.

Some other sources also contribute to the determination of the address of the next microinstruction. The MAR register contains the address of the current microinstruction; some of the bits in MAR, as specified by the current AC value, contribute to the address of the next microinstruction. Also the device contains two flipflops that are normally used to hold status conditions from the processing section. These flipflops can also contribute to the formation of the next address, thus allowing one to perform conditional jumps. The 3001 contains two additional busses, PX and SX, that, depending on the current value of AC, may also contribute to the formation of the next address.

The manufacturer's diagram of the 3001 is shown in Figure 4.24. Note the separation in this diagram of the MA (output) bus into two separate busses, one containing four bits and the other containing five bits. The reason is that the 3001 sequencer was not designed to address a linear, or one-dimensional, control storage; rather, control storage is viewed as a matrix, where each microinstruction resides at a particular row and column location. Thus the 9-bit MA bus specifies two addresses: the row address in the upper five bits, and the column address in the lower four bits. The maximum control-storage size is 32 rows and 16 columns, or 512 words.

Figure 4.25 defines the external connections to the 3001 sequencer. The PX and SX busses, or "primary instruction" and "secondary instruction" busses, can be directly gated into MAR if LD is set to 1. For instance, if the microprogram is currently in the process of decoding the next machine instruction to be executed, the operation code of the machine instruction can be fed directly into the 3001, causing the 3001 to perform up to a 256-way branch.

As mentioned earlier, the AC inputs specify how the next microinstruction address should be calculated and also contain several bits that are to be directly injected into the address.

The major output of the 3001 is the 9-bit MA bus, which specifies the row and column address of the microinstruction to be fetched from control storage. Another output is ISE, which is similar in concept to the VECT output of the 2910; ISE is enabled when a particular branch operation is specified by AC. It is intended for use by the microprogram to signal another device to check for interrupt conditions

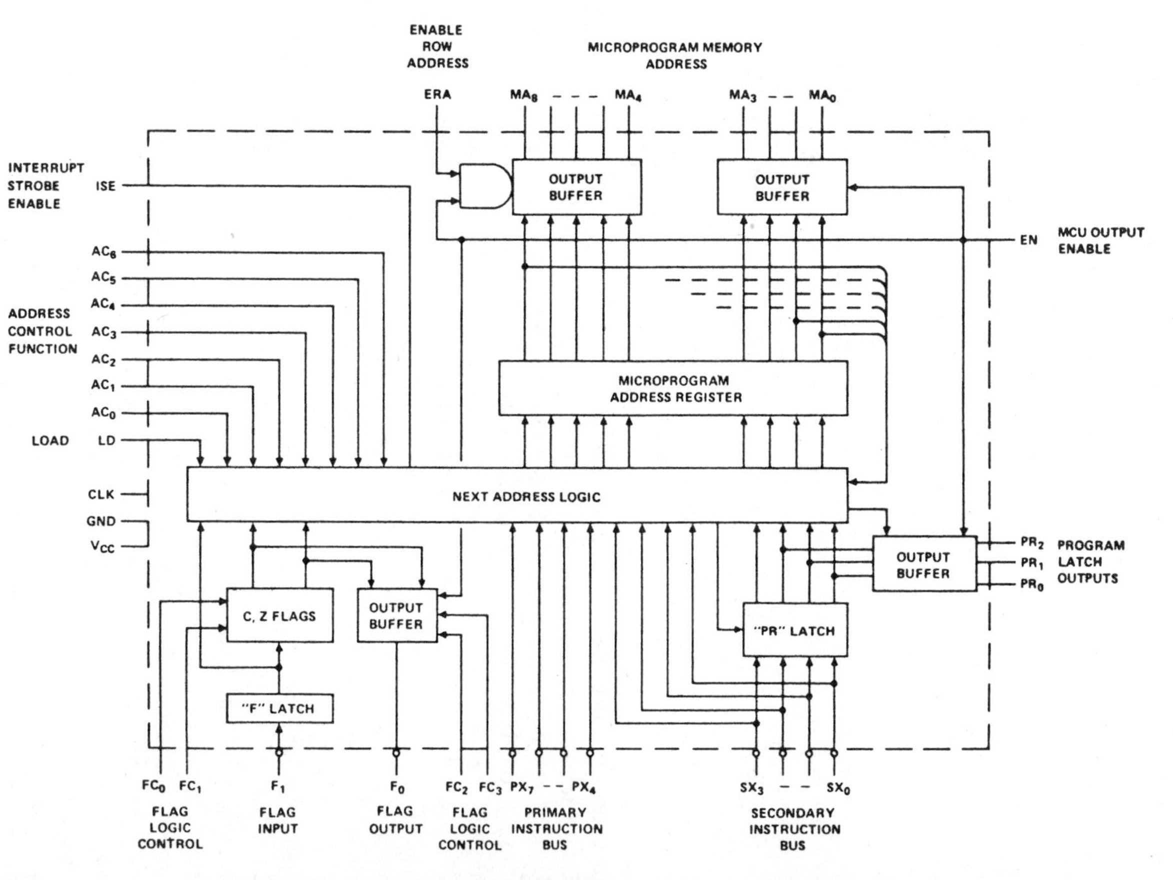

Figure 4.24. Organization of the 3001 (Reproduced with permission of Intel Corp.).

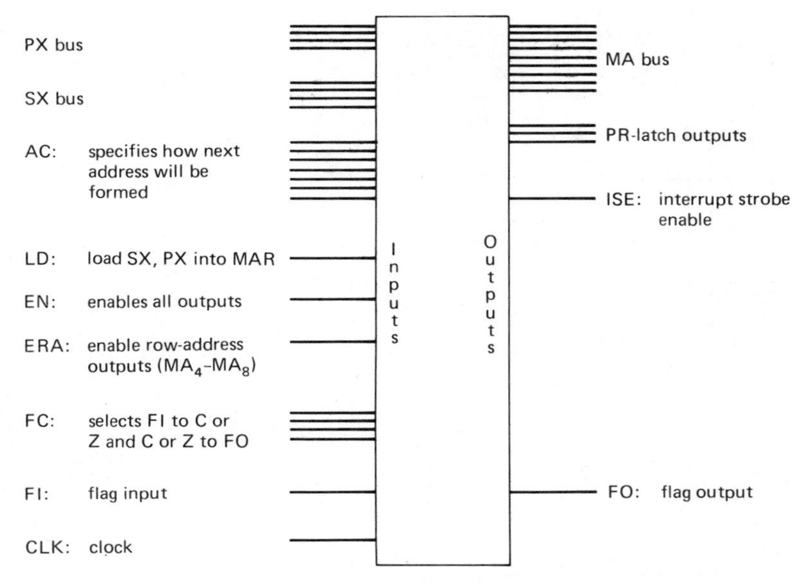

Figure 4.25. 3001 external connections.

when the microprogram is about to initiate the fetching and execution of the next machine instruction.

In addition to being a sequencer, the 3001 contains some flag or status logic that is closely allied with the 3002 ALU/register slice. These interfaces are the connections labelled FC, FI, and FO. They are connected to the carry-output and shift-input pins of the 3002 slices to implement a variety of shift, rotate, and status-checking functions. The FC inputs specify whether the FI input is to be gated to the C and/or Z flipflops in the 3001, or neither. The FC signals also specify whether the FO output should have the value 0, 1, C, or Z.

3001 Functions

Before analyzing how microinstruction branching is performed with the 3001, it is necessary to analyze its unique way of addressing control storage. Control storage is viewed as a two-dimensional array as shown in Figure 4.26. Each microinstruction or location in control storage is specified by both a row and column address. In addition, every location has a "jump set" associated with it. The jump set is the set of control-storage locations that can be addressed or reached from a microinstruction in a particular location. It is important to note that a jump set never contains all locations in control storage; that is, from any given microinstruction in control storage, one can jump to only a limited number of other locations, the most frequent cases being microinstructions in only the current column or row. In fact, in some cases a microinstruction can address only seven other control-storage locations. Given the limited jump sets, and given that there

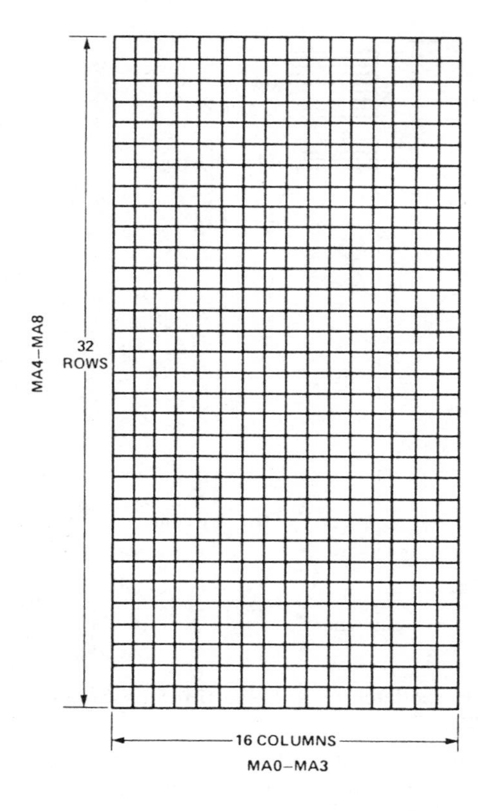

Figure 4.26. Control-storage organization for the 3001.

is no sequencing through control storage (each microinstruction must explicitly point to its successor or successors), one is faced with the combinatorially difficult task of plugging individual microinstructions into control-storage locations such that it is possible for each microinstruction to address its successors.

Figure 4.27 shows how the row and column of the next microinstruction are formed as a function of the seven AC input lines. The AC inputs specify what address should be formed in the MAR register, with one exception. This exception occurs when the LD input is set to 1; in this case, MAR is loaded from the SX and PX busses (the high-order bit of MAR being set to 0).

As shown in Figure 4.27, the first two, three, four, five, or six AC input lines specify the type of jump being performed, and the remaining lines contain bits that are directly gated into MAR. Thus the AC lines specify both the branch operation and part of the address of the next microinstruction. The first four entries in Figure 4.27 specify unconditional branches. (Remember that *every* microinstruction must specify some type of branch, since there is no sequencing to successive control-storage locations.) For instance, the first entry, *jump in current column,* allows a microinstruction to branch to another microinstruction in the current column. In this case, the column-address bits in MAR do not change, and the row address is obtained directly from the five low-order bits of the AC inputs. The

MNEMONIC	DESCRIPTION	FUNCTION							NEXT ROW					NEXT COL			
		AC_6	5	4	3	2	1	0	MA_8	7	6	5	4	MA_3	2	1	0
JCC	Jump in current column	0	0	d_4	d_3	d_2	d_1	d_0	d_4	d_3	d_2	d_1	d_0	m_3	m_2	m_1	m_0
JZR	Jump to zero row	0	1	0	d_3	d_2	d_1	d_0	0	0	0	0	0	d_3	d_2	d_1	d_0
JCR	Jump in current row	0	1	1	d_3	d_2	d_1	d_0	m_8	m_7	m_6	m_5	m_4	d_3	d_2	d_1	d_0
JCE	Jump in column/enable	1	1	1	0	d_2	d_1	d_0	m_8	m_7	d_2	d_1	d_0	m_3	m_2	m_1	m_0
JFL	Jump/test F-latch	1	0	0	d_3	d_2	d_1	d_0	m_8	d_3	d_2	d_1	d_0	m_3	0	1	f
JCF	Jump/test C-flag	1	0	1	0	d_2	d_1	d_0	m_8	m_7	d_2	d_1	d_0	m_3	0	1	c
JZF	Jump/test Z-flag	1	0	1	1	d_2	d_1	d_0	m_8	m_7	d_2	d_1	d_0	m_3	0	1	z
JPR	Jump/test PR-latches	1	1	0	0	d_2	d_1	d_0	m_8	m_7	d_2	d_1	d_0	p_3	p_2	p_1	p_0
JLL	Jump/test left PR bits	1	1	0	1	d_2	d_1	d_0	m_8	m_7	d_2	d_1	d_0	0	1	p_3	p_2
JRL	Jump/test right PR bits	1	1	1	1	1	d_1	d_0	m_8	m_7	1	d_1	d_0	1	1	p_1	p_0
JPX	Jump/test PX-bus	1	1	1	1	0	d_1	d_0	m_8	m_7	m_6	d_1	d_0	x_7	x_6	x_5	x_4

SYMBOL	MEANING
d_n	Data on address control line n
m_n	Data in microprogram address register bit n
p_n	Data in PR-latch bit n
x_n	Data on PX-bus line n (active LOW)
f, c, z	Contents of F-latch, C-flag, or Z-flag, respectively

Figure 4.27. Definition of the AC branch-control inputs (Reproduced with permission of Intel Corp.).

second form allows a microinstruction to jump to any microinstruction in row 0. The third form allows a microinstruction to designate, as its successor, any microinstruction in the current row. The fourth form is an extremely limited form in that the jump set is only seven other control-storage locations. Here the jump is in the current column, but only to a limited number of rows in that column, since the microinstruction can change only the three low-order row address bits. Hence one could name this the "jump in the current column, to a row in close proximity of the current microinstruction" operation.

This form, *jump in column/enable,* also causes the current value in the PR latch to be enabled onto the 3-bit PR-latch outputs. (The PR latch contains four bits, but the PR-latch outputs contain only three bits, a design anomaly.) The purpose of the PR-latch outputs is not clear; for instance, in the Intel SBC 310 High-Speed Mathematics Unit, a design employing the 3001, the PR-latch outputs are unconnected.

The next three jump forms are conditional jumps; they allow a microinstruction to jump to one of two successor microinstructions. The one chosen is dependent on the value of the FI input (the value in the F latch) or the C or Z flipflops. Let us look in detail at the *jump/test F-latch* operation. Here the AC inputs specify the row in which the next microinstruction is contained, but they can specify only one of 16 rows. If the current microinstruction is in rows 0–15, the successors must be in rows 0–15; the same is true for rows 16–31. Notice how the column address is generated. If the current microinstruction is somewhere in columns 0–7 (i.e., $m_3 = 0$), the successor microinstructions must be in columns 2 and 3. If the value of the FI input is 0, the successor in column 2 is addressed; otherwise, the successor in column 3 is addressed. Hence, with this conditional jump, the two successor microinstructions must be in columns 2 and 3 (or columns 10 and 11, if the current microinstruction is in columns 8–15), and they must also be in the same "hemisphere" of rows in which the current microinstruction resides.

By now the reader should be gaining an appreciation of the difficulty that one encounters, when using the 3001 sequencer, in organizing the microprogram in an appropriate way in control storage. This task of allocating microinstructions to specific control-storage locations becomes a major design problem when using the 3001. In fact, it is analogous to the problem of finding a good layout of chips on a printed-circuit or wire-wrap board.

The third group of functions are multiway branches based on the bits in the PR latch. For instance, *jump/test PR-latches* allows one to branch to one of 16 microinstructions. The column is determined by the four bits in the PR latch, and the targets of the branch may be in one of eight rows. The last function, *jump/test PX-bus,* allows one to branch to one of 16 columns, which is specified by the value on the PX bus; these columns may be in one of four rows.

The 3001 contains an output line ISE, or interrupt-strobe enable. This output is similar to the VECT output of the 2910. ISE is raised in only one circumstance; when the AC inputs specify the *jump to zero row* branch function and column 15 (i.e., the AC input is set to 0101111). The intention is that the microinstruction located at row 0, column 15 be the initial microinstruction in the machine-instruction fetch process. Hence, ISE is raised at the beginning of the processing of each machine instruction. Normally, processors handle interrupts (e.g., I/O interrupts) only between the execution of machine instructions. ISE could be used to enable an interrupt controller, which might set the ERA (enable row address) input of the 3001 to 0 (disabling the 3001's row-address outputs) and gate a row address to the row-address lines of control storage, thus forcing a branch in the microprogram to one of several microinstructions that handle the particular interrupt condition detected.

The rather unconventional and inflexible branching functions in the 3001 were motivated by a desire to limit the number of pins on the chip and also limit the number of bits devoted to branching in the microinstruction. Unfortunately this makes the 3001 a particularly unwieldy device to work with, particularly because of the care that must be taken in mapping microinstructions into control-storage row and column locations. In reviewing the branching operations, one can readily see that row 0 and columns 2, 3, 10, and 11 would tend to fill up with microinstructions quickly. (All conditional jumps must end in columns 2, 3, 10, or 11). Experience has shown that, despite the 512-word addressibility of the 3001 sequencer, if one has a design needing over 300 microinstructions, the layout of microinstructions into appropriate control-storage locations becomes an enormously difficult task.

Even if one gets through the assignment process successfully, one is still faced with several other problems. The first is that there are only 128 possible destination pairs for the conditional-jump functions, since all use columns 2, 3, 10, and 11 as their only targets. This implies that even if a microprogram is considerably smaller than 512 microinstructions, one might not be able to fit it into control storage if the microprogram contains a large amount of decision-making logic.

A second serious problem is one of extensibility. Even if one is able to map a microprogram into control storage, it is likely that the exercise will have to be repeated if one wants to add another microinstruction somewhere within the microprogram. Moreover, one may find that a given modification to a microprogram is actually *impossible,* since the possibility exists that the revised microprogram cannot be mapped into control storage, thus invalidating the entire control-section design.

3001 Flag Logic

In addition to the sequencing functions in the 3001, the device also provides a set of flag logic. The logic consists of two flipflops in the 3001, C and Z, which are multiplexed from an input line (FI) and multiplexed to an output line (FO). The four FC input lines control both multiplexers and allow one to explicitly force the FO output to 0 or 1. When coupled to a set of 3002 cascaded ALU/register slices, the CO (carry-out) of the most-significant 3002 slice and the RO (right-shift output) of the least-significant slice are normally connected to the FI input of the 3001, and the FO output of the 3001 is connected to the CI (carry input) pin of the least-significant 3002 slice and the LI (left-shift input) pin of the most-significant 3002 slice. Furthermore, the FC control inputs of the 3001 might be driven by a microorder in the microinstruction. Hence, the microinstruction can force a 0 or 1 carry into the ALU when doing an arithmetic operation, or inject a 0 or 1 into the high-order bit when doing a right-shift operation. Also one can store the shift or carry outputs of the ALU from one machine cycle and use them as inputs in a subsequent machine cycle.

3001 Timing

Figure 4.28 specifies some representative times for Signetics' N3001 sequencer, which is approximately 30% faster than the Intel version (I3001). The minimum clock-cycle time of the N3001 is 60 ns.

THE MC10801 SEQUENCER SLICE

The MC10801 microprogram sequencer is a 4-bit slice in a 48-pin quad inline package; it is an ECL device from Motorola's M10800 ECL family [7]. The MC10801 is a sequencer slice in the sense of the 2909 and 2911. However, the MC10801 is significantly more powerful. Referring to Figure 4.10, the design incorporating 2911 slices and an external counter, several multiplexers, a 29811, and possibly a register (not shown) to contain the status conditions, the MC10801 is a substitute for approximately all of this logic. In other words, this slice provides a set of powerful branching functions (similar to those provided by the 29811/2911 combination), a register that may be used as a counter, and a register and corresponding control logic that can be used to maintain the system's status conditions.

The organization of the MC10801 is shown in Figure 4.29. Compared to the other sequencers discussed so far, this slice is rather complicated. Essentially it contains a 4-word pushdown stack, four additional registers, and an incrementer, as well as one input bus, two output busses, and two bidirectional busses.

From input	Minimum set-up time to clock fall (ns)	Minimum set-up time to clock rise (ns)
LD, AC, FC	7	—
FI	12	—
SX, PX	—	28

From input \ To output (Max. propagation times (ns))	MA	FO	PR, ISE
AC	—	—	32
FC	—	24	—
EN, ERA	26	26	26
CLK	36	36	—

Figure 4.28. Representative N3001 guaranteed times.

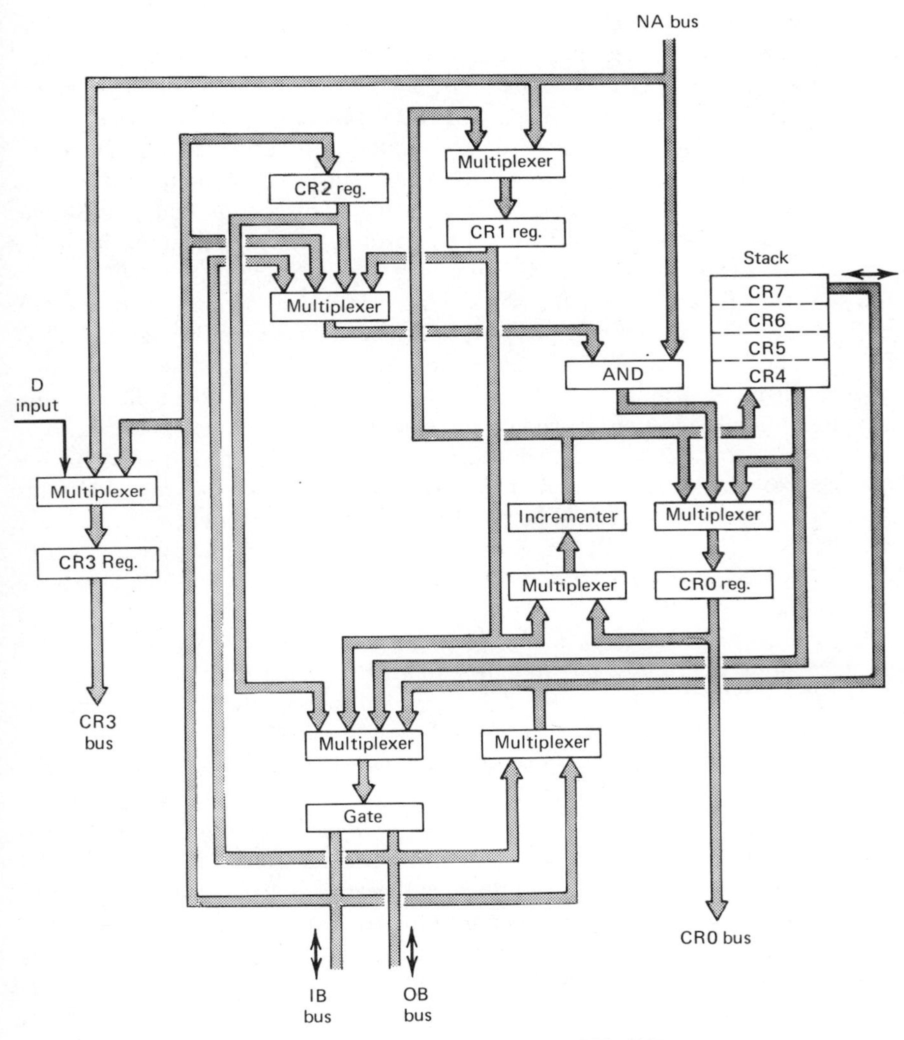

Figure 4.29. Organization of the MC10801.

The device is best understood by analyzing the intended uses of the registers and busses. Register CR0 (across cascaded MC10801s) is intended to hold the address of the current microinstruction; its contents are gated to the CR0 bus, which is normally connected to the address inputs of control storage. The branching functions provided by the slice are described by all possible ways of gating values into the CR0 register. Although these branching operations are discussed later, one can see from the diagram that CR0 may be loaded from the stack, the incrementer, and a set of AND gates. Assuming for the moment that one set of inputs to the AND gates have the value 1, one can see that the NA, IB, and OB busses, and the CR1 and CR2 registers, can be gated into the CR0 register. The NA bus is normally connected to the next-address microorder (MIF) of the current microinstruction. The IB and OB busses may be used to obtain addresses from additional sources, such as mapping PROMs and interrupt controllers.

Register CR1 is intended to be used as a counter for iterations within the microprogram. However, CR1 itself is not a counter; in order to count in CR1, its contents must be routed through the incrementer. This means that, during one clock cycle, one cannot increment CR1 and also increment CR0 (i.e., form the address of the current microinstruction plus one).

As shown in Figure 4.29, CR1 can be loaded from the incrementer or the NA bus. The value of CR1 can be gated to the IB or OB busses, or through the AND feeding the CR0 register. At first glance, the latter doesn't make much sense, since it is difficult to see why one would want to gate an iteration count directly into CR0 (next microinstruction address). However, in addition to its use as an iteration counter, CR1 can also be used as an alternative to the stack, that is, saving the address to which control should be returned after a subroutine call. Using this branching operation, CR0 would be routed though the incrementer, and this value would be routed into CR1, thus saving the address of the next microinstruction in CR1. One would later route the value of CR1 back into CR0 in order to resume control at the original point (i.e., return from the subroutine).

The CR2 register has no particular fixed purpose. Essentially it can be used to hold an address that one wants to gate into CR0 at some later point. The only source for the CR2 register is the IB bus; the value in the CR2 register may be gated to the IB or OB busses and, through the AND, into CR0.

Registers CR4–CR7 are organized as a 4-word pushdown stack for the subroutine branching mechanism. The stack, however, is not a pure pushdown stack, since the low-order register in the stack (CR7) is connected to the IB and OB busses. This was done to allow one to detect the stack-full condition and extend the depth of the stack, if desired, with external logic. When a value is popped from the stack, zeros are stored in the bottom register of the stack. Hence, by reading CR7 via the IB or OB busses, one can determine whether the stack is currently full. By reading out the value of CR7 on the OB bus, for instance, during a push operation, one has the ability, if the stack is full, to move the bottom value in the stack out through the bus to an external stack. When the sequencer performs a pop operation, a logic "1" state is present on the IB and OB busses. One of these busses can then be used to gate information into CR7 from the external stack.

The last register in the MC10801 is CR3, which is intended to be used as a status register. For instance, in a design with three cascaded slices, CR3 allows one to hold up to 12 status conditions in the sequencer. CR3 can be loaded from either the NA or IB bus. In addition, using some of the control inputs to the slice, any single bit in CR3 can be set or cleared from the single D input line. The value in CR3 is continuously available on the CR3 output bus. The status conditions in CR3, as well as any status condition generated externally, can be used to control conditional branches.

One last facility in the MC10801, although not shown in Figure 4.29, is a repeat flipflop (RSQ). RSQ is set when a repetition constant is loaded into CR1 and cleared when CR1 reaches its maximum value (1111). (Unlike the 2910, counting in the MC10801 proceeds upwards; to repeat something *n* times, one loads the two's complement of *n* into CR1.) The RSQ flipflop is used by the branching logic to detect the end of a repetitive operation.

Figure 4.30 is the manufacturer's block diagram of the MC10801.

Figure 4.31 illustrates the pins on the MC10801 sequencer slice. The only pins not shown are four ground connections and four supply-voltage connections (−2.0V and −5.2V).

Most of the connections in Figure 4.31 are either self-explanatory or were mentioned earlier. The four IC lines specify the type of branch to be performed, that is, how the next value for the CR0 register is to be obtained. Given the on-chip status register and the multitude of internal interconnections in the slice, the MC10801 contains a secondary set of function-control signals, the CS lines. Briefly, the nine CS lines have the following functions

CS0–CS3	—specify which bit of the status register (CR3) shall be loaded from D_{in}, which bit of CR3 shall be gated to the XB bus, or whether CR3 should be loaded from the IB or NA bus.
CS4	—specifies whether a status condition on the B line shall be used in determining the state of the current conditional branch.
CS5	—the output-enable signal for the CR0 bus.
CS6–CS8	—specify one of four output sources and to which bus (IB or OB) it should be gated.

The B line is used to gate an external status condition into the slice for branching purposes. Some of the branch operations allow the operation to be decided by the condition CR3 bit AND $\overline{B}$. The single D_{in} input line can be gated, under control of the CS inputs, into a particular bit of the status register. The reset input causes all registers to be reset on a rising clock pulse. However, resetting the stack to zero requires enabling the reset input and providing five clock pulses to CLK.

The $\overline{XB}$, or extender bus, is a two-way connection. Its motivation is similar to that of the Z line on the 2903 ALU/register slice, that is, it is a way of broadcasting status from one cascaded MC10801 slice to another. The need for $\overline{XB}$ is dictated by the placement of the status register on the chip. As mentioned earlier, a design incorporating three slices, for instance, can maintain up to 12 status conditions in the CR3 registers in the three slices. However, if one wishes to base a conditional branch on one of the status bits in one of the CR3 registers, how do the other slices know the value of the bit? This is achieved with the $\overline{XB}$ bus. Control lines CS0 and CS1 select one of four status bits on the slice. If input line CS3 is 0 on a slice, it indicates that $\overline{XB}$ shall be an output and its value will be the status bit selected by CS0 and CS1. If a CS3 value of 1 is sent to a slice, $\overline{XB}$ is designated as an input. Hence, to select a particular status bit on which a branch decision is to be based, the same CS0 and CS1 values are sent to all slices, but a

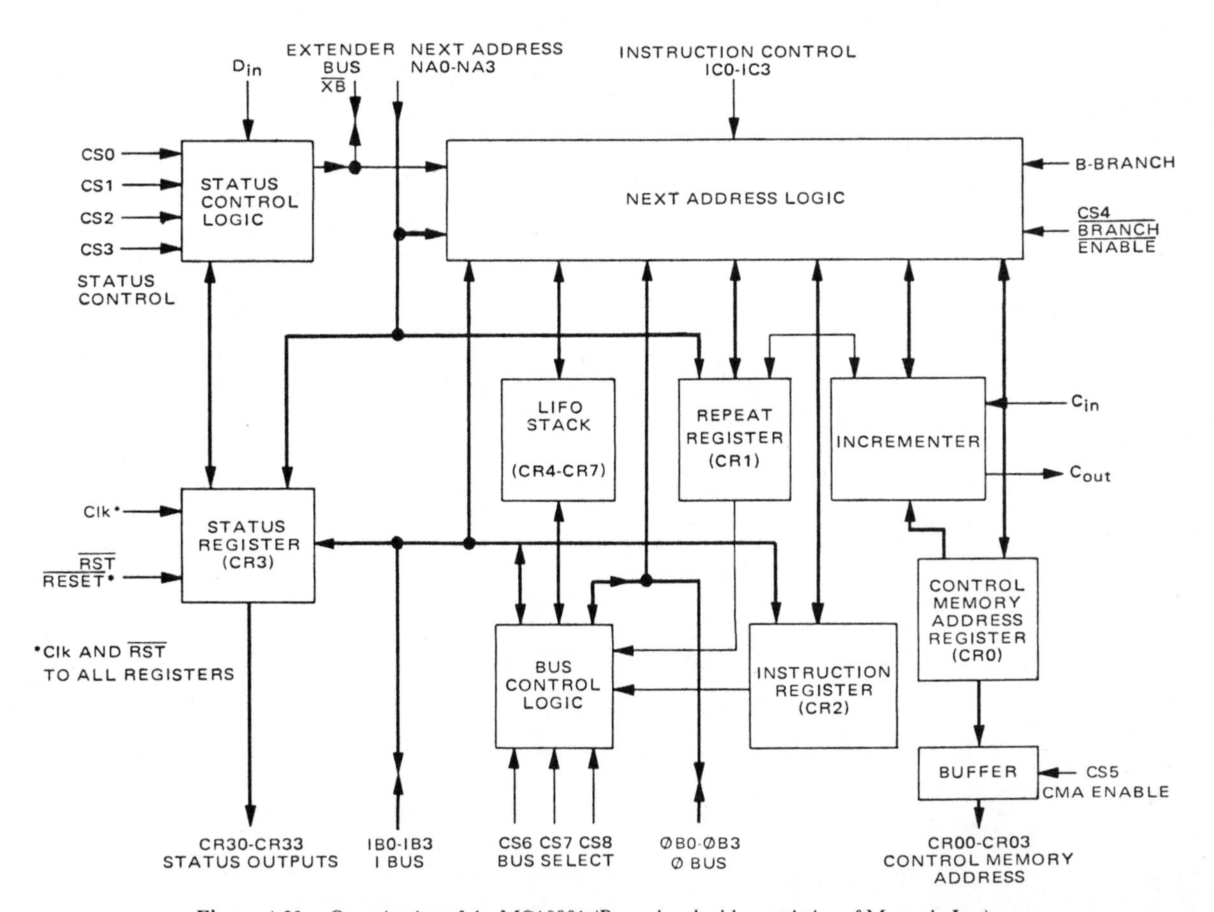

Figure 4.30. Organization of the MC10801 (Reproduced with permission of Motorola Inc.).

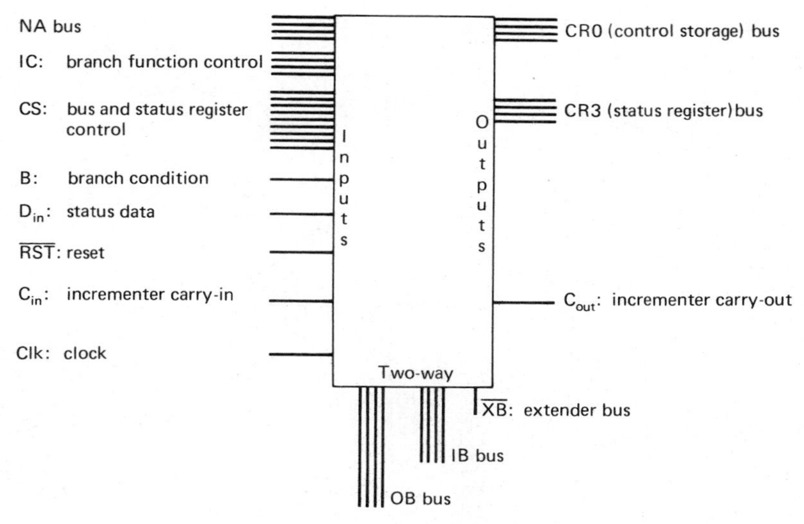

Figure 4.31. MC10801 external connections.

CS3 signal of 0 is sent to only that slice on which the particular status bit is contained. Thus $\overline{XB}$ broadcasts that status bit to the other cascaded slices.

MC10801 Functions

Assuming one needs the full function of the MC10801 slice, the microinstruction would contain a branch-address (MIF) field feeding the NA bus, a 4-bit branch-type microorder feeding the IC inputs, and an 8-bit bus-and-status-control microorder feeding the CS inputs. As described earlier, if one is using the CR3 registers on more than one slice to contain status values, the microinstruction would also contain a field to select the appropriate CR3 register (i.e., a micro-order that is decoded to enable the CS3 input on the proper slice).

The branching operations specified by the IC inputs are defined in Figure 4.32. The *increment* and *jump-to-next-address* operations should require no further explanation. The next four branch types allow one to branch to an address specified by the IB or OB bus, or stored in the CR2 register. Notice that the actual address formed is the designated source ANDed with the value on the NA bus, allowing one to mask the branch address with a value on the NA bus (or vice versa).

As shown in Figure 4.32, three branching operations, *jump to subroutine, return from subroutine,* and *repeat instruction,* have two different effects depending on the value of RSQ (the repeat flipflop). If RSQ is 0, the normal effect of *jump to subroutine* and *return from subroutine* is a normal unconditional subroutine call and return. The MC10801 has iteration capabilities, but they are performed differently from those in other sequencers, such as the 2910. Because the MC10801 has no separate counter (CR1 is just a register), one cannot

MNEM	CODE				DESCRIPTION	RESET	BRANCH OR REPEAT	REGISTER AND FLIP FLOP OUTPUTS[4] V_{OL} ⌐ V_{OH}				
	IC3	IC2	IC1	IC0		$\overline{RST}$	CONDITION[2]	CR0	CR1	CR2	LIFO STACK CR4 – CR7	RSQ[3]
X	X	X	X	X	RESET CONDITION	0	X	0	0	0	"PUSH" CR0 TO STACK	0
INC	1	1	0	0	INCREMENT	1	X	CR0 plus C_{in}	–	–	–	–
JMP	0	0	1	0	JUMP TO NEXT ADDRESS	1	X	NA	–	–	–	–
JIB	1	0	0	0	JUMP TO I BUS	1	X	IB·NA	–	–	–	–
JIN	1	0	0	1	JUMP TO I BUS & LOAD CR2	1	X	IB·NA	–	IB	–	–
JPI	1	0	1	0	JUMP TO PRIMARY INST.	1	X	CR2·NA	–	–	–	–
JEP	1	1	1	0	JUMP TO EXTERNAL PORT	1	X	ØB·NA	–	–	–	–
JL2	0	0	0	1	JUMP & LOAD CR2	1	X	NA	–	IB	–	–
JLA	0	0	1	1	JUMP & LOAD ADDRESS	1	X	NA	CR0 plus C_{in}	–	–	–
JSR	0	0	0	0	JUMP TO SUBROUTINE	1	$\overline{RSQ}+RIN\cdot\overline{XB}=0$	NA	–	–	"PUSH" CR0 TO STACK	–
						1	$\overline{RSQ}+RIN\cdot\overline{XB}=1$	NA	–	–	"PUSH" CR0 plus C_{in}	–
RTN	1	1	1	1	RETURN FROM SUBROUTINE	1	$\overline{RSQ}+RIN\cdot\overline{XB}=0$	CR4	CR1 plus C_{in}	–	"POP" STACK TO CR0	–
						1	$\overline{RSQ}+RIN\cdot\overline{XB}=1$	CR4	–	–	"POP" STACK TO CR0	0
RSR	1	1	0	1	REPEAT SUBROUTINE	1	X	CR0 plus C_{in}	NA	–	–	1
RPI	1	0	1	1	REPEAT INSTRUCTION	1	$\overline{RSQ}+RIN\cdot\overline{XB}=0$	–	CR1 plus C_{in}	–	–	–
						1	$\overline{RSQ}+RIN\cdot\overline{XB}=1$	CR1·NA	–	–	–	0
BRC	0	1	0	1	BRANCH ON CONDITION	1	$\overline{XB}\cdot(CS4+\overline{B})=0$	NA	–	–	–	–
						1	$\overline{XB}\cdot(CS4+\overline{B})=1$	CR0 plus C_{in}	–	–	–	–
BSR	0	1	0	0	BRANCH TO SUBROUTINE	1	$\overline{XB}\cdot(CS4+\overline{B})=0$	NA	–	–	"PUSH" CR0 plus C_{in}	–
						1	$\overline{XB}\cdot(CS4+\overline{B})=1$	CR0 plus C_{in}	–	–	–	–
ROC	0	1	1	1	RETURN ON CONDITION	1	$\overline{XB}\cdot(CS4+\overline{B})=0$	CR4	–	–	"POP" STACK TO CR0	–
						1	$\overline{XB}\cdot(CS4+\overline{B})=1$	NA	–	–	–	–
BRM	0	1	1	0	BRANCH & MODIFY	1	CS4=1	NA	–	–	–	–
						1	CS4=0	CR00=NA0·B CR01=NA1·XB CR02=NA2 CR03=NA3	–	–	–	–

NOTES.
1. X = DON'T CARE STATE
   – = NO CHANGE
2. EQUATIONS APPLY AS SHOWN, WHERE:
   $\overline{RIN}$ = (CR13·CR12·CR11·CR10)
   $\overline{XB}$ = EXTERNAL EXTENDER BUS NODE
   $\overline{B}$ = COMPLEMENT OF BRANCH INPUT
3. RSQ = OUTPUT OF RSR FLIP FLOP
4. ALL REGISTERS AND RSR FLIP FLOP CHANGE STATE ON V_{OL} TO V_{OH} (POSITIVE GOING) CLOCK TRANSITION
5. NEGATIVE LOGIC USED THROUGHOUT

Figure 4.32. MC10801 branching operations (Reproduced with permission of Motorola Inc.).

simultaneously increment, decrement, or test the iteration count and increment the current control-storage address. Hence one cannot easily implement iterative loops with the MC10801. However, there are two alternatives. One is to define the loop as a subroutine, and then call the subroutine iteratively. The other alternative, provided by the *repeat-instruction* operations, is an iterative loop, but the loop can contain only a single microinstruction.

One would implement an iterative loop as a subroutine in the following fashion. First, a microinstruction would specify the *repeat-subroutine* operation. This causes control to be passed to the microinstruction following this one, but it also sets CR1 from the NA bus and sets RSQ. Since CR1, via the incrementer, counts in an upward fashion to 111..., one would place the desired number of repetitions in two's-complement form on the NA bus. The next microinstruction might specify *jump to subroutine*. Here, since RSQ is 1, a branch would be taken to the address specified on the NA bus (the address of the first microinstruction in the subroutine), but rather than pushing the value CR0 + 1 onto the stack, the value in CRO is pushed onto the stack. Hence, when the subroutine returns, it will activate the microinstruction that jumped to the subroutine.

Since RSQ is still 1 when the return operation is performed, CR1 is incremented by 1. When CR1 reaches the value 111..., RSQ is reset. The microinstruction that specifies the *jump-to-subroutine* operation will either push its own address onto the stack, or push the address of the next microinstruction onto the stack (if RSQ = 0). Hence, iterative subroutine calls can serve as a substitute for iterative loops.

The *repeat-instruction* operation works in a similar fashion, where, depending on the value in RSQ, the operation causes the current microinstruction to either branch to itself or to the address specified on the NA bus.

The next three operations represent the conditional branch and conditional subroutine call and return. Notice that the branch condition involves testing three input signals: $\overline{XB}$, CS4, and B. CS4 is normally used to specify whether one wants to base the condition on $\overline{XB}$ or $\overline{XB}$ AND $\overline{B}$. That is, if CS4 = 0, the condition tested is $\overline{XB}$ AND $\overline{B}$; if CS4 = 1, the condition tested is $\overline{XB}$. Recall that $\overline{XB}$ is normally used to transmit the value of the selected status bit to all the cascaded slices. B can be used to signal an external status condition to the slices.

Recall from earlier examples in this chapter that the control-section designs included the useful ability to branch on the true or false status of a condition (e.g., branch on overflow, and branch on no overflow). Although the status register and branching conditions in the MC10801 are rather powerful, they do not easily provide one with this useful facility.

The last branching operation, *branch and modify,* is used to perform a 4-way branch, where the two low-order bits are influenced by the values $\overline{XB}$ and B. This can be disabled by setting CS4 to 1. The normal mode of operation, when using this branching operation, would be to set CS4 on all but the least-significant slice to 1, thus allowing B and $\overline{XB}$ to modify only the low-order bits of the next microinstruction address.

The CS control inputs primarily select a status condition and control some of the busses, particularly the $\overline{XB}$, IB, and OB busses. CS0 and CS1 select one of four bits in CR3 (the status register). In most cases, if CS3 is 0, the negation of the selected status bit is gated to the $\overline{XB}$ bus; if CS3 is 1, the $\overline{XB}$ bus is used as an input. However, under the conditions where the IC inputs specify the *branch-on-condition, branch-to-subroutine,* or *return-on-condition* branching operations, and CS4 is set to 0, $\overline{XB}$ contains the negation of the selected status bit ANDed with the complement of the input on the B bus.

CS0-3 also allow one to selectively load values into the CR3 status register. If CS2 = 0, the D_{in} signal is loaded into the status bit selected by CS0 and CS1. If CS3 is 0 and CS2 is 1, D_{in} is not used. If CS3 = CS2 = 1, register CR3 is loaded with zeros, itself (no change), or the value on the IB or OB bus, the source being dependent on the value of CS0 and CS1.

CS6–8 are used to gate a source to the IB or OB busses when one wants to use one of these busses as an output bus. When IB or OB is not being used as an output bus, the bus drivers are forced to a low-level logic condition to provide for input operations through these busses. CS6 selects IB or OB as the output bus, and CS7 and CS8 select the source (CR1, CR2, CR4, or CR7).

MC10801 Timing

Figure 4.33 lists some representative set-up times and propagation delays for the MC10801. The propagation delays are specified over the operating range of −30 to +85°C; the set-up times are specified at 25°C.

THE 74S482 SEQUENCER SLICE

The 74S482 is a 4-bit-wide microprogram sequencer slice produced by Texas Instruments [8]. It is a Schottky TTL device in a 20-pin package.

From input	Minimum set-up time to clock rise (ns)
NA	28
IC	44
IB, OB	25
CS0-3	35
C_{in}	15
D_{in}	20
$\overline{XB}$	28

Maximum propagation times (ns)					
From input \ To output	CR0	CR3	IB, OB	$\overline{XB}$	C_{out}
IC	—	—	38	24	27
CS0-4, B	—	—	—	20	—
CS5	11	—	—	—	—
CS6–8	—	—	26	—	—
$\overline{XB}$	—	—	36	—	—
C_{in}	—	—	—	—	8
Clk	17	17	32	23	24

Figure 4.33. Representative MC10801 times.

The organization of the 74S482 is illustrated in Figure 4.34. The device is similar to the 2911 slice. The major differences are that the 74S482 has a full adder instead of an incrementer, and one register rather than two. The register holds the address of the current microinstruction. Into this register, one can gate (1) the top value on the stack, (2) the output of the adder, (3) the input (D) bus, or (4) the register (i.e., to repeat the current microinstruction). The A input to the adder can be zero or the D bus; the B input to the adder can be zero or the register.

The manufacturer's diagram of the 74S482 is shown in Figure 4.35, and the external connections are shown in Figure 4.36. Unlike the other devices discussed, the 74S482 has no output-enable input, meaning that the output bus, F, is not a 3-state output.

74S482 Functions

The six S input lines control the 74S482 by controlling the multiplexer, the stack, and the adder. Two of the lines, S_5 and S_6, designate the value that will be gated into the register on a rising clock edge as indicated below.

00—D bus
01—Adder output
10—Top value on stack
11—Register (hold)

Another group of two lines, S_3 and S_4, cause the following change to occur in the pushdown stack on a rising clock edge.

00—no change
01—overlay the top value on the stack with the adder output
10—pop
11—push the adder output onto the stack

The remaining two lines, S_1 and S_2, specify the inputs to the adder, as shown below.

00—D bus + register
01—D bus + 0
10—0 + register
11—0 + 0

The carry input is a third input to the adder. For instance, if $C_{in} = 1$, $S_1S_2 = 10$, and $S_5S_6 = 01$, the adder becomes an incrementer, incrementing the address in the register by one.

Compared with the other sequencers discussed, the unusual aspect of the 74S482 is its adder. Although, at first glance, an adder appears to be more useful

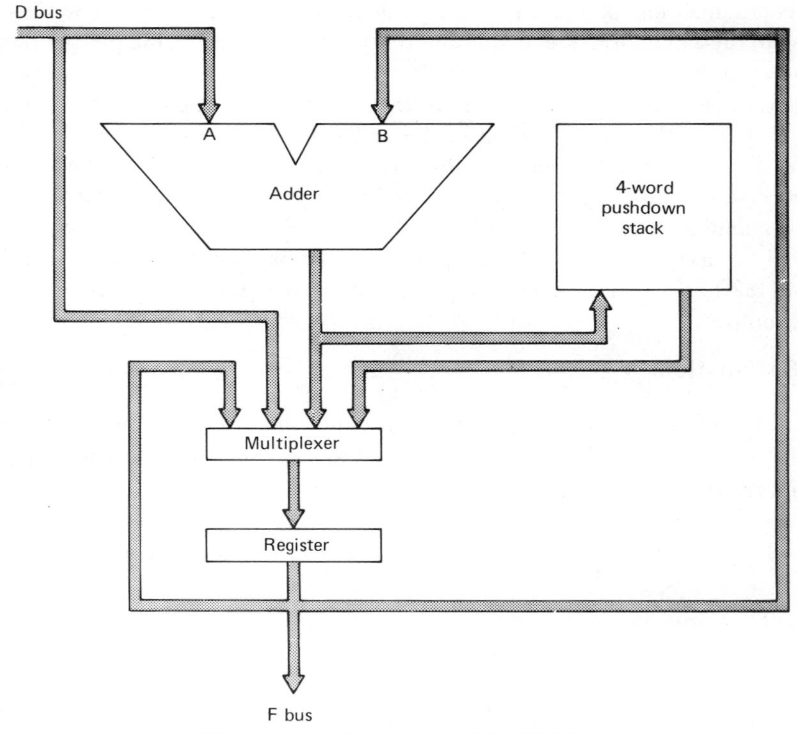

Figure 4.34. Organization of the 74S482.

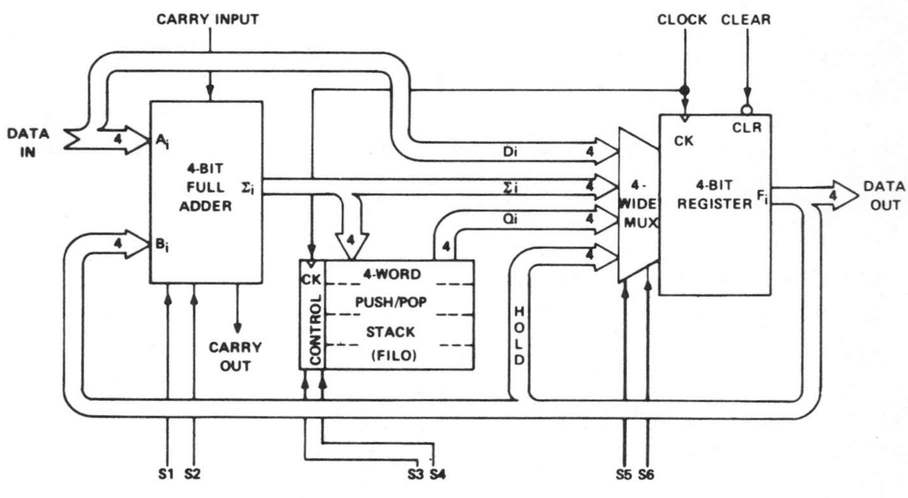

Figure 4.35. Organization of the 74S482 (Copyright 1977 Texas Instruments Inc. Reproduced with permission.).

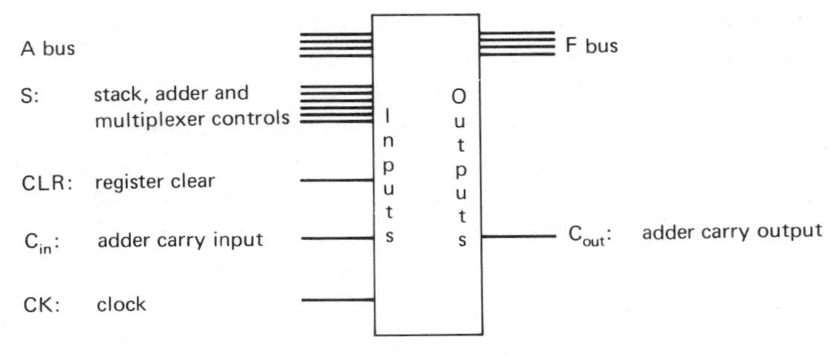

Figure 4.36. 74S482 external connections.

than an incrementer, in practice the adder is normally found to be no more useful than an incrementer. The motivation for the inclusion of the adder was the concept of *indexed* or *relative branching*. A well-known characteristic in both programs and microprograms is that the average *branch distance* (the "distance" between a microinstruction performing a branching operation and the target microinstruction, measured in terms of the difference in their addresses) is small. (This characteristic in programs, also known as *locality of reference*, is one of the reasons why virtual-storage, paging, and cache mechanisms are successful). Because of this, one could possibly shorten the branch-address (MIF) field in the microinstruction, thus saving control-storage space. Rather than specifying an absolute address, the MIF field would specify a control-storage address *relative* to the current address. That is, if a microinstruction at location 932 specified an unconditional branch to the microinstruction at location 939, the MIF field would contain the value 7 rather than 939. The S inputs to the 74S482 would be 000001, which designate the operation

$$\text{register} = \text{D bus} + \text{register}$$

The relative-addressing capability provided by the adder would probably not be used in a 74S482 application. One reason is that it requires additional logic when branching backward (i.e., to a lower-valued address). The logic would have to take the relative address, which is smaller than the control-storage address, and propagate the high-order bit to be able to form a negative input to the adder. Furthermore, relative addressing cannot be used for subroutine calls. When doing a subroutine call, one has to use the adder to increment the current address by one before it is stored on the stack; thus the adder could not also be used during the same cycle to convert the relative subroutine address to an absolute address.

One might try to devise a microinstruction design that employed both absolute and relative addressing. That is, the microinstruction might contain a bit specifying whether the address field is absolute or relative. This bit, among other things, would influence the state of the S_1 and S_2 lines to the 74S482 slices. In a microinstruction specifying relative addressing, the unused bits in the address field (e.g., four, if absolute addresses contain 12 bits and relative addresses contain 8 bits) could be decoded and used for some other purpose. However, this added complexity would probably not be justified.

One situation in which the adder can be useful is implementing multiway branches. If one enabled, for instance, a register, rather than the MIF field in the microinstruction, to the D bus and set S to 000001, one would branch to one of n locations, where n is the maximum value in the register. Using addition for multiway branches, rather than ORing, as with the 2909, means that one does not have to worry about aligning microinstructions at appropriate control-storage address boundaries. Also, n need not be a power of two.

74S482 Timing

Figure 4.37 contains representative set-up times and propagation delays for the 74S482. Speeds of all the sequencers are compared at the end of this chapter.

THE 8X02 SEQUENCER

The 8X02 microprogram sequencer is a 10-bit low-power Schottky device in a 28-pin package. The 8X02 is produced by Signetics [9]. Although it is not one of the better-known sequencers, mostly because it is not a member of a device family, the 8X02 is a simple and powerful sequencer and might be the best choice for a design requiring less than 1024 microinstructions and requiring only limited looping capabilities.

The organization of the 8X02 is shown in Figure 4.38. All busses and registers are 10 bits in width. Like most other sequencers, its central components are a pushdown stack and a multiplexer feeding the output bus. It has two incrementers, allowing the current address to be incremented by 1 or 2. Another unusual feature is the presence of a multiplexer at the input port of the stack, allowing one to push the current address or the current address plus one onto the stack.

From input	Minimum set-up time to clock rise (ns)
A, S_3, S_4	15
CLR, S_5, S_6	0
S_1, S_2	30

From input \ To output (Maximum propagation times (ns))	F	C_{out}
A	—	25
C_{in}	—	18
CLR	20	—
CK	25	—

Figure 4.37. Representative 74S482 guaranteed times.

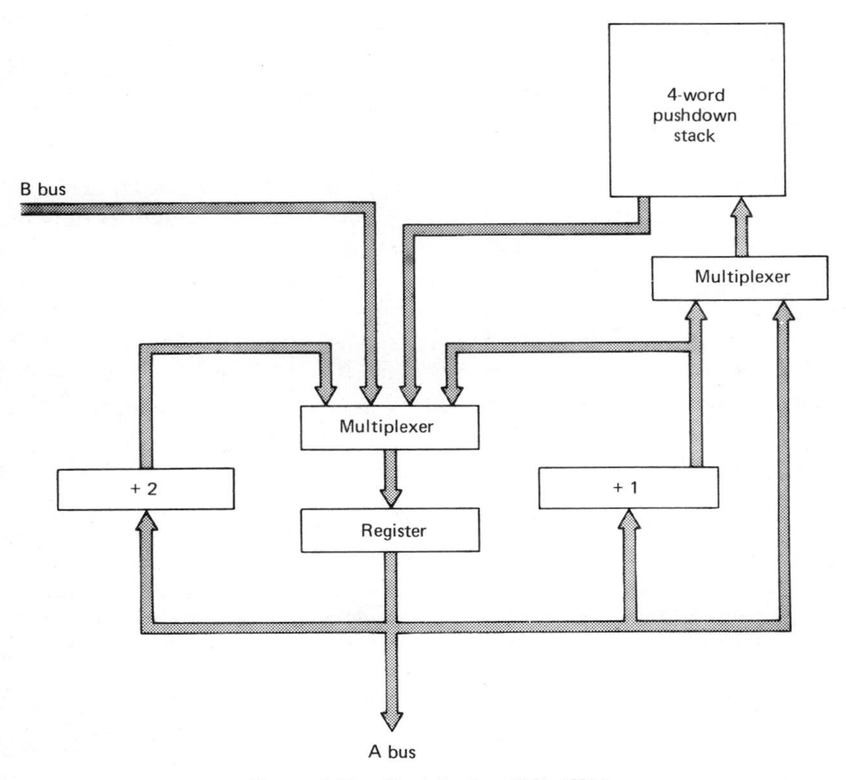

Figure 4.38. Organization of the 8X02.

The external connections of the 8X02 are shown in Figure 4.39. Despite its width (10 bits), the device has only 28 pins (in contrast, the 2910, which is a 12-bit sequencer, has 40 pins). The only output, the A bus, normally feeds the control-storage address inputs. Like most other sequencers, the address output is a 3-state output; it is enabled by the EN input.

The B input bus is normally fed from the address field (MIF) in the current microinstruction. The AC inputs specify the type of branching operation to be

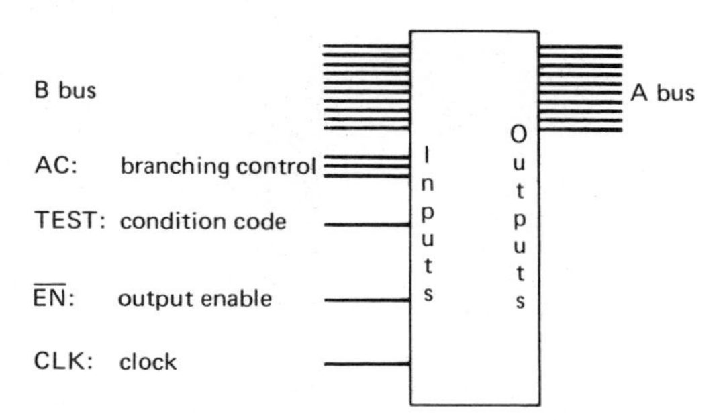

Figure 4.39. 8X02 external connections.

performed. The TEST input supplies a condition (usually the output of a condition multiplexer) to be tested during the current cycle.

8X02 Functions

The three AC lines specify the function to be performed by the 8X02, as shown in Table 4.5. The first operation, *test and skip,* is a useful operation. If the TEST input is 1, the microinstruction following the current microinstruction is skipped. This operation does have the same effect as specifying *conditional branch* and specifying, in MIF, the current address plus two, but *test and skip* does not require use of the MIF field, freeing it for other uses. In a typical design, the microinstruction might have a width of 50 bits. If an 8X02 is being used, the MIF field in the current microinstruction is likely to occupy 10 bits (if addressibility is needed to 1024 control-storage locations). Thus the MIF field occupies 20% of control storage, making it an expensive resource. Because of this, as was discussed earlier, one may want to use this field for multiple purposes, such as an emit field that can be gated into the processing section. However, one cannot, in one microinstruction, use MIF for both purposes (control-storage address and emit value), unless the address happens to equal the emitted value needed, an unlikely and error-prone circumstance. Hence, the fewer microinstructions that use MIF for branching, the more it can be used for other purposes. Since conditionally branching around the next microinstruction is a frequent circumstance, the *test and skip* operation is a welcome operation.

The 8X02 has no internal counter such as the one in the 2910; if iterative loops are needed, one must add an external counter, shown earlier in the chapter. It does have two operations for loops, *conditional branch to loop* and *push for looping,* which use the stack to store the loop address. These operations function slightly differently than the corresponding operations in the 2910, as shown in Figure 4.40. In

TABLE 4.5. 8X02 Branching Functions

AC	Abbre-viation	Branch type	Function TEST=0	Function TEST=1
000	TSK	Test and skip	NA = CA + 1	NA = CA + 2
001	INC	Increment	NA = CA + 1	same
010	BLT	Conditional branch to loop	NA = CA + 1 POP (STK)	NA = STK
011	POP	Return	NA = STK	same
100	BSR	Conditional call	NA = CA + 1	NA = MIF STK = CA + 1
101	PLP	Push for looping	NA = CA + 1 STK = CA	same
110	BRT	Conditional branch	NA = CA + 1	NA = MIF
111	RST	Jump to zero	NA = 000...	same

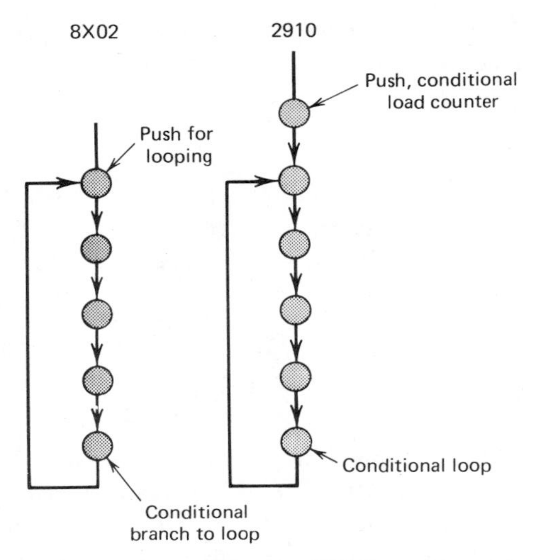

Figure 4.40. 8X02 versus 2910 loop structures.

the 8X02, *push for looping* does not push the address of the next microinstruction onto the stack; it pushes the address of the current microinstruction onto the stack. Similarly, *conditional branch to loop* does not just branch to the address on the top of the stack; it also pops the stack.

The difference between these operations and the corresponding ones in the 2910 is that in the 8X02 the pushing of the loop address onto the stack is done in the first microinstruction in the loop, rather than the preceeding one. This subtle difference has both pro's and con's, meaning that one scheme is not universally better than the other. In the case of the 8X02, one does not have to precede a loop with a microinstruction that pushes the loop address onto the stack, meaning that one can directly branch to, or call, the first microinstruction in a loop. However, this prevents the first microinstruction in the loop from performing a branching operation, such as a conditional branch or subroutine call, which can be done in the 2910. Also, recall that the 2910 has a useful operation for exiting from the middle of a stack-based loop, *conditional jump and pop*. The 8X02 has no corresponding operation. (Also note that loops can be performed without the use of the stack in the 8X02 by using *conditional branch* to branch back to the beginning of the loop.)

The remaining operations in Table 4.5 should be self-explanatory. One small design weakness is the fact that the subroutine-return operation is unconditional. Since the TEST input is not used here, it would have taken very little for the designers of the 8X02 to make this operation conditional, that is, if TEST=0, NA=CA+1. Also, a conditional-return operation would be more useful than the unconditional return in the 8X02. One loses nothing by eliminating the unconditional return, since, as was mentioned in previous sections, it is common practice to permanently define one status condition as *true* (or *false*), allowing one to turn any conditional operation into an unconditional operation. (This would be needed anyway in all designs using an 8X02, since it has no unconditional branch.)

8X02 Timing

The 8X02 is controlled by both the falling and rising clock edges. The falling edge triggers the stack operation; the rising edge stores a value into the register. Representative times for the 8X02 over its operating range are illustrated in Figure 4.41.

THE 67110 SEQUENCER

The 67110, produced by Monolithic Memories, is a 9-bit Schottky TTL sequencer in a 40-pin package [10]. The 67110 was designed to be used with the 6701 ALU/register slice. Like the 3001, the 67110 is not only intended to serve as a sequencer, but also to interconnect the ends of two shifters (the Q and RAM shifters in the 6701 ALU/register slice) and maintain status conditions.

Figure 4.42 shows that the 67110 is organized as four relatively independent sections: branching logic, counting logic, shift logic, and status control. The address-out bus (CRAR) is 9-bits wide, but the address-in bus (N) contains only eight bits. The ninth bit is always generated independently from one of the four status flipflops, the current value on one of the four lines feeding these flipflops, or two other status inputs associated with the shift logic. Hence, all branching functions on an even-odd address-pair relationship. One can branch to a microinstruction whose control-storage address contains a low-order bit of 0, or to the microinstruction beyond it. Although this gives the 67110 considerable branching flexibility, it requires care in laying out the microprogram in control storage (e.g., the insertion of a new microinstruction into the microprogram might throw the even-odd targets onto the wrong boundaries).

The 67110 contains a subroutine stack, but it is only a single register, meaning that one cannot nest subroutines. It also contains a 5-bit counter that can be loaded with the high-order five bits from the N bus. An innovative feature of the

From input	Minimum set-up time to clock fall (ns)	Minimum set-up time to clock rise (ns)
B	—	29
AC	35	90
TEST	28	60

From input \ To output	Max. propagation times (ns): A
EN	35
CLK	40

Figure 4.41. Representative 8X02 guaranteed times.

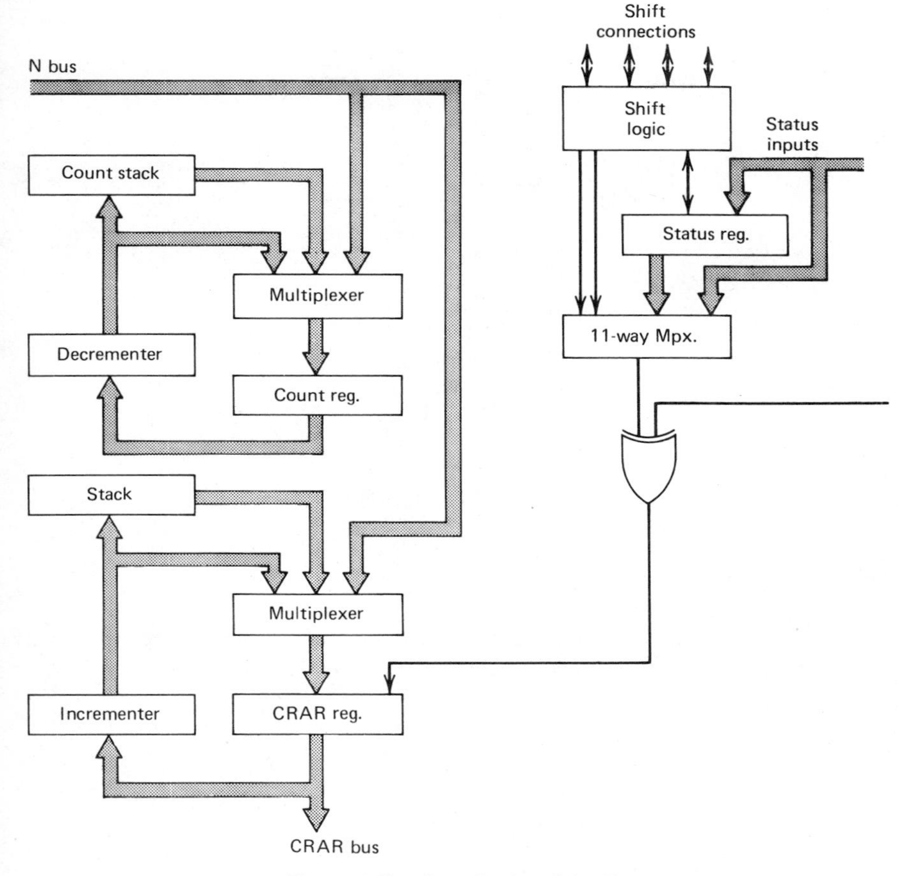

Figure 4.42. Organization of the 67110.

67110 is a stack (one register) for the counter. When one performs a subroutine-call operation, not only is the address of the next microinstruction (high-order eight bits only) loaded into the address stack, but the value of the counter is also saved. This means that one can call a subroutine from within an iterative loop and have the called subroutine also perform an iterative operation. When the subroutine returns, the saved value of the counter is loaded back into the counter.

The external connections of the 67110 are shown in Figure 4.43. The status inputs (and status flipflops) are labelled C (carry), N (negative), V (overflow), and Z (zero), corresponding to the status conditions generated in the 6701 ALU/register slice. Output CRAC is raised when the CRAR register has reached the value 510 or 511. The motivations behind it are to provide a function similar to that of the VECT output of the 2910 and ISE output of the 3001 (e.g., enabling an interrupt controller), and to provide a way to indicate that a 512-word control-storage boundary has been reached, if control storage is larger than 512 words (see the section on addressing large control storages in Chapter 5).

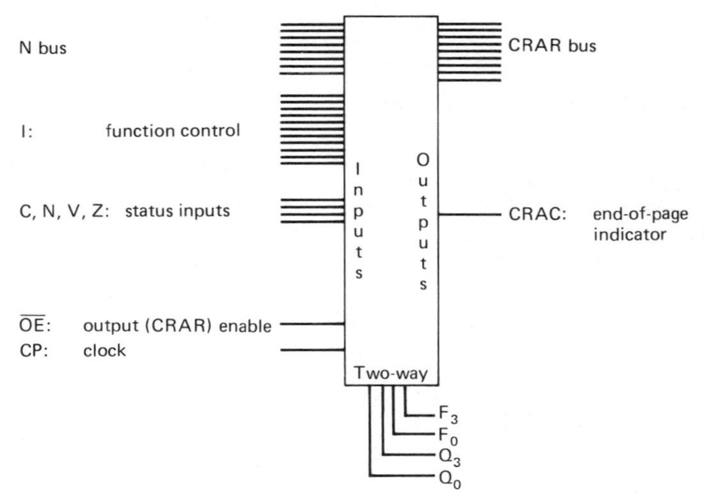

Figure 4.43. 67110 external connections.

67110 Functions

The 10 I function-control inputs are divided into four groups. The first group (three lines) controls the next-address logic. The operations are

Sequence
Sequence and load counter
Repeat
Repetitive subroutine call
Return
Conditional subroutine call on address match
Branch
Subroutine call

Recall that these operations generate only the eight high-order address bits and that the low-order bit is generated independently from the status section. Hence, all the operations above are conditional in the sense that they take one to a microinstruction on an even boundary or the microinstruction beyond that one.

The "conditional subroutine call on address match" is an unusual operation. If the least-significant five bits in the CRAR register (i.e., the low-order five bits of the current microinstruction) are equal to the value in the counter, a subroutine return is performed. Otherwise, a branch is performed. Essentially, this allows one to call a subroutine and dynamically specify, by loading a value in the counter, the last microinstruction in the subroutine (i.e., the microinstruction that will return). However, the utility of this is dubious and it requires the microprogrammer to be more sensitive to the layout of the microprogram in control storage.

A second group of I lines (two lines) controls the shift logic. The four bidirectional shift lines of the 67110 are intended to be connected to the shift pins of the most- and least-significant ALU/register slices (e.g., the RAM and Q shifters in the 6701). The four functions are

1. Double-length left shift, injecting the value of the C flipflop onto Q_0
2. Double-length right shift, injecting the exclusive-OR of the N and V flipflops onto F_3
3. Left rotate (the input on F_3 is gated to F_0)
4. Right rotate (the input on F_0 is gated to F_3)

The definition of the double-length shifts facilitates division and multiplication algorithms in the processing section.

Of the remaining I lines, four control the status logic and one (I_7) is an input to the exclusive-OR gate in Figure 4.42. I_7 allows one to force the low-order CRAR (next address) bit to 0 or 1, or it acts as a polarity control on a status condition feeding the low-order CRAR bit. The four status-control I inputs are defined in Figure 4.44. The last 10 values force one of the status inputs, status flipflops, or Q_0 or Q_3 into the low-order CRAR bit. Some of the other values load the status

CONTROL CODE				ACTION		
I_6	I_5	I_4	I_3			
0	0	0	0	NONE	I_7 TO $CRAR_0$	UNCONDITIONAL BRANCH
0	0	0	1	STORE C	I_7 TO $CRAR_0$	UNCONDITIONAL BRANCH
0	0	1	0	STORE N, V, Z	I_7 TO $CRAR_0$	UNCONDITIONAL BRANCH
0	0	1	1	STORE C, N, V, Z	I_7 TO $CRAR_0$	UNCONDITIONAL BRANCH
0	1	0	0	SHIFT FLAG REGISTER INTO Q_0	I_7 TO $CRAR_0$	UNCONDITIONAL BRANCH
0	1	0	1	SHIFT FLAG REGISTER OUT OF Q_3	I_7 TO $CRAR_0$	UNCONDITIONAL BRANCH
0	1	1	0	INSTANTANEOUS VALUE OF Q_0 TO $CRAR_0$		CONDITIONAL BRANCH
0	1	1	1	INSTANTANEOUS VALUE OF Q_3 TO $CRAR_0$		CONDITIONAL BRANCH
1	0	0	0	STORED VALUE OF C TO $CRAR_0$		CONDITIONAL BRANCH
1	0	0	1	STORED VALUE OF N TO $CRAR_0$		CONDITIONAL BRANCH
1	0	1	0	STORED VALUE OF V TO $CRAR_0$		CONDITIONAL BRANCH
1	0	1	1	STORED VALUE OF Z TO $CRAR_0$		CONDITIONAL BRANCH
1	1	0	0	INSTANTANEOUS VALUE OF C TO $CRAR_0$		CONDITIONAL BRANCH
1	1	0	1	INSTANTANEOUS VALUE OF N TO $CRAR_0$		CONDITIONAL BRANCH
1	1	1	0	INSTANTANEOUS VALUE OF V TO $CRAR_0$		CONDITIONAL BRANCH
1	1	1	1	INSTANTANEOUS VALUE OF Z TO $CRAR_0$		CONDITIONAL BRANCH

Figure 4.44. 67110 status-control inputs.

flipflops. Two values shift the status register (the four status flipflops) by one position, either putting the ejected bit onto line Q_0 or injecting a bit from line Q_3. These unusual operations allow one, in four cycles, to move the values of the status flipflops into the Q register in the processing section, or load the value from the Q register into the status flipflops. This was provided for interrupt handling, where one may want to save the current status in the processor.

67110 Timing

The I inputs must be set up before the falling clock edge, and the status and shift inputs must be set up prior to the next rising edge. After the rising edge, the CRAR outputs are available. Representative guaranteed set-up and propagation times are shown in Figure 4.45.

THE 9408/4708 SEQUENCER

The 9408 and 4708 are 10-bit sequencers produced by Fairchild Camera and Instrument Corp. [11]. The 9408 is an I^3L device (Fairchild's version of I^2L) and is fully TTL compatible. The 4708 is an identical sequencer, but it is a CMOS device.

The organization of the 9408 is shown in Figure 4.46. Like most other sequencers, it contains an address register, an incrementer, and a 4-word stack. It also contains four status flipflops. Noticeably missing is an iteration counter. The multiplexer at the bottom of the diagram allows one to decide whether to place the address register (MPC) on the path from input (BA) to output (A). This associated with the concept of microinstruction pipelining, discussed in Chapter 5. For nonpipelined operation, the A bus is usually driven from the MPC register; for piplined operation, the A bus is usually driven from the input to the MPC register.

The external connections of the 9408 are shown in Figure 4.47. MWB, a 3-bit bus, is a secondary input bus for multiway branches. PLS controls the bottom

From input	Minimum set-up time to clock fall (ns)	Minimum set-up time to clock rise (ns)
$I_{0\text{-}6}$	5	—
I_7	—	16
C, N, V, Z, Q_0, Q_3	—	20
N	—	24

From input \ To output	Max. propagation time (ns): Shift outputs	CRAR	CRAC
$I_{8,9}$	32	—	—
Any shift input	18	—	—
$\overline{OE}$	—	20	—
CP↑	—	35	42

Figure 4.45. Representative 67110 guaranteed times.

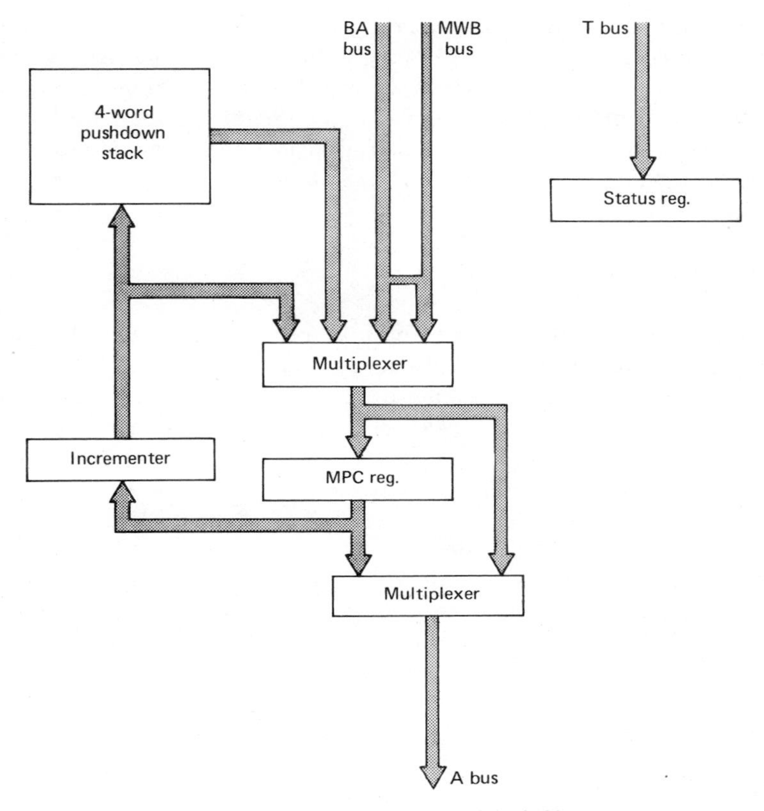

Figure 4.46. Organization of the 9408.

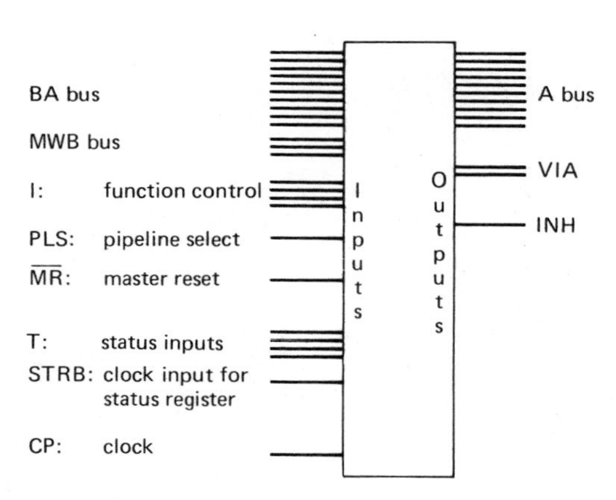

Figure 4.47. 9408 external connections.

multiplexer in Figure 4.46. MR clears the MPC register and the stack. The T bus feeds the four status flipflops. The loading of the status flipflops is independent of the remainder of the device; loading is controlled by the STRB input.

The VIA and INH outputs are similar to the PL, MAP, and VECT outputs of the 2910. They are driven by combinatorial logic decoding the branch (I) inputs and can be used to signal or control other devices when particular branching operations are active.

9408/4708 Functions

The four I inputs give the device 16 branch operations, although the number of unique operations is somewhat smaller (six). Four of the 16 operations are unconditional branches; their only difference is that they set the VIA outputs to four different values. Eight of the 16 operations are conditional branches based on the value (0 or 1) of one of the four status flipflops.

Another branch operation is a multiway branch. Here the output address consists of the seven high-order bits on the BA bus followed by the three bits on the MWB bus. The remaining operations are unconditional subroutine call and return, and sequence. On all operations except sequence and subroutine return, the INH output is 1. Subroutine return and sequence do not use the value on the BA bus (usually driven by a 10-bit branch-address field in the microinstruction). Hence, the designer may wish to use this field for other control purposes in microinstructions specifying a return or sequence operation. The INH output is provided to drive such decoding logic, if it is used.

9408/4708 Timing

Figure 4.48 lists representative times for the 9408 (top value) and 4708. Note that, unlike the times given for most other devices in this book, these times are

From input	Typical set-up times to clock rise (ns)
BA	15 120
I	90 360
STRB	60 240

From input \ To output	A	VIA	INH
BA	28 160	—	—
I	72 360	35 160	35 160
CP	98 392	—	—

Typical prop. times (ns)

Figure 4.48. Representative 9408/4708 typical times.

typical, not worst-case, times. The propagation delays assume that PLS is set such that the A bus is driven from the inputs to the MPC register. The STRB-to-CP set-up time is for the case in which one wants to store status and perform a conditional branch on it in the same cycle. The 4708 times assume a supply voltage of 10 V.

COMPARISONS

In keeping with the style of Chapter 3, this section contains a side-by-side comparison of the sequencers discussed. Table 4.6 compares the physical and organizational characteristics of the sequencers. As was discussed earlier, some of the sequencers perform a second function: the multiplexing of status conditions and, in some cases, the maintenance of status conditions in flipflops. This is reflected in the last four categories in the table.

Table 4.7 compares the branching capabilities of the sequencers. Included in this comparison is the combination of the 29811 and a counter with one or more 2911 slices. Table 4.7 shows, for instance, that the basic slices (2909, 2911, 74S482) have no conditional operations, meaning that external logic must be added to these slices to achieve conditional operations, as was done in Figures 4.5 and 4.10.

The operations in Table 4.7 have been partitioned into eight groups. The first contains the basic sequence and branch operations. The second group contains the repeat operations, that is, operations that use iteration counters. Note that where a particular sequencer does not directly perform a given operation, that operation can usually be implemented via external logic or microprogrammed logic. Also note that only the MC10801 was indicated as having the *repeat-one-instruction* operation. Actually, any sequencer that has a *repeat* operation (e,g., the 2910) can have a one-microinstruction loop, but this requires the use of the next-address (MIF) field in this microinstruction (preventing its use for another purpose in this microinstruction), or it requires that the preceding microinstruction do a *push-and-continue* operation. Hence, the MC10801 has a slight edge for single-microinstruction loops.

The third group represents looping operations, that is, operations that allow one to branch to the address contained on the top of the stack. The fourth group represents operations dealing with microprogram subroutines, and the fifth group is the multiway branch, that is, the presence of an additional bus that can be used to inject bits into the generated address. The sixth and seventh groups contain special operations that, for the most part, are useful for initiating and escaping from loops. The last group indicates whether the sequencer has decoding logic that generates an output signal for the enabling of a secondary address source (e.g., a mapping PROM).

If one simply counts the X's, the 29811/counter/2911 combination and the 2910 have, by far, the most operations. However, this by itself is not a satisfactory measure, since one must weigh the importance of each operation and consider the

TABLE 4.6. Comparison of Sequencer Characteristics

	2909	2911	2910	3001	10801	74S482	8X02	67110	9408
Slice?	yes	yes	no	no	yes	yes	no	no	no
Width (bits)	4	4	12	9	4	4	10	9	10
Technology	LSTTL	LSTTL	LSTTL	STTL	ECL	STTL	LSTTL	STTL	I^3L[6]
Pins	28	20	40	40	48	20	28	40	40
Package	DIP	DIP	DIP	DIP	QIP	DIP	DIP	DIP	DIP
Maximum power (mW)	650	650	1720	1200	2250[4]	700	1000	1270	1000[4,6]
Stack size	4	4	5	0	4	4	4	1	4
Other registers	2	2	2	1	3	1	1	3	1
Address input ports	3	1	1	2	1	1	1	1	2
Address output ports	1	1	1	1	1	1	1	1	1
Address bidirectional ports	0	0	0	0	2	0	0	0	0
Address control inputs[5]	6	6	7	9	9	7	4	3[1]	4
Status inputs	0	0	1	1	6[2]	0	1	6	4
Status flipflops	0	0	0	2	4	0	0	4	4
Status control inputs	—	—	—	4	4	—	—	5[1]	1
Status flipflop outputs	—	—	—	1	4	—	—	1[3]	0

Notes: [1] - plus two more control inputs for shift control.

[2] - includes the XB and B inputs and the 4-bit multipurpose IB bus. The status register can also be parallel loaded from the NA bus, or a single bit can be loaded from the D_{in} bus.

[3] -only one status flipflop (C) can be externally read; it can be gated to one of the shift outputs.

[4] -estimated.

[5] - excluding clock, output-enable, and carry inputs.

[6] - also available as the CMOS 4708, having a maximum power consumption of 15 mW.

TABLE 4.7. Comparison of Branching Capabilities

Operation	2909	2911	29811/ counter/ 2911	2910	3001	MC 10801	74S 482	8X02	67110	9408
Sequence	X	X	X	X		X	X	X	X	X
Unconditional branch	X	X	X		X	X	X		X	X
Conditional branch			X	X	X	X		X	X	X
Two-way branch			X	X						
Repeat			X	X					X	
Repeat using stack			X	X						
Repeat subroutine						X			X	
Repeat until				X						
Repeat one instruction						X				
Uncon. loop using stack	X	X					X			
Con. Loop using stack			X	X				X		
Uncon. subroutine call	X	X					X		X	X
Con. loop using stack			X	X		X		X		
Two-way subroutine call			X	X						
Unconditional return	X	X					X	X	X	X
Conditional return			X	X		X				
Multiway branch	X				X	X				X
Uncon. branch and pop	X	X					X			
Con. branch and pop			X	X				X		
Conditional skip								X		
Relative branch forward							X			
Push and continue	X	X	X	X			X	X		
Pop and continue	X	X					X			
Push and load counter			X	X						
Load counter and continue			X	X		X			X	
Enabling of another address source			X	X	X				X	X

environment in which the sequencer will be used. For instance, the *repeat-until* operation in the 2910 is extremely useful, and the 2910's lack of a *repeat-one-instruction* as implemented in the MC10801 is usually only a minor drawback.

As explained earlier in the chapter, where an unconditional and conditional form of an operation is listed (e.g., subroutine call), it is better to have the conditional form. For instance, the 8X02 has a conditional subroutine call, but only an unconditional return, a slight drawback.

Table 4.8 compares the timing characteristics of the sequencers for three branching operations: conditional branch, sequence, and conditional subroutine call. The times listed were computed from the moment that the branching microorders (e.g., next address, branch type, branch condition) in the current microinstruction are available, to the time that the sequencer's address output is available. For sequencers that do not have internal status-condition flipflops, a

TABLE 4.8. Comparative Sequencer Speeds

	Typical times			Maximum times		
Slice	Cond. branch	Sequence	Cond. call	Cond. branch	Sequence	Cond. call
Am2909/2911	68	29	68	105	43	105
Am2910	45	26	45	78	55	78
N3001	45/58[1]	—	—	60/100[1]	—	—
MC10801	55	45	55	76	72	76
74S482	74	59	74	115	88	115
8X02	103	103	122	135	130	163
67110	35	35	35	70	70	70
9408	98	98	98			
4708	392	392	392			

Notes: [1] - the first value represents testing the C or Z status flag; the second value represents testing an external status condition.

design similar to that of Figure 4.10 is assumed. That is, it is assumed that the status conditions reside in an external register, and that they pass through two multiplexers before reaching the sequencer. Propagation delays of 24 ns typical and 35 ns maximum are assumed through the register and multiplexers. Since the 2909, 2911, and 74S482 do not directly perform conditional operations, the presence of a 29811 or equivalent is assumed in the timings for these devices. Also, where the sequencer is a slice, three cascaded slices are assumed.

The values in the table are revealing because they highlight a difference in sequencer designs. The 2909, 2911, and 2910 appear to be considerably faster than the other sequencers. This is due to the placement of registers along the data path. Referring to Figures 4.1, 4.6, and 4.17, notice that these three sequencers have no register on the path from the address input bus (or stack) to the output bus. This is the primary reason for the apparent speed of these sequencers in Table 4.8. These sequencers do have an address register (MPC), but it appears on the data path *after* the output bus. As mentioned earlier, the incrementing function and the loading of this register are done *in parallel* with the accessing of control storage.

In all other sequencers, the output bus is fed from a register, thus increasing the propagation delay through the sequencers. Furthermore, the incrementing function for the sequence operation is on the critical path. For instance, refer to the diagram of the MC10801 (Figure 4.29). The CR0 register contains the address of the current microinstruction. To perform the sequencing operation, CR0 must be routed through the incrementer, the carry-out signals must ripple through the other slices, and the incrementer output must flow through a multiplexer and back into CR0. In contrast, in the 2911, a sequencing operation simply involves routing the MPC register's value through a multiplexer to the output bus. Hence, this organizational issue—the placement of the incrementer and address register—has a significant effect on the sequencer's speed.

Finally, the following list highlights the relative advantages and disadvantages of each sequencer and sequencer slice discussed.

Advantages	*2909*	*Disadvantages*
Excellent documentation		No counter
OR gates on output bus		Requires external logic to implement conditional operations

Advantages	*2911*	*Disadvantages*
Excellent documentation		No counter
		Requires external logic to implement conditional operations

Advantages	*2910*	*Disadvantages*
12-bit width		Not cascadable
Loop counter		High power consumption
Powerful branch operations		
Repeat-until operation		
Stack-full status output		
Fast		
Excellent documentation		

Advantages	*3001*	*Disadvantages*
Status flipflops (but closely coupled to the 3002)		Limited and unwieldy branching operations
Multiway branch operations		No stack, counter, or incrementer
Fast		Not cascadable, limited to 512 control-storage words
		Microinstruction layout difficulty

Advantages	*10801*	*Disadvantages*
Expandable stack		Has repetitive subroutines, but no repetitive loops
Status-flag register		No polarity test on status flags
Can detect stack overflow		High power consumption
Reset input		

Advantages	*74S482*	*Disadvantages*
Full adder		Output bus is not 3-state
		Requires external logic to implement conditional operations

Advantages	*8X02*	*Disadvantages*
Test-and-skip operation		Slower than most
		Not cascadable
		No loop counter

Advantages	*67110*	*Disadvantages*
Shift logic for shifters in processing section		Limited to one subroutine level
Status flipflops		Even-odd targets of conditional branches
Polarity control for status tests		Not cascadable
Loop counter		Limited width (9 bits)
Loop counter saved on subroutine call		

Advantages	*9408*	*Disadvantages*
Status flipflops		Not cascadable
Reset input		No loop counter
Available in CMOS		Slower than most
		No 3-state output

REFERENCES

1. *The Am2900 Family Data Book.* Sunnyvale, Cal.: Advanced Micro Devices, 1978.
2. *Microprogramming Handbook and Am2910 Emulation.* Sunnyvale, Cal.: Advanced Micro Devices, 1977.
3. D. Mrazek, "Bit-Slice Parts Approach ECL Speeds with TTL Power Levels," *Electronics,* 51(23), 107–112 (1978).
4. *Build a Microcomputer, Chapter II, Microprogrammed Design.* Sunnyvale, Cal.: Advanced Micro Devices, 1978.
5. *Intel Series 3000 Reference Manual.* Santa Clara, Cal.: Intel, 1976.
6. *Introducing the 3000.* Sunnyvale, Cal.: Signetics, 1976.
7. *MC10801 Data Sheet.* Pheonix: Motorola, 1978.
8. *The Bipolar Microcomputer Components Data Book.* Dallas: Texas Instruments, 1977.
9. *Signetics Control Store Sequencer 8X02.* Sunnyvale, Cal.: Signetics, 1978.
10. *57110/67110 Microprogram Controller.* Sunnyvale, Cal.: Monolithic Memories, 1977.
11. *Macrologic Bipolar Microprocessor Data Book.* Mountain View, Cal.: Fairchild Camera and Instrument, 1977.

5

Microinstruction Design

After studying the ALU/register slices of Chapter 3 and the sequencer devices of Chapter 4, one can clearly see that the available bit-slice devices have been designed with microprogrammed control, rather than sequential-logic control, in mind. Indeed, given the advantages of microprogrammed control discussed in Chapter 2, this seems a wise decision.

Given the close coupling between bit-slice logic and microprogrammed control, it is necessary to deviate a bit from bit-slice logic and resume the discussion in Chapter 2, that is, to discuss some of the finer and more-subtle points of microprogrammed control.

This chapter is named *microinstruction design,* as distinguished from *microprogram design*. Microprogram design, which occurs later, entails the structuring of the microprogram itself (for which many principles from the fields of programming and software engineering can be employed) and the determination of the algorithms to be used. Microinstruction design entails not only the specification of the format of the microinstruction (e.g., microorder meanings, format, and layout), but also decisions concerning decoding logic, design of the control section, and system timing. Hence microinstruction design is an integral part of the logic design of the digital system.

Given a design employing bit-slice logic, the microinstruction design process is usually simplified to some extent. For instance, if the processing section is built around the 2901 ALU/register slice and the control section around the 2910 sequencer, certain aspects of the microinstruction can be directly deduced. For instance, it is likely that the microinstruction will have microorders with formats and values identical to the A, B, and I input controls of the 2901, and a microorder directly matching the I control inputs of the 2910. However, it is likely that the design will contain additional, nonbit-slice, devices such as interrupt controls, registers, I/O interfaces, and memory interfaces. Also, it may be the case that the application does not require the full function supplied by some of the bit-slice devices. Hence, there is much more to microinstruction design for a bit-slice-based design than just defining microorders for the control inputs of the bit-slice devices. Also, one must deal with a variety of considerations involving control strategies, system timing, and control-storage addressing.

In general, microinstruction design is an optimization process involving goals concerning system cost, flexibility, and speed. In performing this type of design, one balances such variables as the depth and width of control storage, clocking schemes and speeds, microprogram branching flexibility, and the complexity and overhead of microinstruction decoding logic.

MICROINSTRUCTION PIPELINING

Probably the most useful and powerful design idea in a microprogrammed system is the concept of *microinstruction pipelining* (sometimes referred to as "parallel implementation" as distinguished from "serial implementation" [1,2]). Pipelining is a simple idea, usually requiring a small number of additional components to implement, but it can reduce the cycle time of a system by as much as half.

Although similar in certain respects, microinstruction pipelining should not be confused with machine-instruction pipelining. The latter is a complex technique used in some high-speed processors; the processor is partitioned into several processing stations, such that the processor can perform work on behalf of multiple machine instructions simultaneously. Microinstruction pipelining is simply a technique of allowing the control and processing sections of a processor to operate in parallel, such that the next microinstruction is being addressed and fetched in parallel with the control activities of the current microinstruction.

Figure 5.1 is an overview of the organization of a simple, nonpipelined, microprogrammed processor. The current microinstruction, available on the data-output lines from control storage, is fed directly to the appropriate points in the processing and control sections. Once the status outputs from the processing section are available, they are used by the control section to generate the address of the next microinstruction.

This highly serialized design is depicted in the timing diagram in Figure 5.2. The machine cycle time is CXP+CSA+PXS. Although there may be a small degree of overlap between the processing and control sections (i.e., if the status

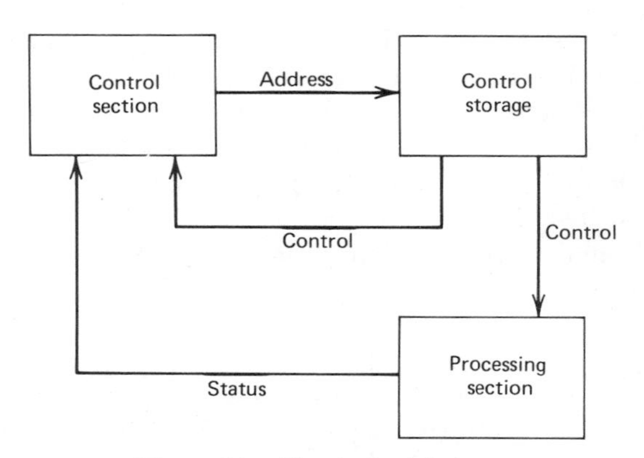

Figure 5.1. Nonpipelined design.

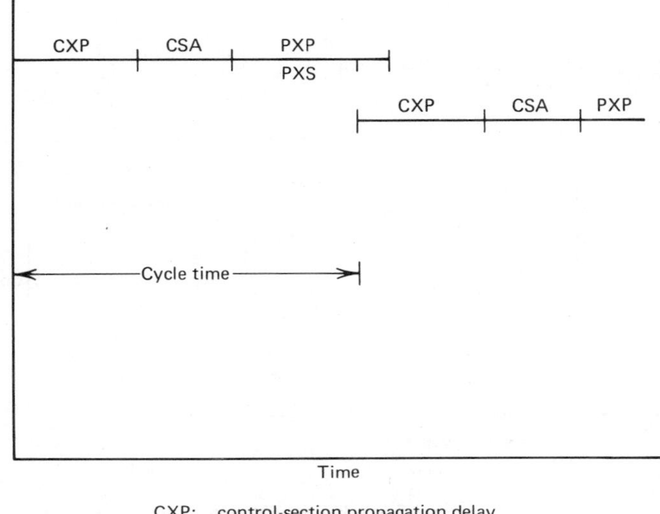

Figure 5.2. System timing for the nonpipelined design.

outputs are available before the completion of the operations in the processing section, or where PXP-PXS > 0), most of the operation of the system is serial, and most of its logic is on the critical timing path (the path dictating what the minimum cycle time can be).

The need for this serialization stems from the conditional-branching operations performed by the control section, since the control section cannot determine the next microinstruction until the status conditions from the current cycle are available. It should be observed that one could place additional logic in the system to observe the branching operation being sent from the current microinstruction to the control section. If the operation is an unconditional one, the logic could cause the control section (but not control storage) to operate in parallel with the processing section for this cycle, thus decreasing the machine cycle time when a conditional-branching operation is not being performed. However, since this timing logic can be complex, and since, in typical microprograms, a high percentage of microinstructions specify conditional-branching operations, this approach is not recommended.

Microinstruction pipelining involves performing control-storage addressing and fetching in parallel with the operation of the processing section, thus shortening the cycle time. The most common implementation of pipelining requires only two changes: the current microinstruction must be held in a register, and the status conditions from the processing section must be held in a register. This is illustrated in Figure 5.3.

Although this form of pipelining is simple to implement, it warrants close atten-

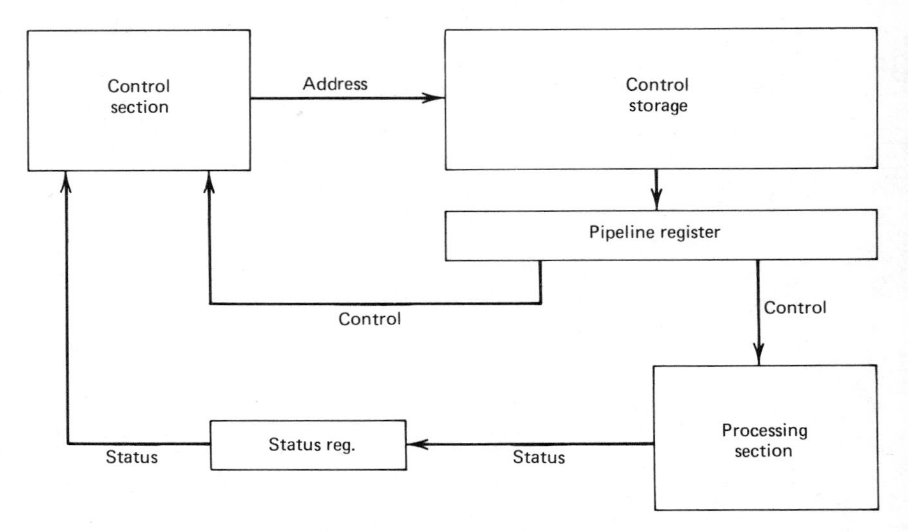

Figure 5.3. Pipelined design.

tion because it changes the semantics of branching operations and involves three machine states: the prior, current, and next machine cycles. View Figure 5.3 in the situation where a microinstruction is held in the pipeline register and is controlling the processing and control sections. Simultaneously, the pipeline register holds the *current* microinstruction, the control section is computing the address of, and fetching, the *next* microinstruction, and the status register holds the status produced from the *previous* cycle. Because of the parallelism, the interpretation of branching operations must change. For instance, a microorder in the current microinstruction specifying "jump if OVR" no longer means "jump if OVR is present in the current cycle;" it means "jump if OVR *was* present in the previous cycle" (i.e., "jump if OVR is present in the status register").

The status register is added to allow the control and processing sections to operate in parallel, but its addition implies that a microinstruction cannot test conditions generated within its own cycle; it tests conditions generated in the last cycle. The pipeline register (usually a set of registers, whose total width equals the width of control storage) allows the control-storage fetch of the next microinstruction to proceed in parallel with the operation of the current microinstruction. The status and pipeline registers usually require no added timing logic; they are triggered by the same clock that triggers the control and processing sections. As described in earlier designs, the low state of the clock causes the control storage to fetch the next microinstruction. The rising clock edge (1) gates values generated in the current cycle into the processing section's registers, (2) gates values generated in the control section into its registers and/or pushdown stack, (3) gates the status conditions generated in the current cycle into the status register, and (4) gates the next microinstruction into the pipeline register.

The pipelining concept splits the design into two paths. One path runs from the control section, through control storage, and into the pipeline register. The second path runs through the processing section into the status register. Since these paths are active simultaneously, the machine cycle time is equal to the longer of the two paths. Figure 5.4 illustrates the system timing under the situation where the addressing path is longer than the processing-section path. Here the cycle time is equal to

$$\text{CXP} + \text{CSA} + \text{PR}$$

Figure 5.5 illustrates the system timing under the situation where the processing-section path is longer than the addressing path. Here the cycle time is equal to

$$\text{PXS} + \text{SR}$$

(assuming that PXP is less than PXS+SR).

As an example, assume the following times

CXP (propagation delay through control section)	= 50 ns
CSA (control-storage access time)	= 65 ns
PXS (propagation delay through processing section to status	= 90 ns
PR,SR (set-up times and propagation delays for pipeline and status registers)	= 15 ns

CXP+CSA+PR = 130 ns and PXR+SR = 105 ns. Hence the machine cycle time can be 130 ns. If a nonpipelined design were used, the machine cycle time would be CXP+CSA+PXS, or 205 ns. As a result the use of pipelining, a rather simple and inexpensive technique, has cut the machine cycle time by over a third.

The drawback of pipelining usually cited is complication of the microprogram.

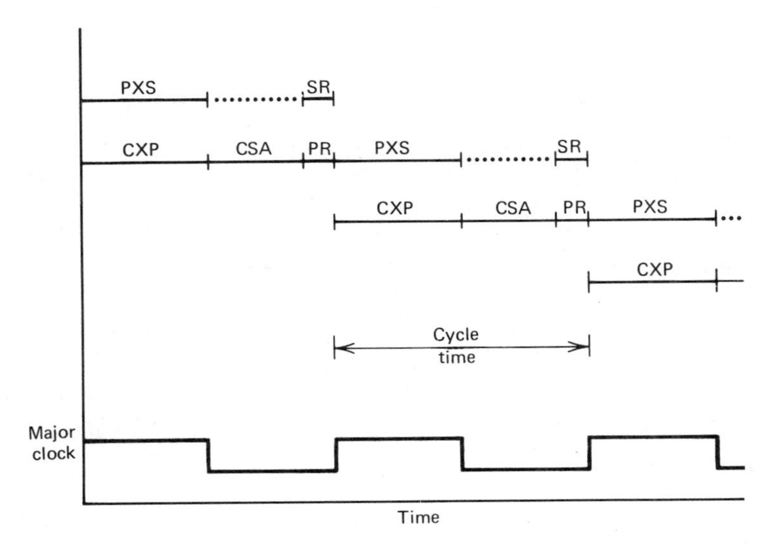

Figure 5.4. Pipelined design timing (control-section time exceeds processing-section time).

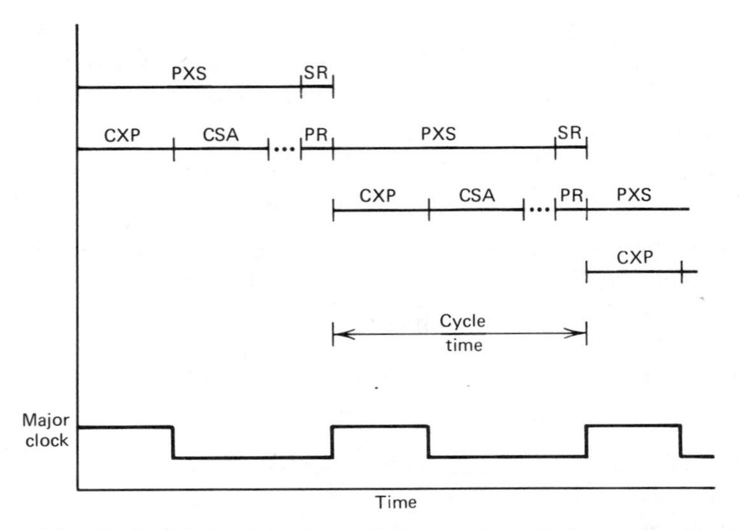

Figure 5.5. Pipelined design timing (processing-section time exceeds control-section time).

In practice this drawback is usually not apparent. One must remember, though, that a conditional branching operation tests conditions established in the *previous* machine cycle, not the current cycle.

One occasional drawback of a pipelined design is that it may cause the microprogrammer to waste a machine cycle now and then. Consider a situation where one desires to test a condition (e.g., "Is the value in one register greater than the value in another register?") and then branch if the condition is present. The organization of the microprogram might be that of Figure 5.6a, where each block represents a microinstruction. In a pipelined design, the comparison and condi-

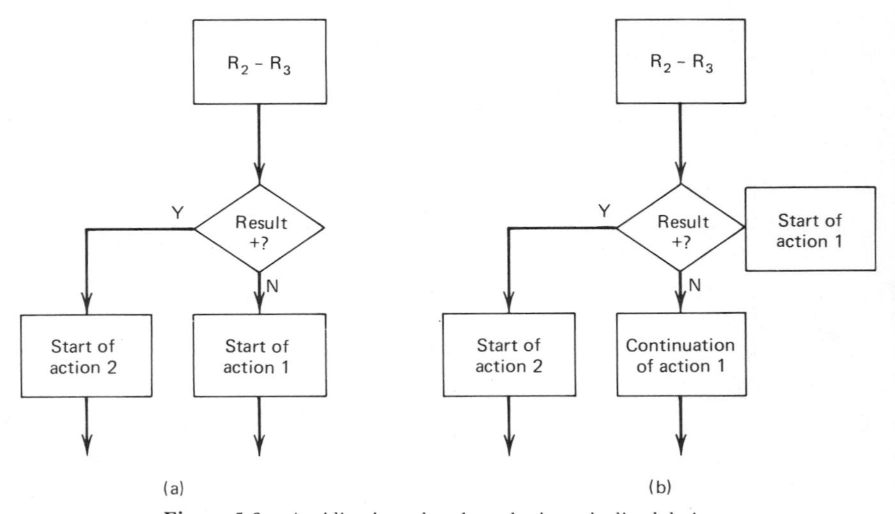

Figure 5.6. Avoiding branch-only cycles in a pipelined design.

tional branch requires two microinstructions, where the two operations could be performed in one microinstruction in a nonpipelined design. Here pipelining seems to be a false savings, since, although the cycles are shorter, more cycles are needed. However, this problem can be alleviated by careful design of the microprogram. For instance, in many situations, additional functions can be performed by the second microinstruction (the one containing the conditional branch); thus the system need not waste cycles by accomplishing only a conditional branch in a microinstruction. Also, as shown in Figure 5.6b, one might predict, when writing the microprogram, the most-probable result of the test and begin some of the work for the most-probable action in the microinstruction containing the conditional branch.

Although the set-up and propagation times of the pipeline or microinstruction register were included as an additional overhead in only the pipelined alternative, it should be noted that such a register is often required in nonpipelined designs to stabilize the control signals and avoid glitches, or spurious control signals.

OTHER FORMS OF PIPELINING

The concept of pipelining involves the placement of two or more registers in the control paths of a design to allow control and processing functions to be performed in parallel. In the previous section, the two registers added to achieve pipelining were the status and pipeline (microinstruction) registers. These allowed the processing section to work in parallel with the control-section-addressing and control-storage-fetching operations.

This pipelining technique is not the only one. Another alternative is shown in Figure 5.7. Here the status register is replaced by a register between the control section and control storage. (Note that many of the sequencers and sequencer

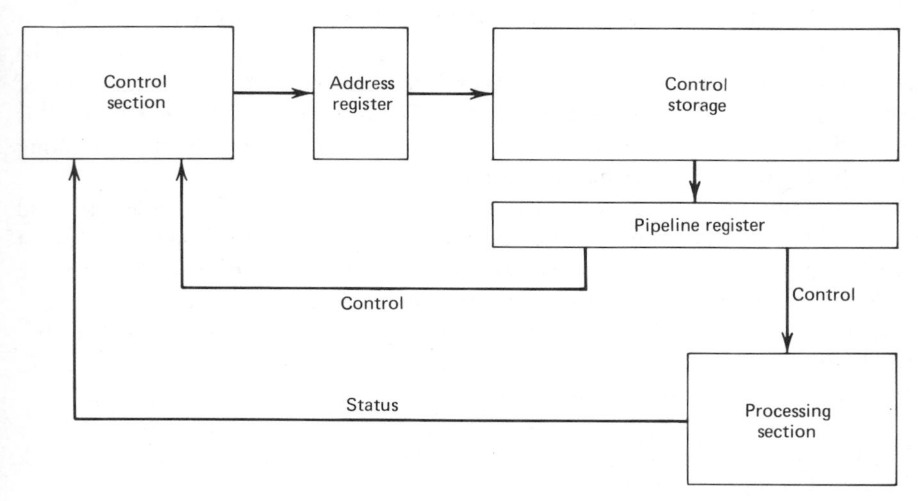

Figure 5.7. Microinstruction/address pipelining.

slices discussed in Chapter 4 have such a register.) With this type of pipelining, the following actions occur in parallel.

1. The clock gates the current microinstruction into the pipeline register, which controls the processing and control sections and eventually causes a generated address to be placed in the address register.
2. The value in the address register at the start of the cycle is used to fetch a microinstruction (the next microinstruction) from control storage.

Hence, a conditional-branching operation is controlled by the status generated in the current machine cycle, but it does not influence the address of the microinstruction in the next cycle (cycle n+1); it determines the address of the microinstruction for cycle n+2 (i.e., two cycles away). In other words, if a microinstruction states "branch to location X if condition Y is true," and Y is true in this cycle, the microinstruction at location X becomes not the next microinstruction fetched, *but the microinstruction fetched after the fetch of the next microinstruction.*

With the organization of Figure 5.7, the machine cycle time can be seen to be the greater of

$$\text{PXS} + \text{CXP} + \text{AR1}$$

and

$$\text{AR2} + \text{CSA} + \text{PR}$$

where AR1 is the set-up time of the address register, AR2 is the propagation delay from the output-enable input, and the other factors have their definitions of the previous section. Using the sample times of the previous section and AR1 = 5 ns, AR2 = 15 ns, the cycle time is the greater of 145 ns and 95 ns, or 145 ns. Hence for these sample times, this is a poorer form of pipelining.

This form of pipelining differs from the form in the previous section, because it places the control-section path on the processing-section path, rather than on the control-storage path. It results in a faster cycle time only when the access time of control storage is disproportionately large. For instance, if CSA = 100 ns, this form of pipelining yields a cycle time of 145 ns, where the cycle time resulting from the earlier pipeline technique would be 165 ns.

Because the times used in the previous section are typical of most designs, and because of the difficulty in understanding conditional branching with this form of pipelining (the address formed determines the address of the microinstruction two cycles later), this form of pipelining is infrequently used.

A third form of pipelining retains the address and status registers, but eliminates the microinstruction register. This form is illustrated in Figure 5.8. Here the following actions occur in parallel.

1. The address register is enabled, causing a microinstruction to be fetched, which in turn controls the processing and control sections.

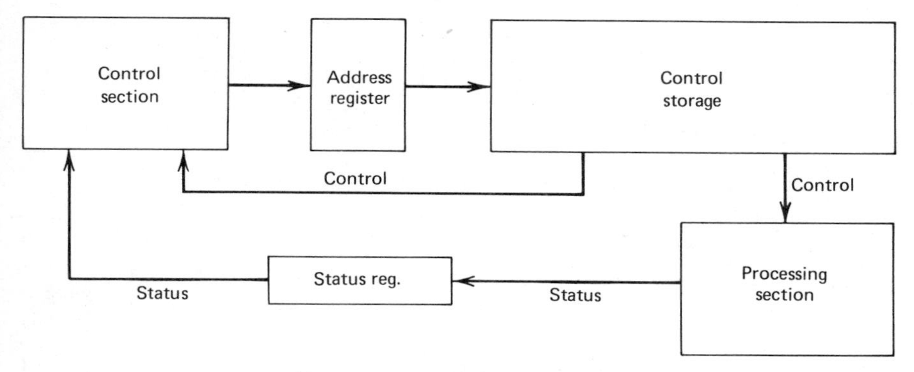

Figure 5.8. Address/status pipelining.

2. The control section operates in parallel with the processing section, using the status in the status register to generate a new value for the address register.

In this form, as in the first form, a conditional-branching operation uses status from the previous cycle to form the address of the microinstruction for the next cycle. The cycle time is the greater of

$$\text{AR2} + \text{CSA} + \text{PXS} + \text{SR}$$

and

$$\text{AR2} + \text{CSA} + \text{CXP} + \text{AR1}$$

Note that the control-storage access time appears in both paths. Using the sample data from before, the minimum cycle time is the greater of 185 ns or 135 ns, or 185 ns. In general, this approach is always slower than the first approach discussed in the previous section. Its only advantage is that it is slightly cheaper, since the address register is normally smaller than a microinstruction (pipeline) register.

The last alternative combines all previous techniques; that is, it creates three parallel paths by employing microinstruction, address, and status registers. Hence, it could be termed *double pipelining*. The structure of this approach is illustrated in Figure 5.9.

This form of pipelining implements the following parallel actions

1. The current microinstruction is gated into the pipeline register, thus controlling the control and processing sections.
2. The control section, using the status register, generates a new value for the address register.
3. The current value in the address register is used to fetch a microinstruction from control storage.

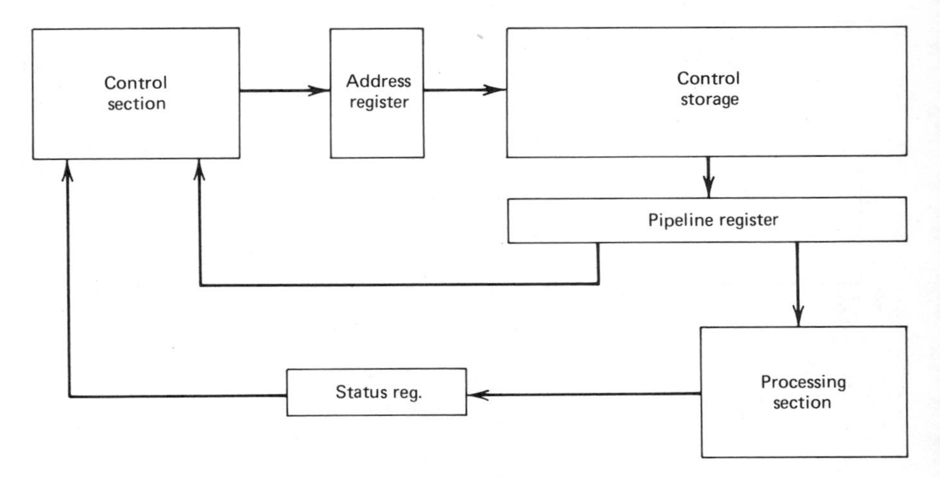

Figure 5.9. Microinstruction/address/status pipelining.

Hence information related to four machine cycles is present simultaneously

1. The status register holds information generated in cycle n−1.
2. The pipeline register holds the microinstruction for cycle n.
3. The address register holds the address of the microinstruction for cycle n+1.
4. The control section is generating the address of the microinstruction for cycle n+2.

A microinstruction thus uses status information from the previous cycle (cycle n−1) to generate the address of the microinstruction *for the cycle after the next cycle* (cycle n+2). (The address of the microinstruction for the next cycle had been generated by the microinstruction in the cycle prior to the current cycle.)

The minimum cycle time can be seen to be the greater of

$$PXS + SR$$

and

$$CXP + AR1$$

and

$$AR2 + CSA + PR$$

Using the sample times, the cycle time is the greater of 105 ns, 55 ns, and 95 ns, or 105 ns. Hence, this type of pipelining achieves its high speeds by decoupling the speeds of the processing section, control section, and control storage. Its logic expense is not much higher than the other approaches. However, the difficulty in implementing microprograms in such a system, because of the confusion surrounding the cycle in which branching operations take effect, makes this approach an infrequent choice. The microinstruction/status pipeline approach (the first one described in the previous section) is the recommended approach for most designs.

PIPELINE PREDICTION

Another approach to microinstruction pipelining is a probabilistic one. The addressing and fetching of the next microinstruction is performed in parallel with the operation of the processing section, making a guess about the "most probable" next microinstruction. If, later in the machine cycle, the guess proves incorrect, the processor must delay itself to address and fetch the correct microinstruction.

In a design with pipeline prediction, the branching operations in a microinstruction are dependent upon status created within the same cycle, and the operations affect the address of the microinstruction for the next cycle. Hence, the semantics of the branching operations are the same as those in a nonpipelined design.

Figure 5.10 illustrates a variety of timing situations. The first case represents the nonpipelined design, where the addressing and fetching of the microinstruction is not overlapped with processing. The second case represents the typical pipelined approach (pipelining with a microinstruction register and status register). The last two cases in Figure 5.10 represent the prediction approach. Although the branch operations are dependent upon status generated in the current cycle, the addressing and fetching of the next microinstruction is performed in parallel with the control of the processing section. If the status conditions generated later in the machine cycle prove the predicted next microinstruction incorrect (e.g., the design guesses that all conditional-branching operations will result in a branch, and the condition being tested turns out to be false), the machine cycle is delayed until the correct microinstruction is fetched.

As an example, assume a design with sequence, unconditional-branch, and conditional-branch operations. Assume that the parallel addressing and fetching logic is designed under the assumption that conditional branches will result in a branch being taken. If 25% of the microinstructions processed specify conditional

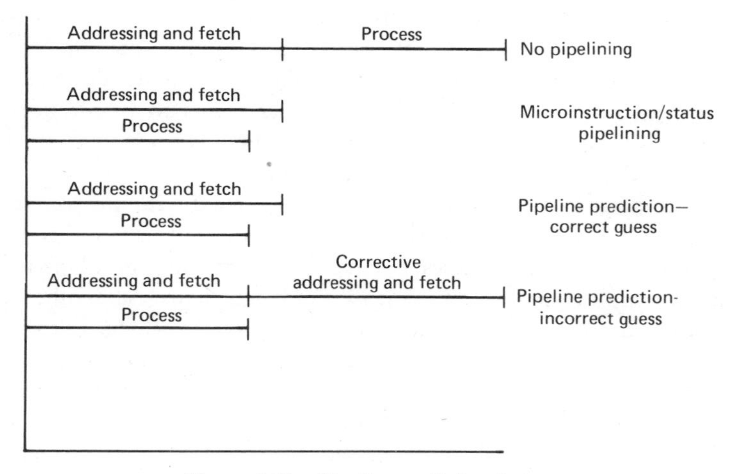

Figure 5.10. Pipeline-prediction timing.

branches, then, at the minimum, 75% of the parallel addressing and fetching processes will produce the correct result. Furthermore, if 70% of the conditional branches result in the branch being taken, only in 7.5% of the machine cycles must the microinstruction addressing and fetching operation be repeated. Assuming that a nonpipelined cycle takes 200 ns and a correct-prediction cycle takes 120 ns, one has reduced the average cycle time to 126 ns.

Rather than blindly making a universal guess about the outcome of all conditional branching operations, one can improve the odds by recognizing that some conditional branches in the microprogram are very likely to result in branches, and others are unlikely to result in branches. In other words, the microprogrammer can usually make such statements as "yes, the conditional branch in this particular microinstruction will almost always result in a branch" and "no, the conditional branch in that microinstruction is rarely taken." This knowledge can be exploited by adding a bit to each microinstruction, a bit that dynamically informs the control section of the guess it should make. For instance, the value 1 might mean "assume the branch is likely" and the value 0 means "assume the branch is unlikely." When the microprogram is being written, the microprogrammer studies the circumstances of each microinstruction containing a conditional branching operation and sets the "branch-predictor" bit accordingly.

Compared to the recommended microinstruction/status pipelining technique, pipeline prediction can achieve close to the same cycle time. In fact, it can lead to a faster system, since traditional pipelining occasionally results in some microinstructions that can accomplish nothing else but a conditional branch. However, pipeline prediction requires additional logic, plus the complexity of having two different machine cycle times.

VARIABLE CYCLE TIMES

The cycle time of a processor is usually determined by locating the longest path through its logic and using this worst case as the minimum possible value of the cycle time. Unfortunately, the longest path often exceeds all other paths by a considerable margin, and one often discovers that this path is infrequently activated. Hence, although many machine cycles could be terminated much earlier, the cycle time is dictated by a worst-case situation.

As an example, Table 3.7 indicates that the maximum propagation delay through the Am2901A for most cases is 119 ns, but if a shift is being performed on the ALU output, the path requires 147 ns. Unless one had no plans to perform shifts, the cycle time would be dictated by the longer path. As another example, the F=0 status output is the latest status output to become stable. Even if one was not interested in the F=0 status in a particular cycle, the simplest system design would require one to wait until all status outputs had stabilized. As a third example, Table 4.8 indicates that the controller designed using the Am2910 has a maximum propagation delay of 55 ns for sequence operations, but 78 ns on conditional branches. If the operation involves decrementing the counter, the delay can be even longer.

Rather than letting these worst-case paths dictate the machine cycle time, the machine cycle can be made variable in duration [3]. As a simple example, assume that a 150 ns machine cycle is sufficient for most cycles, but the longest logic paths require a cycle time of 200 ns. One can design the system with two machine cycle times—150 ns and 200 ns. Normally, the 150 ns cycle is used. When the current microinstruction specifies functions that require the longer logic paths, the current machine cycle is continued for an extra 50 ns.

Doing this requires knowledge of whether the current microinstruction requires a 150 or 200 ns cycle. Three possible approaches are

1. Designing logic that analyzes the microorders in the current microinstruction. If the microorders specify a function that requires a longer machine cycle (e.g., a microorder specifies a shift operation), the final clock pulse is delayed an extra 50 ns.
2. Adding a bit to each microinstruction that indicates whether a longer cycle is needed. As each microinstruction is written, the microprogrammer analyzes it and sets the bit accordingly. The timing logic examines this bit, rather than analyzing the microorders.
3. Adding a bit to the microinstruction as in the above alternative, but rather than having the microprogrammer set the bit, have the microprogram assembler analyze the microorders in each microinstruction and set the bit accordingly.

Alternatives 2 and 3 have the expense of an added bit in each word in control storage. Furthermore, alternative 2 is quite error-prone; if the microprogrammer sets the bit incorrectly, the error can be quite difficult to diagnose. On the other hand, depending on how sophisticated one chooses to get, alternative 1 can involve a substantial amount of logic. Hence, alternative 1 is generally recommended if the logic needed to detect a "longer-cycle" microinstruction is simple; otherwise, alternative 3 is recommended.

One does not want to implement this concept using a variable-speed oscillator, since this results in timing instabilities. Instead, one usually employs an oscillator that runs considerably faster than the machine cycle, and defines the machine cycle as being a given number of oscillator cycles. In the example above, one might employ an oscillator with a 50 ns cycle. Normally, a machine cycle consists of three clock cycles. However, when a longer cycle is needed (i.e., the "longer-cycle" bit is set in the current microinstruction, or logic decoding the microinstruction sees a need for a longer cycle), the machine cycle consists of four clock cycles.

RESIDUAL CONTROL

To this point, all the control signals needed (e.g., for the processing and control sections) have been indicated as emanating from the current microinstruction. However, in certain instances, it is useful to be able to have a microinstruction store control signals in a register, and have these control signals used in later cycles. This design technique is called *residual control.*

Consider a processor that has two machine instructions that AND and OR two variable-length strings of bytes in main storage. Given that the two machine instructions are identical except for the operation performed on the bytes, one is inclined to want to use a common microprogram subroutine or a common set of microinstructions for both, the motivations being to save control-storage space and avoid the cost of writing an almost-identical set of microcode for both instructions. However, this would require a microprogram loop with a test in the middle of the loop to determine whether to AND or OR the bytes obtained from storage. This test would add one or more cycles of overhead to each iteration of the loop.

An alternative is the addition of a residual-control register to the design. Assume the ALU is controlled by a 3-bit microorder. The bits might be defined as

000 Add
001 Subtract
010 AND
011 OR
100 Exclusive-OR
101 Exclusive-NOR
110 Pass left input only
111 Residual control

The value 111 in the microorder specifies that the control signals for the ALU will not come from this microorder; they will come from the residual-control register. Of course, a method must be provided to gate a value into the residual-control register.

Many opportunities for sharing microinstructions now exist. For instance, the microcode corresponding to the "AND bytes" instruction would set a 010 in the residual-control register and branch to a shared sequence of microinstructions (or call a subroutine). The microcode corresponding to the "OR bytes" instruction would set a 011 in the residual-control register and branch to, or call, the same point. Within the shared loop, rather than specifying a distinct ALU operation, 111 ("use residual-control value") would be specified, thus causing the current ALU function to be determined by the value in the residual-control register.

Although residual control, when incorporated into a design, is usually specific to a particular component or function (e.g., controlling the ALU), there is no reason why the concept cannot be generalized, perhaps to all functions. The general residual-control register might be as wide as the widest microorder, and all microorders in which it makes sense might have a value that specifies "use the control value in the residual-control register as if its value were present in this microorder." This would allow one to create residual-control values for the specification of destination registers, ALU operations, memory operations, branch addresses, branch operations, and so on.

Like most everything else, residual control also has drawbacks. Obviously, it requires additional logic. Also, it makes the microprogram more difficult to understand because, when examining a microinstruction specifying a residual-control operation, one cannot immediately discern the effect of that microinstruction. Last, in terms of saving control-storage space, residual control can be self-defeating in that more bits of information are needed in the microinstruction. For example, since one of the eight microorder values is devoted to specifying the residual-control ALU function in the simple example above, the residual-control capability for the ALU requires three eighths of a bit in each control-storage word. Also, one needs enough bits in the microinstruction to control the loading of the residual-control register (e.g., a 1-bit microorder that specifies "load the value of this other field (possibly a multi-use field) into the residual-control register").

MICROORDER ENCODING

As discussed in Chapter 2, the encoding of control information in the microinstruction is usually motivated by two factors: reducing the width of the microinstruction (hence the size of control storage) and reducing the possibility of coding meaningless or erroneous microinstructions (e.g., specifying two functions that are truly mutually exclusive). Considerable work has been done in the area of optimal encodings of microinstructions [4]. Unfortunately, most of the techniques developed are not of practical use in that they are oriented toward obtaining optimal solutions rather than practical engineering solutions. Furthermore, fast and dense control-storage components are now available, making concerns about control-storage space requirements less critical than they were in the past.

Determining the degree of encoding is a tradeoff process. High degrees of encoding save control-storage space, but they increase the cost of decoding logic and may result in slower cycles times because of the added delays through the decoding logic. Hence, one usually sees only limited amounts of encoding in extremely fast systems and in systems with a small number of control-storage words.

Fortunately, the use of bit-slice components reduces the need for encoding considerations, since, in most bit-slice devices, the control signals are already encoded and the decoding is done within the device. For instance, the branching operations of the 2910 sequencer are encoded into four bits and decoded within the chip. In a design employing the 2910, one would usually define a 4-bit microorder for branch operations, and define its values to be identical to the I inputs specified for the 2910. One might only be interested in encoding here if the microprogrammer needed only a limited number of the 2910's branching operations (e.g., four). In this case, a 2-bit branching-operation microorder could be defined, and decoding logic would be needed to map these four values into four of the 16 I input combinations of the 2910. Although one has saved two bits in each control-storage word by doing this, it must be weighed against (1) the loss in branching flexibility, (2) the cost of the decoding logic, and (3) the delay added by the decoding logic.

As a simple example, assume that we are designing a system which requires an ALU to perform the following functions

X + Y
X − Y
X AND Y
X OR Y

and that, for the sake of discussion, we decide to use one or more 74S181 ALUs. The 74S181 performs a large number of functions. Four S inputs select the arithmetic or logical function, the 1-bit M input distinguishes between an arithmetic and logical function, and the C (carry) input must be set to 0 for the addition operation and 1 for the subtraction operation. Hence, the 74S181 requires six control inputs, but since we require only four functions, we might wish to encode these into a 2-bit microorder.

Figure 5.11 shows the required decoder, which might be a high-speed PROM (four 6-bit words) or discrete logic. If the latter, the designer should carefully assign the values of the microorder (add, subtract, AND, and OR) to minimize the decoding logic needed.

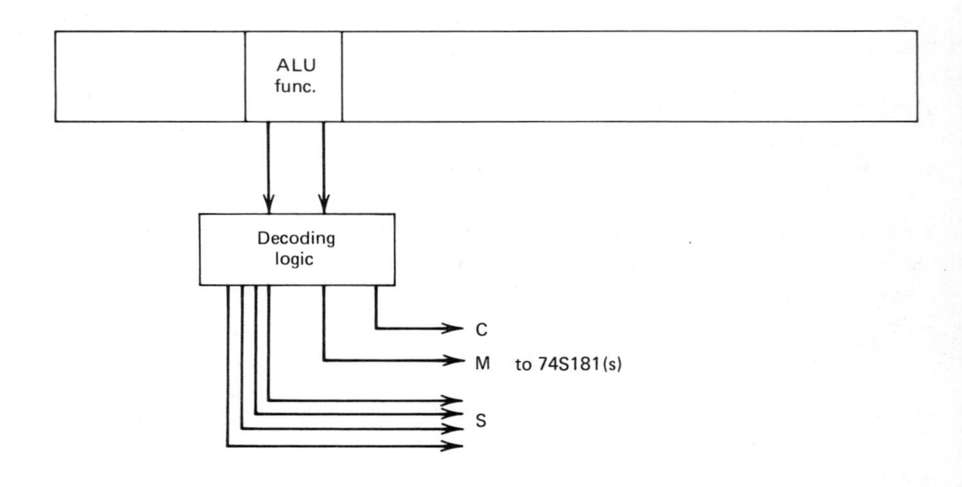

Truth table for the decoder

ALU func.	Meaning	'181 inputs		
		S	M	C
00	A + B	1001	0	0
01	A − B	0110	0	1
10	A or B	1110	1	X
11	A and B	1011	1	X

Figure 5.11. A microorder decoder.

The type of encoding described above and in Chapter 2 is known as *single-level* or *direct encoding*. Here the sole input to a decoder is a single microorder. Another type of encoding, as illustrated in Figure 5.12, is known as *multiple-level* or *indirect* encoding. In Figure 5.12a, some of the decoders have a secondary input; a special microorder controls the decoders. This "decoder-control" microorder allows other microorders to have two or more completely different interpretations. This may be useful if several infrequently used microorders are needed. Rather than widening the microinstruction to include all these microorders, the decoder-control microorder specifies which microorder is present in a shared field in each microinstruction. This technique is illustrated later in the chapter.

As shown in Figure 5.12b, the decoders can be controlled by one or more external signals. As a simple example, assume that a processor is being designed, and that the processor itself must control the I/O devices (i.e., there are no independent I/O processors or channels). Assume that certain microorders are applicable only when the processor is acting as a processor (i.e., interpreting machine instructions) and that certain microorders are applicable only to I/O control. Rather than placing all these microorders in the microinstruction, they may be overlaid. That is, certain fields in the microinstruction have two different interpretations, depending whether the processor is in the "CPU" state or "I/O" state. A flipflop holds the current state of the processor (CPU or I/O control) and this flipflop is fed to the decoders as a secondary input.

To repeat, the degree of encoding employed requires careful analysis, since higher degrees of encoding result in additional logic costs and the overhead of propagation delays through the decoding logic. In light of the latter, one technique is to employ little or no encoding on microorders that appear in the critical timing paths, and employ higher degrees of encoding on microorders not on the critical timing path.

Another form of microinstruction encoding, making use of two control storages, is discussed in a later section in this chapter.

PRE- AND POST-PIPELINE DECODING

Until this point, nothing has been said about the location of microinstruction decoding logic, in particular, in relation to the pipeline register, if one is used. However, by the appropriate placement of the decoding logic, one may be able to shave 5 to 10% from the machine cycle time.

The traditional placement of the decoding logic is shown in Figure 5.13a. Here the pipeline register has the same width as control storage. At the beginning of the machine cycle, the current microinstruction in the pipeline register is decoded, resulting in signals being sent to the control points of the system. However there is no requirement that the pipeline register hold the form of the microinstruction as it appears in control storage. In fact the microinstruction can be decoded into indi-

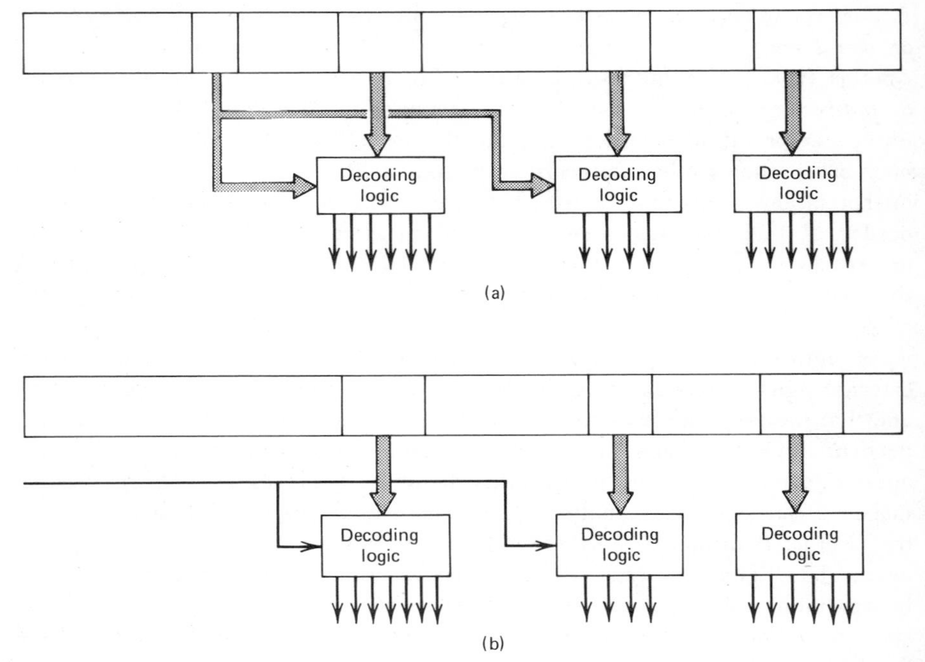

Figure 5.12. Two forms of indirect encoding.

vidual control signals before it is sent into the pipeline register as is shown in Figure 5.13b. Doing so requires a wider pipeline register, but the advantage may be a decrease in machine cycle time in certain situations. Figure 5.13b moves all the decoding logic into the timing path of the control section rather than the processing section. Hence, if the critical path through the control section and control storage is shorter than that through the processing section, and if the decoding

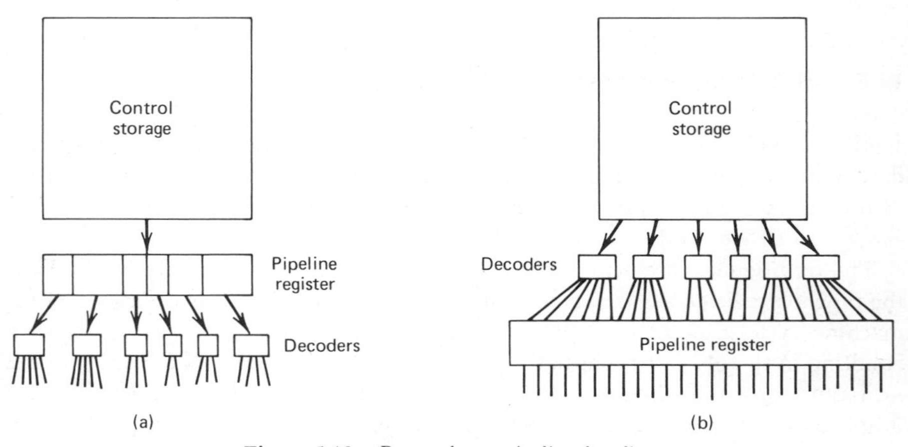

Figure 5.13. Pre- and post-pipeline decoding.

needed for the processing section is substantial, the longest path (that through the processing section) may be shortened, thus reducing the machine cycle time.

This is illustrated in Figure 5.14. Here two situations are illustrated. The first is where the path through the control logic (microinstruction sequencing logic and control-storage access) is the longer path. The second situation is where the path through the processing section is the longer path. For both situations, post-pipeline decoding and pre-pipeline decoding are considered.

The symbols used in Figure 5.14 are the same as those used earlier. However, where microorder decoding time was not explicitly considered in the previous discussions of pipelining, it is broken out here. PD represents the propagation delay through decoding logic for the processing section and CD represents the propagation delay through decoding logic for the control section (e.g., delays through one or more status-condition multiplexers).

First we shall consider the case where the control path is longer than the processing path, making the assumption that pipelining with a status and microinstruction register is being used. In this case the minimum machine cycle time is given by CD+CXP+CSA+PR, assuming that post-pipeline decoding is being used. One sees that moving all the decoding logic into the control path by placing it between control storage and the pipeline register has no beneficial effect on timing. If pre-pipeline decoding is used here, the minimum machine cycle time is given by CXP+CSA+PR, plus the maximum of PD, CD. Hence in this case pre-pipeline decoding has no time advantage; in fact it will be slower if PD is greater than CD.

Next consider the situation where the path through the processing section is the longer path. If post-pipeline decoding is being used, the minimum machine cycle time is PD+PXS+SR. In this case it may be advantageous to place the decoding

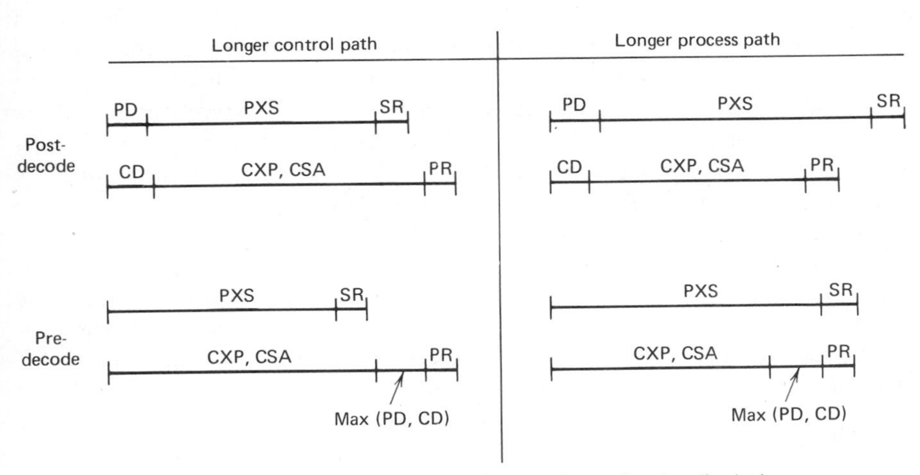

PD: critical timing path through processing-section decoding logic
CD: critical timing path through control-section decoding logic

Figure 5.14. The effects of pre- and post-pipeline decoding.

logic between control storage and the pipeline register, as the decoding-time component (PD) can be removed from the minimum machine cycle time. If pre-pipeline decoding is used here, the minimum machine cycle time is given by the greater of

$$PXS + SR$$

and

$$CXP + CSA + PR + \max(PD,CD)$$

In almost all cases, this should result in a lower minimum machine cycle time. Hence, if the processing-section path is the longer path, and there is some decoding logic on this critical path, the machine cycle time can be reduced by placing the decoding logic between control storage and the pipeline register. The tradeoff is that this requires a pipeline register of perhaps considerably more width.

In certain situations, it may be advantageous to mix these two approaches by doing some of the decoding before the load into the pipeline register and other decoding at the output of the pipeline register.

ADDRESSING LARGE CONTROL STORAGES

By cascading a number of bit-slice sequencers, one can address a control storage of any size. However, a situation may arise where one wants to use a fixed-length sequencer (e.g., 2910 or 8X02) but one needs more control storage than the sequencer can address [5]. This section illustrates a few techniques for doing so.

For purposes of illustration, assume that one determines that a particular 10-bit sequencer will be used, but more than 1024 control-storage words are needed; assume that the design needs 4096 words of control storage. Figure 5.15 illustrates one way of accomplishing this. Here it is assumed that the microinstruction contains a 12-bit branch-address field, suitable for addressing all 4096 words. The low-order 10 bits of the branch address become inputs of the sequencer, and the sequencer outputs feed the low-order 10 bits of the control-storage address inputs. The high-order two bits of the branch-address field in the microinstruction are gated to the two high-order control-storage address inputs.

This allows one to address all of control storage, even though the sequencer is only 10-bits wide. However, this is not without a penalty; it requires special considerations when writing the microprogram. For instance assume the sequencer contains a pushdown stack for managing control-storage addresses during subroutine calls and returns. The stack is only 10-bits wide, meaning that if one calls a microprogram subroutine in a different 1024-word area ("quadrant" or page) in control storage, the subroutine will return to the wrong location because the sequencer is not capable of remembering the entire 12-bit control-storage address. This means that, when using this technique, one must avoid subroutine calls across 1024-word boundaries, or one must employ some care in doing so (which is discussed later in the section).

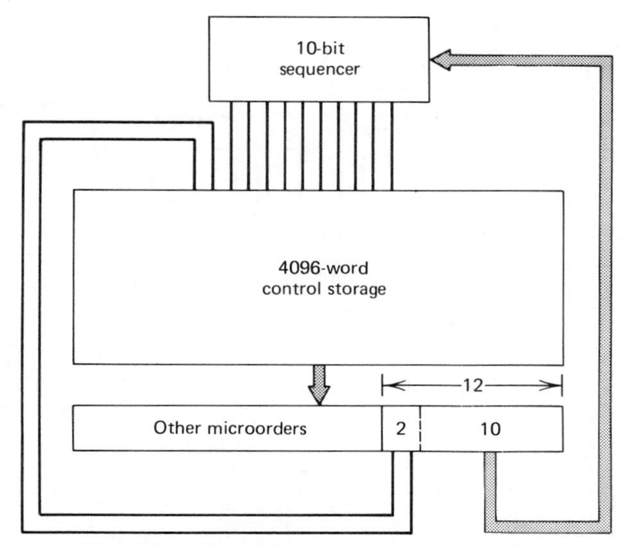

Figure 5.15. Generating 12-bit addresses with a 10-bit sequencer.

Given this problem, one is likely to organize the microprogram into four distinct sections, each section residing in a 1024-word page of control storage. The partitioning would be done such that the most-frequent branching operations are branches to other microinstructions in the current page, and that branches from one page to another will be infrequent. This implies some redundancy in the microinstructions, since it is highly likely that the two high-order bits in the branch-address field in the next microinstruction will be the same as those in the current microinstruction. Another technique can be employed to eliminate these "page address" bits from the branch-address field. This is shown in Figure 5.16.

In Figure 5.16, the high-order control-storage address inputs do not come from the current microinstruction; they come from a page register (a 2-bit register in this example). Here the branch-address field has been shortened to 10 bits, and the normal circumstance is for a branching operation to address, as its target, a microinstruction in the current 1024-word page of control storage. However, one needs the ability to branch from one page to another, which is accomplished by the ability to alter the page register. Here a 1-bit microorder, A, has been added to the microinstruction. Assume that another microorder B exists in the microinstruction and that this microorder, perhaps, has multiple purposes. If the A microorder has the value 1, the first two bits of the B microorder are loaded into the page register, thus causing the next microinstruction to be fetched from a different page.

This concept removes the page-address bits from the microinstruction by substituting the page register. The technique is similar to the idea of residual control discussed earlier. With this technique, one still has the subroutine-branching problem mentioned above. In addition, one must be cautious in writing the micro-

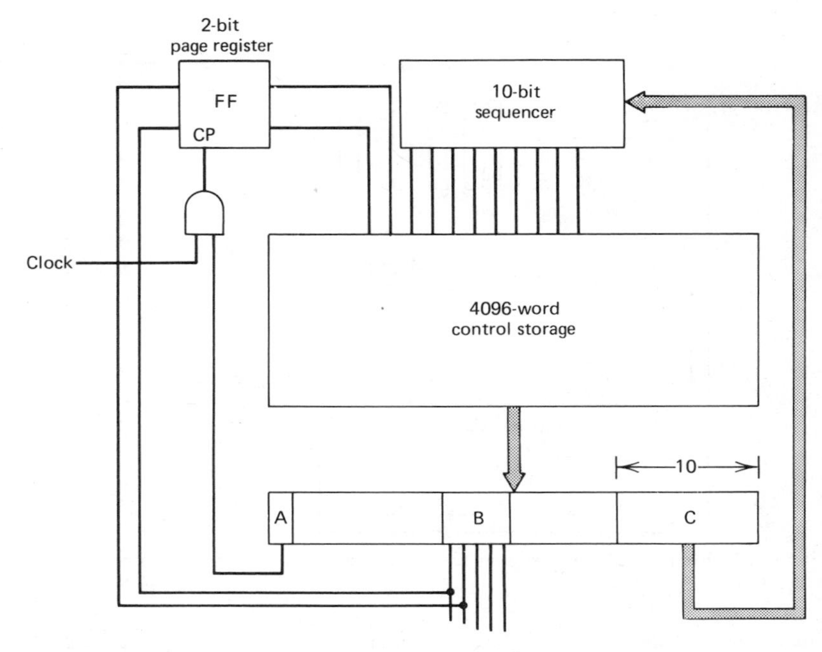

Figure 5.16. Generating larger addresses via a page register.

program such that a sequence operation is not performed in the last microinstruction in a page.

To illustrate the problem with subroutine calls, assume we are dealing with the design in Figure 5.16. Assume that the microinstruction at location 4F7 (hexadecimal) wishes to call a subroutine beginning at location 142. Location 4F7 is in page 1; location 142 is in page 0. In the microinstruction at location 4F7, one would specify the call operation, setting the 10-bit branch-address field to 142 and setting the page register to 00. However only the 10 low-order bits of the address of the next microinstruction in control storage (the one at 4F8) would be stored in the sequencer's stack. Hence the value 0F8, not 4F8, would be stored in the stack. When the subroutine eventually performs a return operation, it will transfer control to the wrong point.

One can manage this problem in several ways. An obvious solution is to place a copy of each subroutine in each page of control storage in which it is needed. In some situations this solution may be acceptable, although it does increase the control-storage requirements. A second solution is to provide surrogate representations of subroutines that need to be called from multiple pages. Each surrogate would consist of two microinstructions and would reside in the page in which the subroutine resides.

For instance, assume one has a subroutine named XYZ in page 0, and that this subroutine must be called from microinstructions in pages 1, 2, and 3. In addition to writing the microprogram for subroutine XYZ, one will also write, in page 0, three 2-microinstruction sequences.

Examining one of them, the first microinstruction will call XYZ and the second microinstruction will specify a return operation, setting the page register to a particular value. For instance, let the 2-microinstruction routine that serves as a surrogate for calls to XYZ from page 1 be called XYZ1. The first microinstruction in XYZ1 will simply be a call to XYZ; the second microinstruction will contain a return operation and set the page register to 1. Likewise, the second microinstruction in XYZ2 will set the page register to 2 upon returning, and so on. Hence, if a microinstruction in page 1 wishes to invoke subroutine XYZ, it should call XYZ1, setting the page register to 0. XYZ1 will, in turn, call XYZ, XYZ will eventually return to XYZ1, and then XYZ1 will return to the calling point in page 1.

Although this does allow one to call subroutines in different pages, the drawbacks are two additional machine cycles incurred when an out-of-page subroutine is invoked, and an additional entry placed on the sequencer stack when this occurs.

A third alternative involves the addition of logic to the control section to stack the current value of the page register when a subroutine is called. For instance, a 2-bit by 4-word LIFO stack might be added, as well as some logic that decodes the branching-operation microorder. When the microorder specifies an operation that causes an address to be pushed onto the stack in the sequencer, the current value of the page register is pushed onto the auxiliary stack. When the branching-operation microorder specifies an operation that removes a value for use from the sequencer's stack (e.g., subroutine return), the top value in the added stack is popped into the page register. Hence this stack is acting as an extension to the stack in the sequencer.

One might also wish to use the idea of a page register even if the sequencer being used is large enough to address the entire control storage. The motivation may be to reduce the length of the microinstruction by shortening the branch-address field within the microinstruction.

This is illustrated in Figure 5.17. In this example, control storage contains 1024 words, and a 10-bit sequencer (sufficient to address the entire control storage) is employed. However, the page register was added so that the branch-address field in the microinstruction could be shortened from 10 bits to 8 bits. Thus, in this example, control storage is divided into four 256-word pages. A microinstruction can branch freely within its own page. If a microinstruction needs to branch to a microinstruction outside of the current page, it must alter the value in the page register. The page register, rather than being directly gated to control storage, is gated to the two high-order bits of the sequencer's address inputs. This approach does not have the subroutine problem described earlier, since the stack within the sequencer is wide enough to remember the full control-storage address. However one must still be careful to partition the microprogram

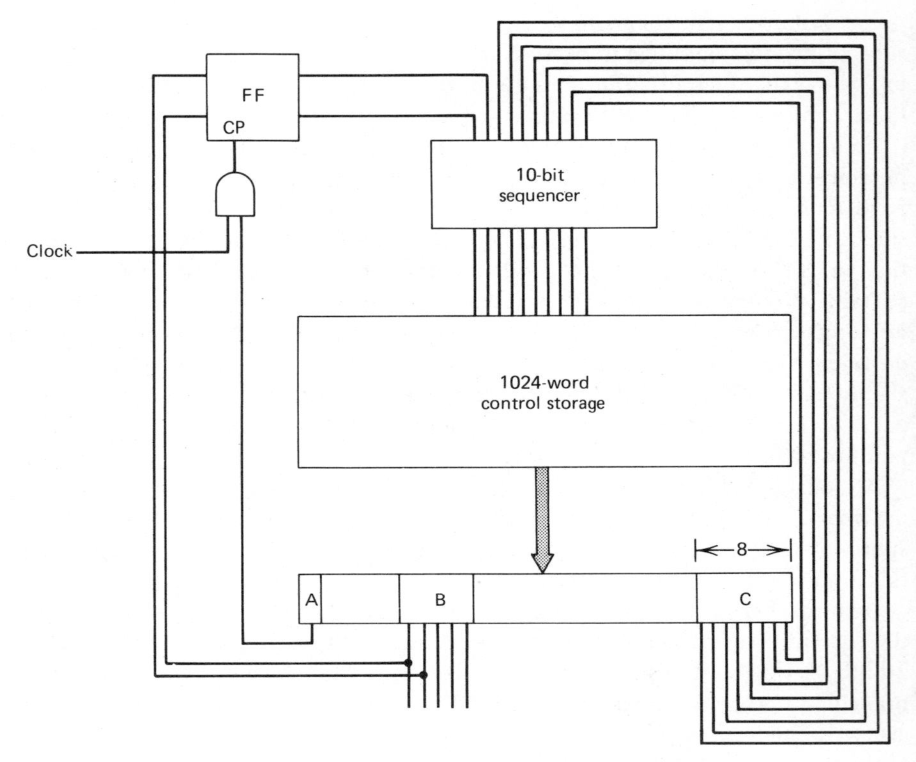

Figure 5.17. Use of a page register to reduce microinstruction size.

into four distinct sections such that interpage branches are infrequent. Also, one must not make the mistake of thinking that a microinstruction at the end of a page can perform a sequence operation to get to the first microinstruction on the next page. (It could do so, providing it also changes the page register.)

HORIZONTAL VERSUS VERTICAL MICROINSTRUCTIONS

Even for people well versed in the concept of microprogramming, a point of significant confusion is the use of the terms "horizontal microinstruction" and "vertical microinstruction." In fact this confusion is very much related to the difficulty in the definition of microprogramming as discussed in Chapter 2. In addition to sorting out the distinctions between these two types of microinstruction design and looking at their implications on system design, this section will return to the problems surrounding the definition of microprogramming and illustrate how they are related to the notions of horizontal and vertical microinstructions.

As a basis for starting, all microinstruction examples presented to this point are microinstructions in the horizontal category.

One source of confusion is the descriptive terms themselves. The terms "horizontal" and "vertical" have an curious origin; they were selected to depict an "artist's conception" of the shape of control storage in machines having either type of design. In a machine with a horizontal-type microinstruction, the control storage, relatively speaking, tends to be wide and shallow; hence the term "horizontal" was applied. In a machine with a vertical microinstruction design, control storage tends to be relatively narrow (short words) and deeper (more words).

The first thing to realize is that the terms vertical and horizontal do not imply that one is faced with a binary decision; rather, they are used to describe designs toward the ends of a continuous spectrum of designs. This is illustrated in Figure 5.18. Notice here that the spectrum has been drawn with a boundary on the left, but no defined boundary on the right. This lack of a recognized boundary at one end of the spectrum is associated with the definition problem of microprogramming, which is discussed later.

One can label the spectrum in at least two ways. One way to view the spectrum is as a measure of the degree to which the microinstruction can explicitly and directly control the actual logic devices in the system. At one end of the spectrum, each bit in the microinstruction controls a logic device within the system. This was the property of the design of the initial microprogrammed machine in Chapter 2. Such a machine would be termed a "highly horizontal" design. However, as we saw in Chapter 2 and in this chapter, one can employ various degrees of encoding on the microinstruction. As the microinstruction becomes more and more encoded, the microprogrammer becomes more and more divorced from the actual logic devices in the system; he is faced with an "abstract machine" created by the decoding logic. As one moves further to the right, one can envision a microinstruction design that is so highly encoded that its definition is quite abstract and independent from the data flow and logic devices of the system. Designs in this area of the spectrum are called vertical designs.

One can see, perhaps, how this is related to the definition of microprogrammed control. In Chapter 2, microprogrammed control was defined in terms of the ability, through stored-program logic, to exercise explicit and direct control over the logic devices in the system. This ability is reduced as one moves toward the

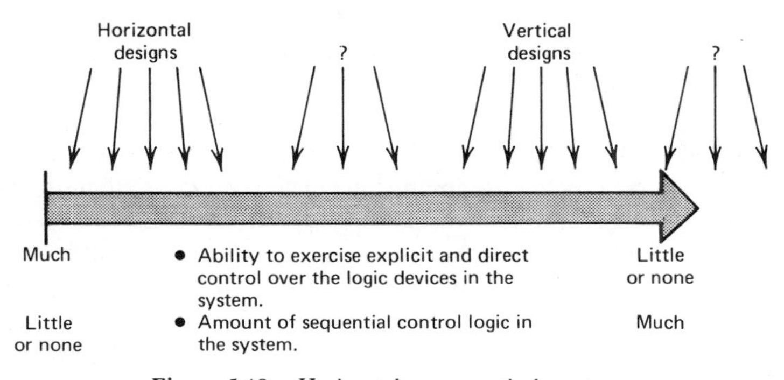

Figure 5.18. Horizontal versus vertical spectrum.

right in the spectrum. Hence, in a highly vertical design, one could quarrel about whether the machine is even a microprogrammed machine, or whether the things called highly vertical microinstructions are in fact software.

Another way of labelling the spectrum is as a measure of the amount of sequential control logic in the machine. This may seem an unusual measure, since microprogrammed control was defined as an *alternative* to sequential logic control. However one should realize that the two concepts may be mixed. In a machine with a highly horizontal design, there is little or no sequential control logic, except for the control-storage device itself. As one moves toward the right, the amount of sequential control logic normally increases. Again, one is faced with the problem of establishing the point at which there is so much sequential control logic that the machine cannot be considered to be a microprogrammed machine. As an example, the IBM S/370 Model 138 processor has a highly vertical microinstruction that is decoded and interpreted by a significant amount of discrete combinatorial and sequential logic. Because of this, one could argue that this machine is not a microprogrammed machine; in fact one could argue that it does not even have the S/370 architecture. One could say that it is a sequential-logic-controlled machine with a foreign architecture, and that the S/370 architecture was implemented by writing a machine-language program on the foreign architecture that, in turn, interprets programs encoded to the S/370 architecture.

Figure 5.18 also defines a gray area between the horizontal and vertical designs. For lack of a better term, designs in this area are often called "diagonal" microinstruction designs, indicating that they share characteristics of both horizontal and vertical designs. The question mark at the right side of the spectrum highlights the confusion over whether to label such designs vertical-microinstruction designs or non-microprogrammed designs with a sequential control network.

Horizontal and vertical designs are often typified by four characteristics. The first is the physical appearance of the microinstruction. A horizontal microinstruction contains a large number of independent microorders which exercise control over individual parts of the data flow. A highly vertical microinstruction contains relatively few fields. Usually, only one of the fields is a control field, and the other fields contain data or register addresses. A highly vertical microinstruction usually triggers a predefined sequence of events. For instance, consider a vertical microinstruction containing three fields. The first field might be used as an operation designator and the second two fields as operand designators. One value of the operation designator might be defined as "take the value of the memory word addressed by the register specified in the second field and add it to the value of the register specified in the third field." This implies that some amount of sequential control logic exists to control the implied sequence of events, that is, fetching a value from storage and then adding it to a register.

Such a microinstruction looks suspiciously like a machine instruction, thus pointing out the problem of defining precisely the difference between highly vertical microinstructions and machine instructions. Figure 5.19 illustrates a few of the many types of vertical microinstructions in the Burroughs B1700 processor.

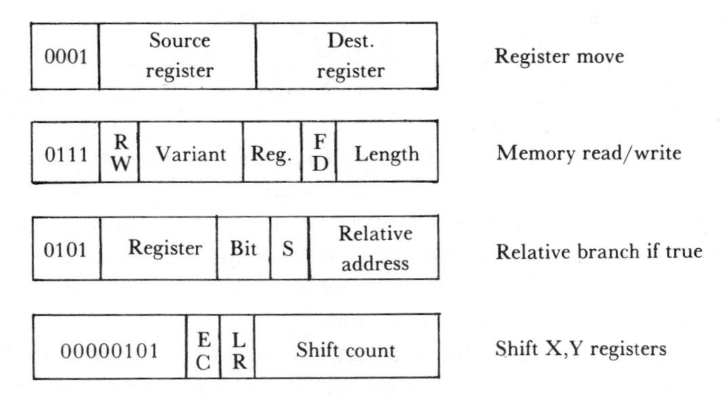

Figure 5.19. Vertical microinstructions in the Burroughs B1700.

Another way of differentiating between horizontal and vertical microinstructions is their length. In most designs, a horizontal microinstruction may typically be from 30 to well over 100 bits in length; a vertical microinstruction may typically be from 15 to 30 bits long.

Another distinguishing characteristic is the degree of parallelism that can be achieved within a machine cycle. A horizontal microinstruction exhibits a high degree of parallelism because its microorders control many simultaneous operations. A vertical microinstruction usually involves little or no parallelism within the machine cycle; instead it initiates a single sequence of events.

Another characteristic usually, but not always, identified with both forms of designs is the way in which sequencing or branching operations are represented. In a horizontal microinstruction, as has been the case with all microinstructions discussed thus far in the book, branching operations are specified by one or more microorders in each microinstruction. That is, every microinstruction performs a branching operation (although, in many cases, it might merely be the sequence operation, meaning fetch the next microinstruction from the next location in control storage). In a vertical design, branching operations are usually not performed in parallel with the control operations. Rather, each microinstruction simply sequences to the next microinstruction in control storage. Whenever a branching operation is needed, it is performed in a separate cycle. That is, the definition of the operation field in a vertical microinstruction might contain several values that specify branching operations. In a vertical microinstruction that specifies a branching operation, the other fields typically specify a branch address.

Although there are many exceptions to this rule in existence, a rule of thumb is that higher-speed systems usually have horizontal designs, and slower-speed systems usually have vertical designs. This occurs for three reasons. First, a vertical microinstruction is more highly encoded, meaning that there is more overhead in the microinstruction decoding process. Second, the large number of microorders in a typical horizontal microinstruction means that a large number of operations can potentially be performed in parallel in a single machine cycle.

Third, because of the separate branching microinstructions in most vertical designs, the vertical machine wastes a machine cycle whenever a branch is needed; in the horizontal machine, branching operations are performed in parallel with control and processing operations.

On the other hand, microprograms are usually easier to develop for a machine with a vertical microinstruction, since this form of microprogramming is quite similar to software programming. Because of this, and because vertical designs tend to stray from the true meaning of microprogramming, horizontal designs are often referred to as "hard" microprogrammed machines, and vertical designs are known as "soft" microprogrammed machines.

Another difference between the two is the amount of control storage needed. One would expect that a vertical machine would usually require less control-storage space for a given microprogram, given its more highly encoded microinstruction. Although this seems to be the case, one experiment has indicated that there was no significant difference in the total number of control-storage bits needed [6]. The experiment involved selecting a problem and microprogramming its solution on four existing horizontal and vertical machines. However, the experiment is not conclusive, since it was not an "apples-to-apples" comparison; it involved comparing horizontal machines with completely different vertical machines. If one took a given machine data flow and then designed both a horizontal microinstruction format and a vertical microinstruction format, the microprogram written to the vertical design should take less total space.

One last aspect of vertical and horizontal microinstructions needs clarification. Some system designs exist that claim to have *two* levels of microprogrammed control; the lowest level of control is perhaps a horizontal design and this interprets another level of stored program logic which is often called vertical microcode. This organization is illustrated in Figure 5.20. Notice that the lowest level of microprogrammed control interprets the level above it by controlling the hardware beneath. Although the level above is often called "vertical microcode" because it,

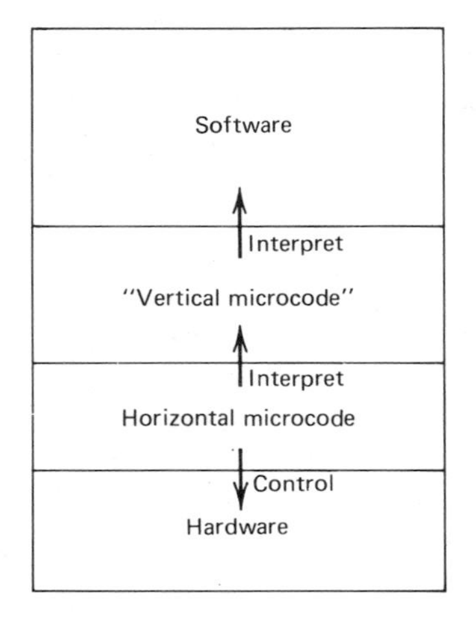

Figure 5.20. Example of misuse of the term "microcode."

in turn, interprets a higher software level, this is a misuse of the term microcode or microprogram. Returning to the definitions in Chapter 2, it makes no sense to call a layer of a system a microprogram layer if it in turn is interpreted by a microprogram. If a microprogram is interpreting a higher set of control logic, this control logic is defined to be software, or machine instructions. Hence, the second layer in Figure 5.20 is software, and one is simply faced with a system having multiple layers of interpretive software, an attribute that most existing software systems have to some degree.

In contrast, there are microprogrammed systems with one level of microprogram control, but two levels of control storage, the subject of the next section.

TWO-LEVEL CONTROL STORAGES

The section on microinstruction encoding alluded to at least one more technique that reduces the amount of control storage needed in a design. This is the technique of an indirectly addressed, or two-level, control storage.

Before examining this technique, it might be helpful to review the reasons why one is often interested in reducing the size of control storage. Using memory technologies available in 1980 as an example, the densest memory component available for use in control storages (assuming an access time of well under 100 ns is needed) is a 4096-bit chip. Given a control storage in a medium- to large-scale processor of perhaps 2048 100-bit words, one would need 50 chips for the control storage alone, a sizable investment in terms of component costs, physical board space, and power consumption. (Actually, since most memory chips have a width of one bit, 100 chips might be needed.)

Space-saving techniques are concerned with the elimination of redundancy. When one looks at control storage redundancy on a macroscopic scale, one could say that the subroutine concept is a redundancy remover. A motivation for employing microprogram subroutines is to avoid repeating a sequence of microinstructions that perform a given function in multiple places in control storage. The concept of microorder encoding looks at redundancy in a microscopic manner. This encoding, for instance, takes advantage of the fact that mutually exclusive control signals can be encoded in a smaller number of bits.

There is an intermediate type of redundancy that may be tackled by the designer. For instance, even if microprogram subroutines are employed and microorders are highly encoded, one might discover that some control-storage words have values identical to other control-storage words. Perhaps a microinstruction specifying something like "gate registers A and B to the ALU, perform an addition operation, gate the ALU output to register B, initiate a memory read operation, and sequence to the next microinstruction" appears 10 times within the microprogram. Perhaps a subroutine ABC has been written, but control storage now contains four identical microinstructions that specify "set R5 to zero and call ABC." The idea of an indirectly addressed, or two-level, control storage is concerned with the elimination of this type of redundancy.

To implement this concept successfully, one must be faced with the situation where identical microinstructions appear frequently in control storage. Figure 5.21 will be used to illustrate the concept. Assume the design has a 100-bit microinstruction and that approximately 2048 microinstructions are needed. The first alternative is simply to define a control storage containing 2048 100-bit microinstructions. The second alternative is to provide two control storages, perhaps a 9-bit control storage of 2048 words, and a 100-bit control storage containing considerably fewer words, perhaps 512 as shown in Figure 5.21.

The 9-bit control storage can be considered as the place in which the microinstructions logically reside. However, rather than place control signals in the 9-bit word, each word specifies an address of a microinstruction in the second control storage. Hence all microinstructions in the second control storage will be unique and many words in the 9-bit control storage will point to the same microinstruction in the second control storage. Rather than having redundant (identical) 100-bit microinstructions, one ends up with redundant 9-bit addresses, thus saving a considerable amount of space.

The logic needed to achieve this is straightforward. The control section first fetches a word from the 9-bit storage and then uses its value as the address used to fetch an actual microinstruction from the 100-bit storage.

Given the example above, one sees a significant savings in the total amount of control storage. In the first alternative, 204,800 bits are needed. With the two-level concept, only 69,632 bits are needed. The obvious disadvantage is speed, since the control section must perform two sequential memory fetches to eventually acquire the microinstruction for the next machine cycle. However, if pipelining is used, and depending on the relative speeds of the control and processing sections and control storage, this concept may add little or no overhead.

Branching operations are performed in the following way. Each 100-bit microinstruction specifies a branch operation and possibly a branch address. However, all branches are performed with respect to the 9-bit control storage. For

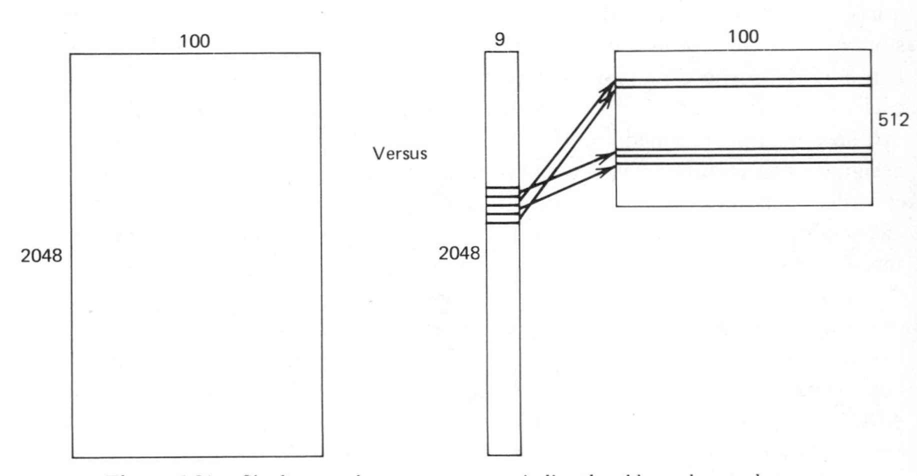

Figure 5.21. Single control storage versus an indirectly addressed control storage.

instance, if a microinstruction specifies the sequence operation, the operation does not mean "fetch the next microinstruction from the next location in the 100-bit control storage;" it means "fetch the value of the location in the 9-bit memory beyond the location currently being addressed." If a microinstruction specifies a branch to an address, the address specified is for a word in the 9-bit control storage.

Terminology often applied to these memories is "micromemory" for the address memory (the 9-bit memory in this example) and "nanomemory" for the memory containing the actual microinstructions.

A variant of this technique adds more information to the address memory with the motivation of reducing the number of words needed in the microinstruction memory. For instance, it is often the case that large numbers of microinstructions are *almost* identical, except for the values of one or two microorders. If the address memory can provide certain parameters (e.g., register specifications) for use in the current machine cycle, then a larger number of nonunique microinstructions can be found, thus reducing the size of the microinstruction memory.

The Nanodata QM-1 machine can be used as an illustration [7]. The machine contains two control storages. One control storage has an 18-bit word, where seven bits specify an address in the other control storage and the remaining 11 bits can specify one or two values for use by the microinstructions in the second memory. The other control storage has a 360-bit word. Rather than placing one microinstruction in one word, each word in this nanomemory contains a 72-bit field of residual-control information, which is active whenever this word is active, and four 72-bit microinstructions. When a word in the first memory is selected, it causes the machine to activate the residual-control information and the first microinstruction in the selected nanomemory word, as well as make the 11 bits of information in the first memory word available for use by the microinstruction (e.g., as register designators). This microinstruction can then sequence to the subsequent microinstruction in the current nanomemory word, branch to one of the other microinstructions in the current nanomemory word, or branch to the first microinstruction in another nanomemory word. The structure of the two memories is shown in Figure 5.22.

USING MAIN STORAGE AS THE CONTROL STORAGE

Another way of minimizing the number of bits of control storage is to eliminate it altogether. Given that the processor has a main storage in which machine-language programs and their data are stored, there is no reason why the microinstructions themselves cannot be stored here. This does not reduce the total number of storage bits required, but it does reduce the cost, since main storage is slower and is considerably cheaper. The IBM 370/145 and the Burroughs B1800, to name a few examples, store their microprograms in main storage.

One obvious drawback is that the approach becomes unwieldy if the microinstruction size does not correspond to the main-storage word size; one would probably not do this if the size of the microinstruction is larger than the

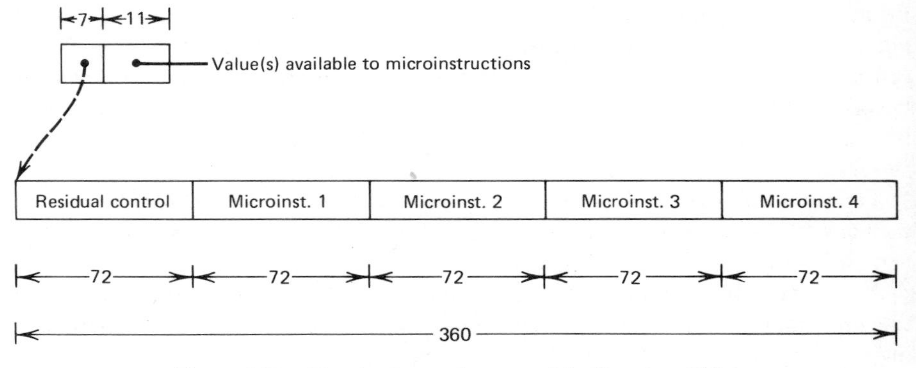

Figure 5.22. Two-level control storage of the Nanodata QM-1.

size of words in main storage, since this would entail several storage accesses to obtain a microinstruction. Hence, this approach is normally used only on small machines with vertical microinstruction designs (i.e., on machines likely to have small microinstructions). The other obvious problem is one of performance, since main storage is usually considerably slower than a control storage. However, this can be overcome, as is done in the Burroughs B1800, by the use of a cache: a small, high-speed buffer memory. Rather than reading only a single microinstruction from main storage, a block of main storage (i.e., a set of microinstructions) can be fetched and stored in the cache. Hopefully, there is a high probability that the microinstruction needed for each machine cycle will be in the cache. This hope is based on the expectation of locality of reference in the microprogram, that is, the hope that microinstructions tend to address, as their successors, microinstructions in the immediately surrounding area. A machine with a microinstruction cache might have two machine cycle times: a fast cycle that is used when the next microinstruction is found in the cache, and a considerably slower cycle when the next microinstruction must be fetched from main storage.

Although this concept has not been studied fully, it seems to have drawbacks that would eliminate it from consideration for most designs. In addition to the extra overhead incurred when a microinstruction must be obtained from main storage, one must also contend with the problem of memory interference, that is, microinstructions as well as machine instructions and their operands are being fetched from a single memory device. Another drawback is cost, since the use of a cache normally implies a small associative memory to search for items in the cache. A third drawback is the complexity added because of the cache and the need for two highly different cycle times.

A CASE-STUDY DESIGN

As a final illustration of some of the ideas in this chapter, the microinstruction design of an actual processor will be examined. The processor itself will not be illustrated here. Briefly, however, it contains a 24-bit data path, a 4-bit data path used for decimal and logical operations, a pair of memory-address registers, a pair of memory-data registers, and a microprogrammed control section. The 24-bit data path is largely constructed from six cascaded Am2903 ALU/register slices. The control section employs an Am2910 microprogram sequencer and an Am29803 multiway branch control unit. The control storage contains 8096 words; a 1-bit page register exists in the control section to generate the high-order control storage address bit.

The microinstruction design is illustrated in Figure 5.23. Because of the large number of words of control storage, an attempt was made to minimize the size of the microinstruction without sacrificing function. This was accomplished by defining four formats of microinstructions with the format specified by the first two bits of the microinstruction. Thus indirect encoding of the microorders is done; the inputs to many of the decoders are not only the microorder, but also the first two bits of the microinstruction. For instance, the full capability of the 2903 slice is needed, but placing full control of the 2903 in a single microinstruction would lead to a wider microinstruction because of the large number of control inputs to the 2903. Hence the following was done. The 2903's extended functions (e.g., normalization, multiplication, division) are needed infrequently. Rather than adding four bits to each microinstruction to control the 2903 extended functions (control inputs I_5-I_8), the extended functions can only be performed by a format-3 microinstruction (bits 0–1 = 11). As another example, the multiway branching capability of the 29803 is needed in the design, but again it is something that is done rather infrequently in the microprogram. Rather than placing a control microorder for the 29803 in all microinstructions, it exists in only the format-2 microinstruction (bits 0–1 = 10).

Basically, the format-0 microinstruction allows one to perform a branching operation, manipulate the 24-bit data path, and initiate a memory operation. Bits 2–13 normally specify a branch address; that is, they feed a 12-bit address to the 2910 sequencer. This field can also be used to feed a constant value into the RC counter in the 2910. If this field is not being used in a microinstruction to specify a branch address or a count, an emit value can be placed in this field, and other microorders control the gating of this value into the data path.

The three microorders in bits 15–23 specify a branch operation. Bits 16–19 specify which of the 2910 branch operations is to be performed. This microorder is sent intact to the control inputs of the 2910. However, since one of the 2910's operations was not needed in this design, one value of this microorder is decoded and causes a customized branching operation to be performed. Bits 20–23 are meaningful during conditional branching operations; the microorder selects one of 16 conditions to be tested. The 1-bit microorder in bit 15 is the polarity control, specifying whether the condition is to be tested for true or false. Status/microinstruction pipelining is used in this design.

The next microorder (bits 24–27) specifies the input to be gated into the left port of the ALU in the 2903s. Since the processor has several registers in addition to those in the 2903s, this field is not sent directly to the A-register input of the

	0 1	2–13	14	15	16–19	20–23	24–27	28–31	32–34	35–36	37–39	40–43	44–46	47	48
24-bit data path control	00	Branch address or count or emit	Memory wait	Branch T or F	Branch type	Branch condition	ALU_L source	B register select	MDR1 (P0–P5) select	ALU_R source	ALU function	ALU shift and destination	Memory request	MARX	MDRX
Super branch	10				P register select	Multiway branch control or compare value			Branch type / Y control						
4-bit data path control	01				Branch type	Branch condition	$DALU_L$ source	$DALU_R$ source	DALU function	C control	MAR select for count / MAR ± 1	DALU destination			
Extended operations	11						ALU_L source	B register select	(unused)	Addr. prefix	(unused) / PIU sig	ALU extended function	Extended memory request		

Figure 5.23. Microinstruction design example.

2903 slices. Rather, if this microorder has a value in the range 0000–1001, it specifies one of the first 10 registers in the 2903s as one ALU source. For values 1010–1111, it causes certain external registers, or the emit field, to be gated into the DA bus of the 2903s, and causes this bus to be selected as the left ALU source by the 2903s.

The microorder in bits 28–31 is sent directly to the B-register address inputs of the 2903s. The microorder in bits 35–36 selects the other ALU input source. There are four possibilities: the register specified as the B register, the Q register, the value of the emit field in the current microinstruction (12 zeros followed by bits 2–13 of the current microinstruction), or the value 6 (a wired value fed into the 2903s over the DB bus).

The next microorder, in bits 37–39, specifies the function to be performed by the ALU in the 2903s. Omitting the 2903's extended operations, the 2903 ALU performs 16 functions specified on four input lines, but this microorder contains only three bits. The reason is that only 8 of the 16 ALU functions were needed in the design. Hence decoding logic exists between this microorder and the 2903 to convert the 3-bit values into the 4-bit control values needed for the 2903. (Although this may seem a costly way to save one bit, remember that each microinstruction bit saved is a savings of 8096 bits of control storage.)

Bits 40–43 in the format-0 microinstruction contain a microorder that controls the shifters within the 2903 slices, as well as the routing directions for the output of the ALU. Since the ALU output can be routed from the 2903s' Y bus into several external registers, this 4-bit microorder does not directly correspond to the four ALU destination-control signals (control lines I_5-I_8) of the 2903. Some of the 2903 shift-and-destination functions are not used; these unused values are decoded to direct the routing of the 2903 Y-bus output.

The next microinstruction format (bits 0–1 = 10) is quite similar to the first format. The major difference is that the two branch-control microorders are decoded differently. Bit 32 causes one of two branching modes to be selected. One mode, a customized operation, is not relevant here; the other is the multiway branch provided by the Am29803. When this type of branch is specified, bits 20–23 of the microinstruction, rather than being decoded as a branch-condition selector, become the four control signals needed by the 29803. The microorder in bits 16–19, rather than being interpreted as a branch operation to be sent to the 2910, is used to select one of 15 4-bit registers in the processor. When a multiway branch is specified, the value of the selected register is gated to the T inputs of the 29803. Since the 2910 sequencer does not have OR gates on its output (like the 2909), four OR gates exist in the low-order four bits of the address bus between the 2910 and control storage. The 29803 outputs are connected to these OR gates.

The design contains logic to force the 2910 to perform an unconditional branch when the multiway-branch operation is being used, and to deactivate (to zero) the 29803 outputs at all other times.

As mentioned previously, the processor also contains a 4-bit data path. This data path requires a substantial number of control signals. Again, rather than placing them in a single, wide microinstruction, they are isolated in a third format of the microinstruction (the first two bits = 01). This microinstruction can initiate a 2910 branching operation, as was the case in the first format, control the 4-bit data path, and initiate memory requests. The microorders in the first two formats that applied to the 24-bit data path are decoded differently; they control the 4-bit data path. For instance, two microorders specify the inputs of the 4-bit ALU (one of 14 registers, zero, or the low-order four bits of the emit field). A microorder in bits 32–34 specifies one of eight functions for that ALU, and a microorder in bits 40–43 specifies the destination of the ALU output. The 4-bit data path contains a carry flipflop associated with the ALU; this flipflop can be explicitly set or reset by the microorder in bits 35–36. Two other microorders allow one to increment or decrement one of the two memory address registers.

A set of operations that are needed infrequently within the microprogram were combined into a fourth format of the microinstruction. Here the microorder in bits 40–43 specifies which of the extended functions in the 2903 is to be performed. As mentioned earlier, control storage was partitioned into two pages of 4096 words each, because the 2910 can only address 4096 words. In this processor, branches from one page to another occur infrequently, and when they do occur, this type of microinstruction (format-3) must be used. The microorder in bits 35–36 is used to set the single-bit prefix or page register to 0, 1, or unchanged. The format-3 microinstruction can also perform a 2910 branch operation and control some additional, more exotic, memory operations.

The machine cycle time is normally 150 ns. However, it is extended to 180 ns when logic detects a 2903 shift operation or a 2910 branch operation that involves decrementing its counter.

REFERENCES

1. S. R. Redfield, "A Study in Microprogrammed Processors: A Medium Sized Microprogrammed Processor," *IEEE Trans. on Computers,* **C-20**(7), 743–750 (1971).

2. A. K. Agrawala and T. G. Rauscher, *Foundations of Microprogramming: Architecture, Software, and Applications.* New York: Academic, 1976.

3. G. F. Muething, Jr., "Microprogramming Techniques to Achieve Maximum Performance," *Digest of Papers, Spring Compcon 78.* New York: IEEE, 1978, pp. 72–74.

4. T. Agerwala, "Microprogram Optimization: A Survey," *IEEE Trans. on Computers,* **C-25**(10), 962–973 (1976).

5. P. Chu and V. Coleman, "Expanding the Addressing Capability of your Microprogram Sequencer," *EDN,* **23**(11), 137–143 (1978).

6. R. Johnson and R. E. Merwin, "A Comparison of Microprogramming Minicomputer Control Words," *Digest of Papers, Fall Compcon 74,* New York: IEEE, 1974, pp. 161–166.

7. G. Frieder and J. Miller, "An Analysis of Code Density for the Two Level Programmable Control of the Nanodata QM-1," *Proc. of the Tenth Annual Workshop on Microprogramming.* New York: ACM, 1977, pp. 26–32.

8. A. D. Robbi, "Microcache: A Buffer Memory for Microprograms," *Digest of Papers, Fall Compcon 72.* New York: IEEE, 1972, pp. 123–125.

6

Other Bit-Slice and Support Devices

Although ALU/register slices and microprogram sequencers embody a considerable amount of the logic needed in a digital system, a substantial number of auxiliary or support devices are needed to complete the design. Some of this auxiliary logic has been packaged into LSI components, sometimes as bit slices. This chapter surveys these auxiliary components and discusses some of the more-interesting ones in detail.

BIT-SLICE FAMILIES

Most manufacturers of bit-slice devices have defined families containing MSI and LSI support components, components that are usually needed to complete a design. Although one can mix components from a variety of families, one usually sees a high degree of design consistency among devices within a family.

Advanced Micro Devices originated the popular 2900 family. Some of the major components in this family are

2901 ALU/register slice
2903 ALU/register slice
2904 Status and shift controller
2909 Microprogram sequencer slice
2910 Microprogram sequencer
2911 Microprogram sequencer slice
2913 Priority interrupt expander
2914 Vectored priority interrupt controller
2925 Clock generator and driver
2930 Program controller slice

2940 DMA address generator
2942 Programmable timer/counter
29803 16-way branch controller
29811 Next-address controller

The 2900 family also includes a carry-lookahead generator and a variety of bus transceivers, registers, multiplexers, decoders, and memories.

The 3000 series, originated by Intel, contains the following components

3001 Microprogram sequencer
3002 ALU/register slice
3214 Interrupt controller

as well as a carry-lookahead generator, latch, and bus transceiver.

The high-speed ECL M10800 family from Motorola contains the following components

10800 ALU/register slice
10801 Microprogram sequencer slice
10802 Timing controller
10803 Memory interface slice
10806 Dual-port register array
10808 Multibit shifter

as well as bus transceivers and ECL/TTL translators.

The Fairchild 9400 Macrologic family contains the following major components

9401 CRC generator/checker
9403 FIFO buffer memory
9404 Data path switch
9405 ALU/register slice
9406 Pushdown stack and controller
9407 Memory interface slice
9408 Microprogram sequencer
9410 Register array

The 9400 family, except the 9401, is available in CMOS as the 4700 family. The 4700 family also contains a programmable bit-rate generator (4702).

The Fairchild F100220 ECL family contains the following devices

F100220 ALU/register slice
F100221 Multiple-function network

F100222 Dual-access stack
F100223 Programmable interface unit
F100224 Microprogram sequencer

THE 2930 PROGRAM CONTROL UNIT

Most conventional computer architectures of today embody a variety of addressing modes in their machine instructions, including indirect addressing, indexed and based addressing through registers, and addressing relative to the current value of the program or instruction counter. Hence, there is often a small amount of arithmetic to be performed in locating the operands of a machine instruction being executed. Also, when fetching a machine instruction from storage, the processor must usually perform arithmetic on the program counter (i.e., increment it).

If the processor consists of only a control section and processing section, the above functions must be performed in the processing section, leading to a highly serial design. If additional speed is needed, a design similar to that of Figure 6.1 might be employed. Here the design is expanded to include a third section: a program/memory controller. Depending on the design, the controller may perform some or all of the following functions

1. Serve as one or more memory address registers for the processing section.
2. Fetch the next machine instruction from main storage.
3. Resolve the operand addresses of the next machine instruction.
4. Fetch the operands from storage.
5. Process all branch-type machine instructions.
6. Fetch and buffer, for the processing section, the next n machine instructions.

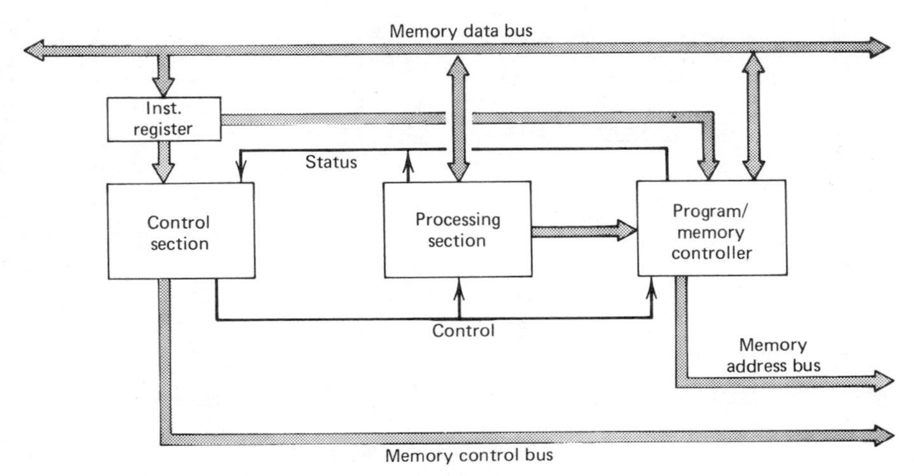

Figure 6.1. Processor organization with a program/memory controller.

The Am2930 program control unit [1] was designed to serve as a program/ memory controller. It is a cascadable 4-bit slice in a 28-pin package. As shown in Figure 6.2, the 2930 consists of two registers (PC and R), an adder, an incrementer, and a 17-entry pushdown stack.

2930 External Connections

The external connections of the 2930 are shown in Figure 6.3. The five I inputs control the multiplexers and the stack logic. The IEN input, when high, inhibits all registers except R from changing state, but it does not affect the combinatorial logic and the outputs produced by the I inputs. RE, when low, causes the R register to be loaded from the D bus. The CC input is used to gate a condition to be tested into the 2930.

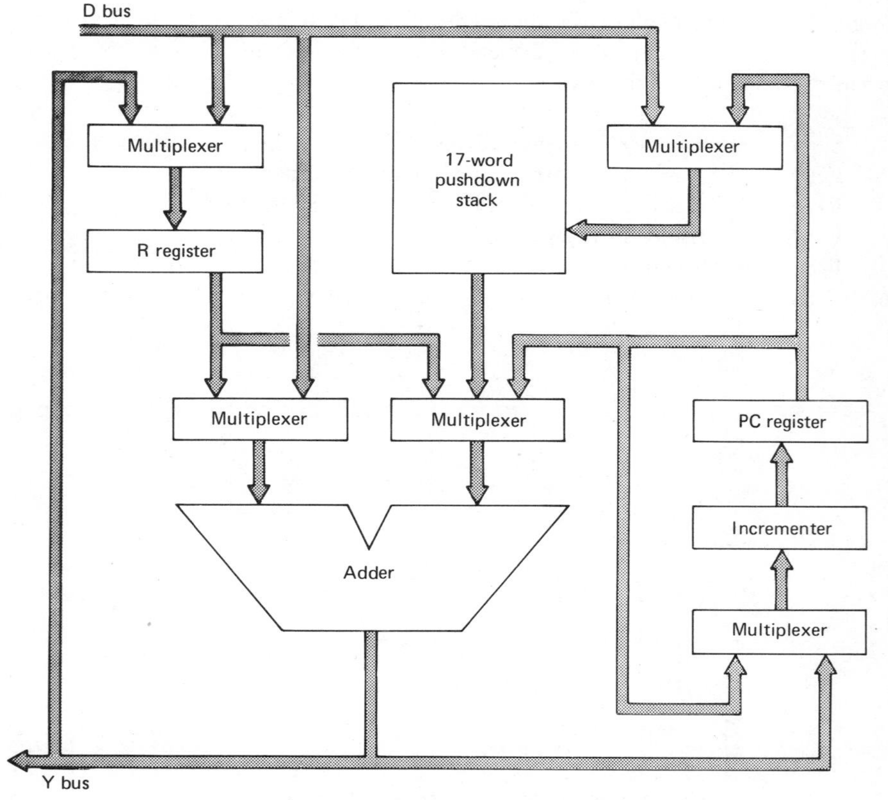

Figure 6.2. Organization of the 2930.

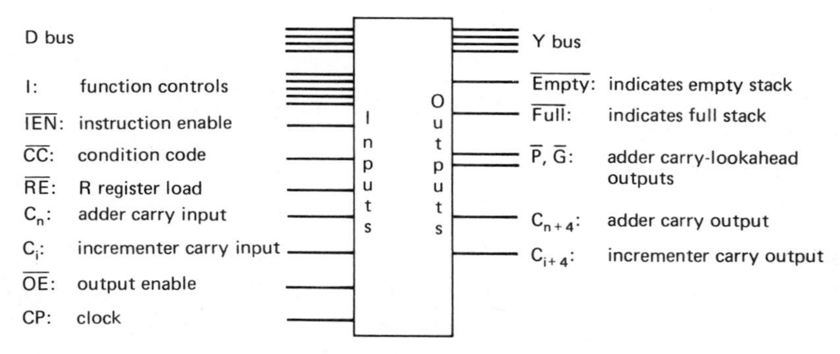

Figure 6.3. 2930 external connections.

Both the adder and incrementer have carry inputs and outputs, and carry-lookahead outputs are provided for the adder.

2930 Functions

The 2930's I inputs give it 32 functions. When taken with the CC and RE inputs, the total number of unique functions is 46.

When the I_4 control input is 0, the 2930 performs an unconditional function. Most of these functions involve the selection of an output (Y bus) value and incrementing the value in the PC register (if $C_i = 1$). The possible Y outputs are

PC
R
D
$R + D + C_n$
$PC + D + C_n$
$PC + R + C_n$
$S + D + C_n$

where D is the value on the input (D) bus and S is the value on the top of the stack. Four of the remaining functions gate PC to the Y bus, increment PC by C_i, and (1) load D into R, (2) push PC, (3) push D, or (4) pop the stack. One function ($I_{4-0} = 01101$) pops the stack onto the Y bus and increments PC by C_i. Another ($I_{4-0} = 01111$) gates PC into Y and does not increment PC.

When $I_4 = 1$, the 2930 performs conditional operations. If $\overline{CC}=1$, PC is gated to the Y bus and PC is incremented by C_i. If $\overline{CC}=0$, one of 16 operations is performed. Six of these are labelled "conditional jumps;" they gate the following to the Y bus

R

D

0

$R + D + C_n$

$PC + D + C_n$·

$PC + R + C_n$

and gate the Y-bus value plus C_i into the PC register. Another six are conditional subroutine jumps; they have the same effect as the conditional jumps, but they also push the value of PC onto the stack. Two others are return operations; they pop the stack and gate either S or $S+D+C_n$ to the Y bus and to PC. The 15th operation simply gates PC to the Y bus. The last operation, "suspend," involves no state changes and causes the Y bus to be held in the high-impedence state.

As shown in Figure 6.1, the 2930 is intended for use as a memory interface and as a device to handle branch instructions in the machine-language program being interpreted. The PC register is intended to be used as the instruction counter (address of the next machine instruction). The auxiliary R register, because of its independent enable input (RE), can be used as a memory address register by the processing section. The adder allows one to implement relative branching in the machine-language program (e.g., "branch to the current instruction plus N") and indexed branching (e.g., "branch to the address in register M plus N"), where N is the value of a field in the branch instruction and M is a register in the processing section. The CC input allows one to implement conditional branch instructions. The stack is provided for the implementation of a subroutine mechanism in machine-language programs. In most applications, the Y bus is connected to the memory address inputs and the D bus is fed from the processing section and the instruction register.

Since the 2930 requires a large number of control inputs (10) from the microinstruction, and since the 2930 is not normally used every cycle, its control inputs might come from multipurpose fields, rather than dedicating distinct microorders to the 2930. The IEN input allows one to disable the 2930 when it is not being used.

Another possible application of the 2930 is as a microprogram sequencer.

Critique

Despite its apparent sophistication, the 2930 has several drawbacks, each serious enough to preclude its use in most applications. The first is the stack. The intention of the stack is the implementation of subroutine call and return instructions in the machine-language instruction set. However, virtually all architectures that use a stack for this purpose (e.g., PDP-11, 8080) require that the subroutine stack be held in main storage so that the stack is addressable by machine instructions. One reason is that the stack is used in these architectures for more than holding return addresses; its entries are often used for passing arguments to subroutines and for addressing. Another reason is that one must be able to switch stacks if the processor is being switched among different programs. Hence it is probable that if the 2930 is being used as a memory/program controller, no use will be found for its stack.

Another problem is the PC incrementer. The intention was that PC would be used to hold the address of the current machine instruction, and when the instruction is fetched, PC would be incremented by one and thus point to the next machine instruction. Unfortunately, this is useful only if the memory is word addressable and each instruction occupies one word. The majority of machine architectures have byte-addressed memories, where a machine instruction occupies several, and sometimes a variable number of, bytes.

Third, the 2930 is not particularly fast. For instance, a few typical (not maximum) times are

I input to Y output	—60 ns
I input to G,P outputs	—50 ns
C_n input to Y output	—25 ns

For use as a microprogram sequencer, the 2930 is considerably slower than the 2910 and lacks the invaluable counter and some of the branch operations of the 2910. As a memory/program controller, the designer has another alternative: an ALU/register slice, such as the 2901A. Not only is the 2901A considerably faster, but its register array and multifunction ALU provide more flexibility as a program controller.

THE 9407 PROGRAM CONTROL UNIT

The 9407 is a Schottky TTL program control unit bit-slice device in a 24-pin package [2]. It is also available as the 4707, a CMOS device.

The organization of the 9407 is shown in Figure 6.4. It consists of an adder, four registers, and two output busses. When the 9407 is used as a program/ memory controller, the registers are typically used for a program counter (address of the current or next machine instruction), stack pointer (address of an area of main storage being used as a pushdown stack for subroutine management), and a general memory address register. Since the registers, as well as the adder, can be gated onto the output bus, one has the option of incrementing the program-counter value before or after it is gated onto the output bus. The two output busses allow one to connect the 9407 to both the memory address and data busses.

Figure 6.5 illustrates the external connections of the 9407. Note that it has no carry-lookahead outputs from the adder.

9407 Functions

The function to be performed by the 9407 is specified by the four I inputs. Each combination of I inputs selects an output for the X bus; eight combinations also cause two registers to be modified. The available functions are

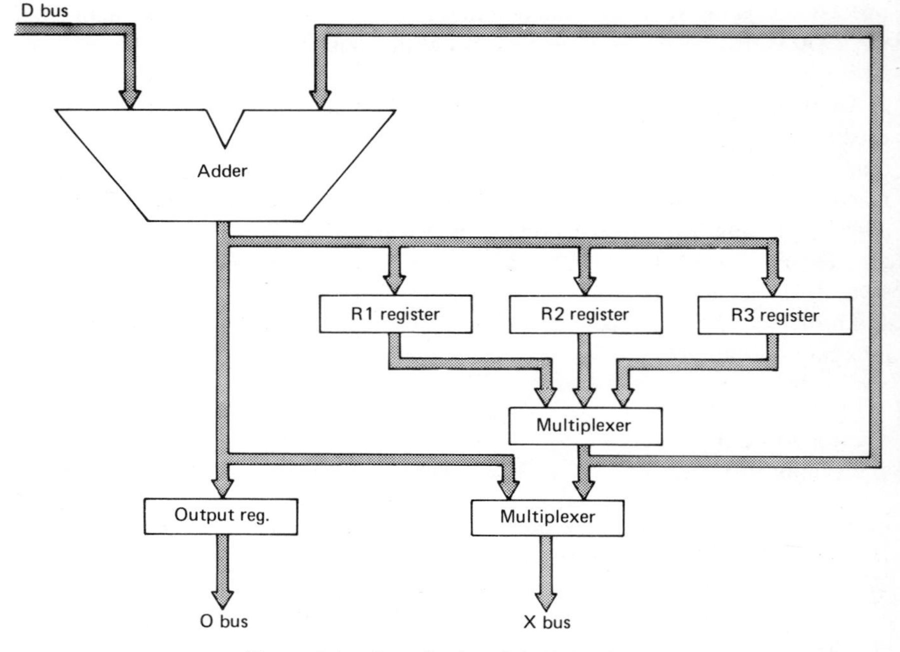

Figure 6.4. Organization of the 9407/4707.

X = R0
X = R1
X = R2
X,R0 = R0 + D + CI
X,R1 = R0 + D + CI
X,R2 = R0 + D + CI
X,R1 = R1 + D + CI
X,R2 = R2 + D + CI
X,R0 = D + CI
X,R1 = D + CI
X,R2 = D + CI

For all functions involving an adder operation, the value is also stored in the output register if the $\overline{\text{EX}}$ input is 0. (The number of functions above is not 16, since not all 16 combinations of the I inputs specify unique functions.) One of the I input lines controls the X-bus multiplexers; that is, it determines whether the X bus will be fed from the adder or the selected register. For instance, if I=0000, the X-bus output is R0; if I=0001, the X-bus output is R0+D+CI. Hence the fourth I input line can be used like the CC input of the 2930, that is, as a status input to implement a conditional-branch machine instruction.

One of the big differences between the 2930 and the 9407 is the absence of a pushdown stack. A companion device, the 9406 program stack, is a 16-word by 4-bit pushdown stack with the necessary control logic. However, as was noted in the previous section, an on-chip pushdown stack is often of no use in a program/memory controller.

On the surface, the 9407 is considerably faster than the 2930. For instance, the typical propagation delay from I to X is 37 ns, versus 60 ns for the 2930. However, the lack of carry-lookahead outputs in the 9407 make it slower than the 2930 for widths of 12 bits or more.

THE MC10803 MEMORY INTERFACE

The most powerful program/memory controller slice currently available is the MC10803 from Motorola [3]. The MC10803 is an ECL 4-bit slice in a 48-pin package.

The organization of the MC10803 is shown in Figure 6.6. Like the other devices in the M10800 family, the MC10803 is typified by an extraordinary number of data paths and a large number of external busses, many of which are bidirectional.

The heart of the 10803 is an ALU, which can perform the addition, subtraction, AND, OR, exclusive-OR, left-shift, and right-shift operations. The ALU also contains AND logic on its B input port, allowing the output of the multiplexer to be ANDed with the value on the P bus before entering the B port. The slice contains six registers. Two (MAR and MDR) are intended to be used as memory-address and memory-data registers. The other four reside in a single-port register array. As shown, one register (register 0, or PC) is separately addressable.

The 10803 contains five 4-bit external busses. Three busses (I, O, and D) are bidirectional. Two busses (A and D) pass through complementers. The A and D busses are intended to serve as the memory address and data busses. The I and O busses are intended for connections to other sections of the system (e.g., the processing section). The P input bus can be used to gate constant values (e.g., from an emit field in the microinstruction) into the slice. Figure 6.7 is an illustration of how the MC10803 might be configured with other devices within the system.

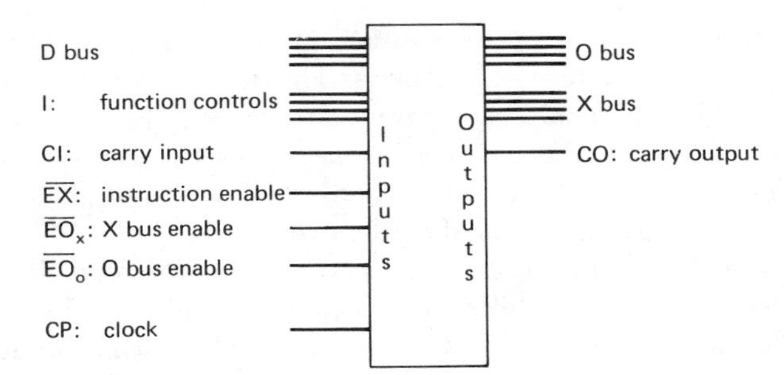

Figure 6.5. 9407/4707 external connections.

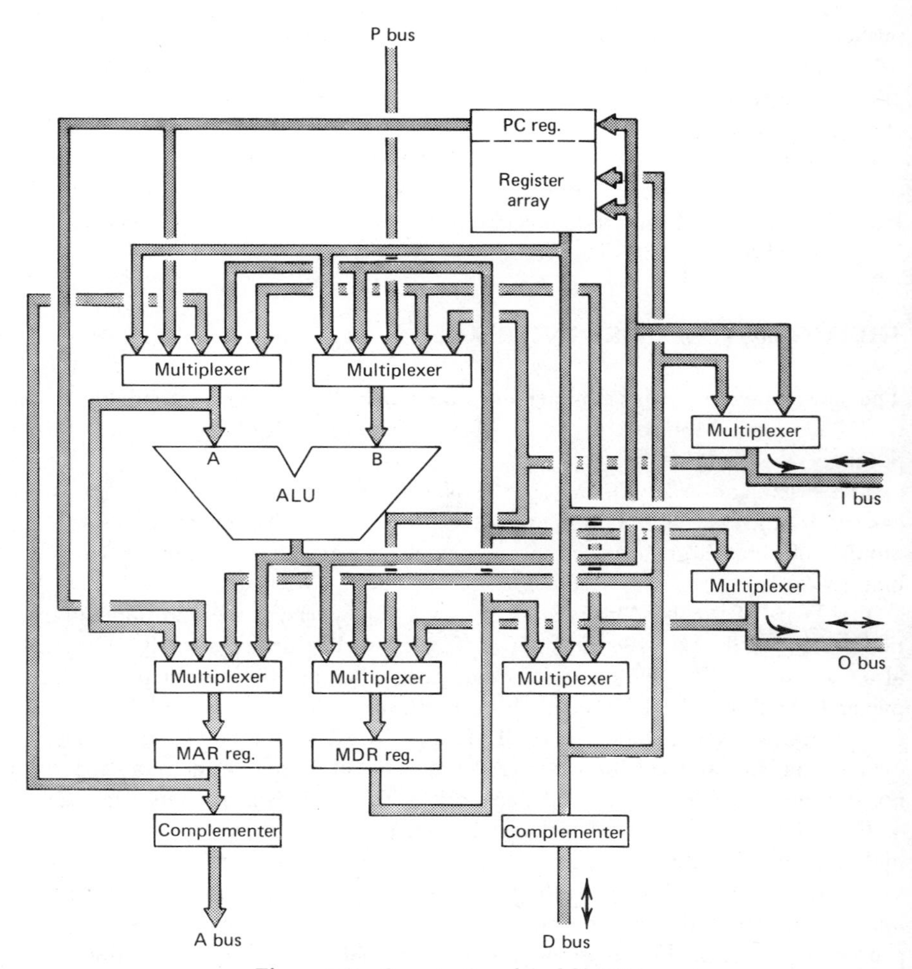

Figure 6.6. Organization of the MC10803.

MC10803 External Connections

Figure 6.8 illustrates the 40 active connections of the 10803 slice. The other eight pins are used for supply voltages and ground. The need for the large number (15) of function-control inputs should be apparent after viewing Figure 6.6. The carry-lookahead outputs have a dual use: producing signals for carry-lookahead logic, and indicating status (zero ALU output and overflow). One of the MS inputs specifies which set of outputs is desired on the two pins.

The 10803 also makes dual use of its carry input and output pins. When the ALU is used as a right or left shifter, the C_{in} and C_{out} pins are used as the shift connections. Because of this, both pins are bidirectional. When a ripple carry con-

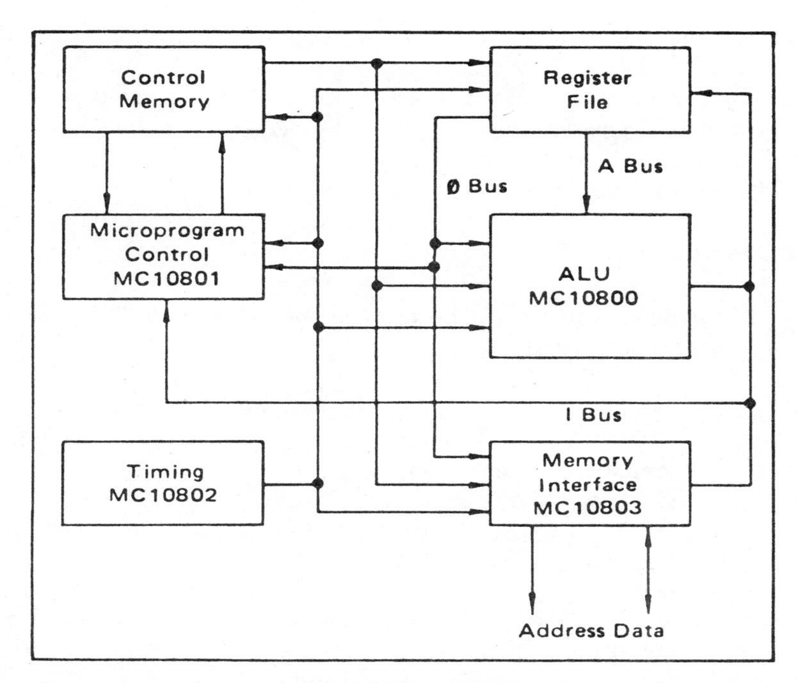

Figure 6.7. Processor organization with M10800 parts (Reproduced with permission of Motorola Inc.).

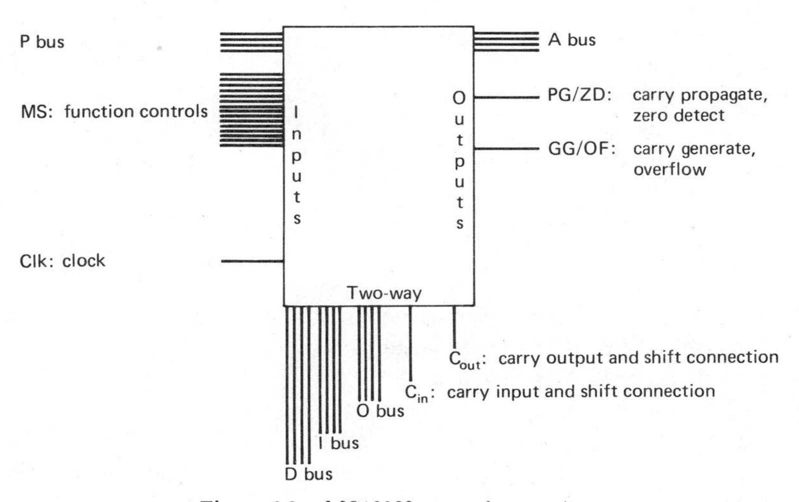

Figure 6.8. MC10803 external connections.

nection is used, no special considerations are needed. However, if a carry-lookahead configuration is used, one must provide buffers between the C_{in} and C_{out} pins of adjacent slices to permit a direct transfer, during a right shift, of C_{in} to C_{out}. (Left shifts are performed by the ALU by simply adding the input value to itself.) Since it is likely that right shifts will not be needed in most applications of the 10803, one can often forget this consideration.

MC10803 Functions

The 15 MS input pins control the functions performed by the 10803. The uses of the MS inputs are summarized below.

MS_{0-3}	gates a value onto the D, I, or O busses, or gates a value into the MDR register, MAR register, or a register in the register array.
MS_4	output-enable control for the A bus.
MS_{5-11}	gates the ALU output to the MAR register, a register in the register array, and/or the PC register (register 0 in the array).
MS_{6-8}	selects the ALU operation.
MS_9	with MS_{6-8}, selects the ALU operation. Also specifies whether carry-lookahead or status outputs are to be produced.
MS_{10-11}	selects the ALU inputs.
MS_{12-13}	selects a register in the register array.
MS_{14}	controls the complementers in the A and D busses.

To examine one set of control inputs in detail, Figure 6.9 defines the control of the ALU inputs and operation. The following symbols are used in the figure

OB	—O bus
IB	—I bus
DB	—D bus
DR	—MDR register
AR	—MAR register
ASL	—arithmetic shift left
LSL	—logical shift left
ASR	—arithmetic shift right
LSR	—logical shift right
V	—overflow status
PC	—register 0 in the register array
RF	—selected register in the register array
R3	—high-order ALU result bit

Note that some of the operations first AND the value on the P bus with a selected input before entering the ALU's B port. Hence one can use the P bus for bit-masking operations.

MS 9 10 11	MS 6 7 8: 0 0 0	MS 6 7 8: 0 0 1	MS 6 7 8: 0 1 0	MS 6 7 8: 0 1 1	MS 6 7 8: 1 0 0	MS 6 7 8: 1 0 1	MS 6 7 8: 1 1 0	MS 6 7 8: 1 1 1
	SUB G_G=V P_G=Z_D	ADD G_G=V P_G=Z_D	ASL G_G=V P_G=Z_D	POINTER G_G=G P_G=P	AND C_{out}=R3 P_G=Z_D	OR C_{out}=R3 P_G=Z_D	ASR C_{out}=R3 P_G=Z_D G_G=1	EOR C_{out}=R3 P_G=Z_D
0 0 0	ØB Minus IB·P	ØB Plus IB·P	ØB	ØB Plus P	ØB·(IB·P)	ØB + (IB·P)	ØB	ØB ⊕ (IB·P)
0 0 1	ØB Minus DR·P	ØB Plus DR·P	DR	DR Plus P	ØB·(DR·P)	ØB + (DR·P)	DR	ØB ⊕ (DR·P)
0 1 0	RF Minus ØB·P	RF Plus ØB·P	RF	RF Plus P	RF·(ØB·P)	RF + (ØB·P)	RF	RF ⊕ (ØB·P)
0 1 1	RF Minus DR·P	RF Plus DR·P	AR	PC Plus P	RF·(DR·P)	RF + (DR·P)	AR	RF ⊕ (DR·P)
	G_G=$G_G P_G$=P_G	G_G=$G_G P_G$=P_G	LSL G_G=$G_G P_G$=P_G	RELATIVE G_G=$G_G P_G$=P_G	C_{out}=R3 P_G=Z_D	EORP C_{out}=R3 P_G=Z_D	LSR G_G=1 P_G=Z_D	MODIFY G_G=$G_G P_G$=P_G
1 0 0	ØB Minus IB·P	ØB Plus IB·P	ØB	PC Plus IB·P	ØB · P	ØB ⊕ P	ØB	AR Plus P
1 0 1	ØB Minus DR·P	ØB Plus DR·P	DR	PC Plus DR·P	DR · P	DR ⊕ P	DR	AR Plus DR·P
1 1 0	RF Minus ØB·P	RF Plus ØB·P	RF	PC Plus ØB·P	RF · P	RF ⊕ P	RF	AR Plus ØB·P
1 1 1	RF Minus DR·P	RF Plus DR·P	AR	PC Plus RF·P	AR · P	AR ⊕ P	AR	AR Plus RF·P

Figure 6.9. 10803 ALU functions (Reproduced with permission of Motorola Inc.).

The large number of function-control inputs and data paths give the 10803 a large set of functions. In some cases, one can alter as many as three registers in one cycle. For instance, one combination of control signals allows one to gate PC into the MAR register, gate the value (or its complement) on the D bus into a register in the array, and gate the ALU result of PC plus P into the PC register. Because of this flexibility, the 10803 also has utility as a microprogram sequencer, peripheral controller, and even a processing-section slice.

The MC10803 is the densest of the M10800 series of bit-slice devices, having approximately 600 gates, compared to the 550 gates in the MC10801 sequencer slice and 350 gates in the MC10800 ALU-register slice.

Some typical propagation delays through the 10803 are

$MS_{7\text{-}11}$ to PG,GG —25 ns
$MS_{0\text{-}3}$ to DB,OB,IB —14 ns
P to IB via the ALU —17 ns

THE 2914 PRIORITY INTERRUPT CONTROLLER

Another large functional area of a processor that has not been discussed yet is the handling of interrupts, such as interrupts from peripheral devices or process-control devices. Interrupt handling usually requires considerations of

1. Being able to handle interrupts from many sources.
2. Being able to prioritize interrupts and, when several interrupts occur simultaneously, being able to select the one of highest priority.
3. When handling an interrupt, being able to disable interrupts of lower priority.
4. Being able to selectively enable or disable all interrupts.
5. Being able to use an encoded representation of a particular interrupt to do a multiway branch to a point within the microprogram.
6. Being able to clear interrupts.

One can see that interrupt handling is a nontrivial process, and that it can result in a considerable amount of logic.

The Am2914 interrupt controller provides the functions above and more in a single 40-pin package [1,4]. The 2914 is a powerful device, allowing one to use a single component for interrupt control, replacing perhaps 25 or more SSI and MSI components.

Figure 6.10 depicts the organization of the 2914. Its principal input is a set of eight interrupt lines (P), and its principal output is a 3-line encoded representation of the interrupt to be handled (V). It includes an 8-bit interrupt register into which interrupt signals are gated, an 8-bit mask register allowing one to selectively mask (disable) interrupts, and a 3-bit status register which can hold a value representing the lowest enabled interrupt level. The 2914 also contains a

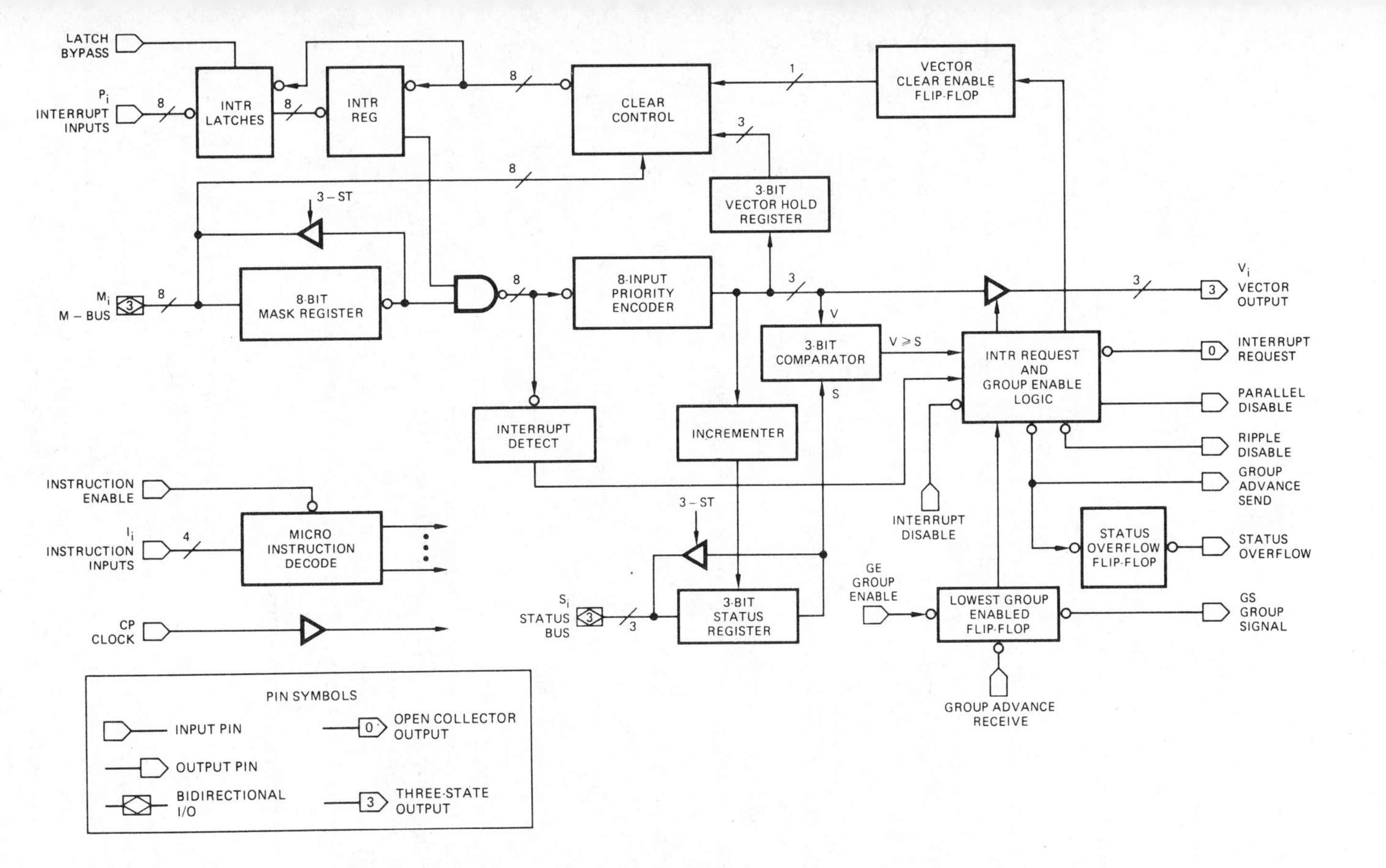

Figure 6.10. Organization of the 2914 (Copyright 1978 Advanced Micro Devices, Inc. Reproduced with permission of copyright owner.).

considerable amount of logic for cascading several 2914s together to handle a larger number of interrupts (the flipflops at the right side of the diagram). This cascading logic will be ignored for the moment and discussed at the end of the section.

To describe the logic within the 2914, the discussion will begin at the left side of Figure 6.10 and proceed toward the right. This corresponds roughly to the flow of interrupt information through the 2914.

Interrupt signals enter the 2914 from the 8-bit P bus, pass through a set of latches, and enter the interrupt register on the rising edge of the clock input to the 2914. A low (0) value in the register is considered to be an interrupt. One has the choice of disabling the latches, thus causing the 2914 to respond to level-indicated interrupts, or to enable the latches so that they can catch low pulses, thus causing the 2914 to respond to edge-indicated interrupts.

The 8-bit mask register, which can be loaded from an external bus, can be used to indicate the subset of interrupts to which the system is currently willing to respond. A 0 in a position in the mask register indicates that the corresponding interrupt is enabled. The values in the mask and interrupt registers are ANDed and sent to a priority encoder. If multiple interrupts are present and are enabled by the mask register, the priority encoder selects the highest-valued interrupt, encoding it as a 3-bit value.

Four things happen with this encoded interrupt value (called an *interrupt vector*). First, it is compared with the value in the status register. The status register, which can be loaded from an external bus, indicates the lowest-valued interrupt to which the system is currently willing to respond. If the interrupt vector is greater than or equal to the value in the status register, the interrupt-request output is signalled. Second, the interrupt vector is placed on the vector-output (V) 3-state bus, allowing external control logic to enable and read the interrupt vector. Third, the vector is placed in the vector-hold register, allowing the system to later clear the interrupt. Fourth, when the external logic reads the interrupt vector from the V bus, the interrupt vector is incremented and placed in the status register. This automatically sets the 2914 in a state such that it will respond to only interrupts that are higher than the interrupt currently being processed.

The status and mask registers allow one to design a system with priority interrupts, nonprioritized interrupts, or a combination of both.

2914 External Connections

Figure 6.11 depicts the input, output, and bidirectional pins of the 2914. The P bus is used to gate interrupt signals into the 2914. The four I signals specify one of 16 functions to be performed; normally these signals emanate from the current microinstruction. The IE input can be used to tell the 2914 to ignore the I inputs, thus allowing one to place the I inputs in a multifunction field within the microinstruction. The LB input is used to enable or disable the interrupt latch, as mentioned earlier. The remaining inputs (except for CP) are associated with the cascading of 2914s.

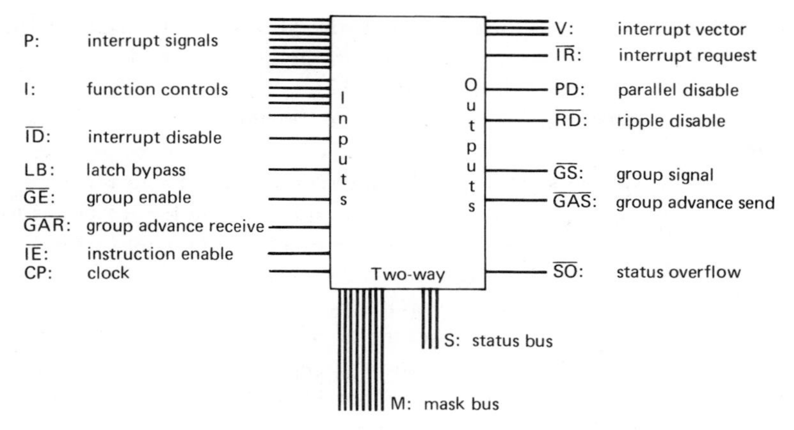

Figure 6.11. 2914 external connections.

The 2914 contains two bidirectional busses. The M bus can be used to gate a value into the mask register, read the current value in the mask register, or specify, to the interrupt-clear logic, a set of interrupts to be cleared. The S bus is used to gate a value into, or read, the status register.

The major outputs are the V bus (containing the encoded value of the interrupt selected) and IR, a status output indicating whether the 2914 has selected an interrupt. The SO output specifies whether the status register, as fed from the incrementer, has overflowed. Overflow would occur if the highest-priority interrupt has occurred. In a noncascaded design, the SO output is usually connected to the ID input, causing the 2914 to respond to no further interrupts while the highest-priority interrupt is being processed.

Typically, the V bus feeds the address inputs of a PROM containing branch addresses within the microprogram. The PROM might be enabled, for instance, by the VECT output of a 2910 sequencer, thus allowing the microprogram, at selected points, to branch to one of eight locations within the microprogram to handle the interrupt.

2914 Functions

The I inputs allow one to perform 16 functions. These functions involve setting and reading the mask and status registers, enabling the V bus (reading the selected interrupt vector), and clearing interrupts. The 16 functions are summarized below.

Master clear	clears all registers and flipflops.
Clear all interrupts	clears the interrupt register and latch.
Clear interrupts from mask register	for each bit in the mask register with the value 1, the corresponding bit in the interrupt register and latch is cleared.

Clear interrupts from M bus	for each line on the M bus with the value 1, the corresponding bit in the interrupt register and latch is cleared.
Clear interrupt, last vector read	clears the bit in the interrupt register and latch associated with the last interrupt vector read.
Read vector	enables the interrupt vector formed by the 2914 onto the V bus. Also, this value plus one is loaded into the status register, the interrupt vector is placed in the vector-hold register, and the vector-clear-enable flipflop is set.
Load status register	loads the status register from the S bus and the lowest-group-enabled flipflop from the GE input line.
Read status register	enables the status register onto the S bus.
Load mask register	loads the mask register from the M bus.
Read mask register	enables the mask register onto the M bus.
Set mask register	sets all bits in the mask register to 1, thus inhibiting all interrupts.
Clear mask register	resets all bits in the mask register to 0.
Bit clear mask register	for all lines on the M bus with the value 1, clears the corresponding mask register bits.
Bit set mask register	for all lines on the M bus with the value 1, sets the corresponding mask register bits.
Disable interrupt request	resets the interrupt-request-enable flipflop, thus disabling all interrupts.
Enable interrupt request	sets the interrupt-request-enable flipflop, thus enabling all interrupts (subject to the values in the mask and status registers).

Cascading the 2914

The 2914 also contains a considerable amount of logic for the cascading of multiple 2914 controllers to handle more than eight interrupt levels. Basically, the problems to be solved were that the status register must be logically expanded and the 2914s must be capable of communicating with one another to resolve the priority of simultaneous interrupts.

Going back to Figure 6.10, the lowest-group-enabled (LGE) flipflop is used in a cascaded design to indicate which 2914 contains the lowest priority level which will be accepted. The vector-clear-enable flipflop (VCE) indicates if the last interrupt vector read was from this 2914. It is needed during interrupt-clearing operations to cause the proper 2914 to clear its interrupt.

Going back to Figure 6.11, three input lines and four output lines are associated with cascading. Their purposes are

$\overline{\text{ID}}$ (input)	When 0, inhibits the interrupt-request output and gates a 0 to the RD output.
$\overline{\text{GE}}$ (input)	The load-status function gates this input into the LGE flipflop.
$\overline{\text{GAR}}$ (input)	The read-vector function uses this, with other signals, to load the LGE flipflop. GAR on the 2914 with the lowest-priority interrupts must be 0.
PD (output)	PD is 1, when LGE is 0 or an interrupt request is generated.
$\overline{\text{RD}}$ (output)	RD is 0 when the ID input is 0, LGE is 0, or an interrupt request is generated.
$\overline{\text{GS}}$ (output)	GS is the output of the LGE flipflop.
$\overline{\text{GAS}}$ (output)	During a read-vector function, this output is 0 when the highest-priority (7) interrupt is being read from this 2914.

The 2914 can be configured in two ways—ripple cascade and parallel cascade. In the ripple cascade mode, the RD output of a 2914 is connected to the ID input of the 2914 representing the next-lowest-priority set of interrupts. This has the effect of suspending all lower-priority 2914s when an interrupt occurs. One can also use standard carry-lookahead generators to accomplish this; when this is done, the PD output, instead of RD, is used.

When cascading multiple 2914s, one must expand the status and interrupt-vector busses. For instance, with four 2914s providing 32 interrupt levels, one needs 5-bit status and interrupt-vector busses. An auxiliary device provided for this is the Am2913 priority interrupt expander [1]. The 2913 is an 8-to-3 encoder with a 3-state output. The attribute that makes the 2913 unique is its five inputs that control the 3-state output. These inputs can be directly connected to the I (function-control) inputs of the 2914s, thus making the 2913 transparent to the microprogram (i.e., no additional control signals are needed for the 2913).

Figures 6.12 and 6.13 show, in two parts, the use of two 2913 expanders and eight 2914 controllers to create an interrupt-control section for up to 64 interrupts. In Figure 6.12, a 2913 is used to create an interrupt vector of six bits. The 2913 encodes the RD outputs of the 2914s to create the upper three bits in the interrupt vector. The 2913's controls are connected to the 2914 I controls such that the 2913's output is enabled during a read-vector operation.

Figure 6.13 illustrates another 2913 for the expansion of the status bus. The 2913's controls are connected to the 2914 I controls such that the 2913's output is enabled during a read-status operation. Since the status bus is bidirectional, a 3-to-8 decoder is also used. Note the use of the GE input and GS output. Since these are connected to the LGE (lowest-group-enabled) flipflop, the high state of one LGE flipflop in the 2914s corresponds to the upper three bits on the status bus. The GAS and GAR connections are made to set the LGE flipflop in the next 2914 when the status register in one 2914 overflows (i.e., when the highest-priority interrupt within a 2914 occurs).

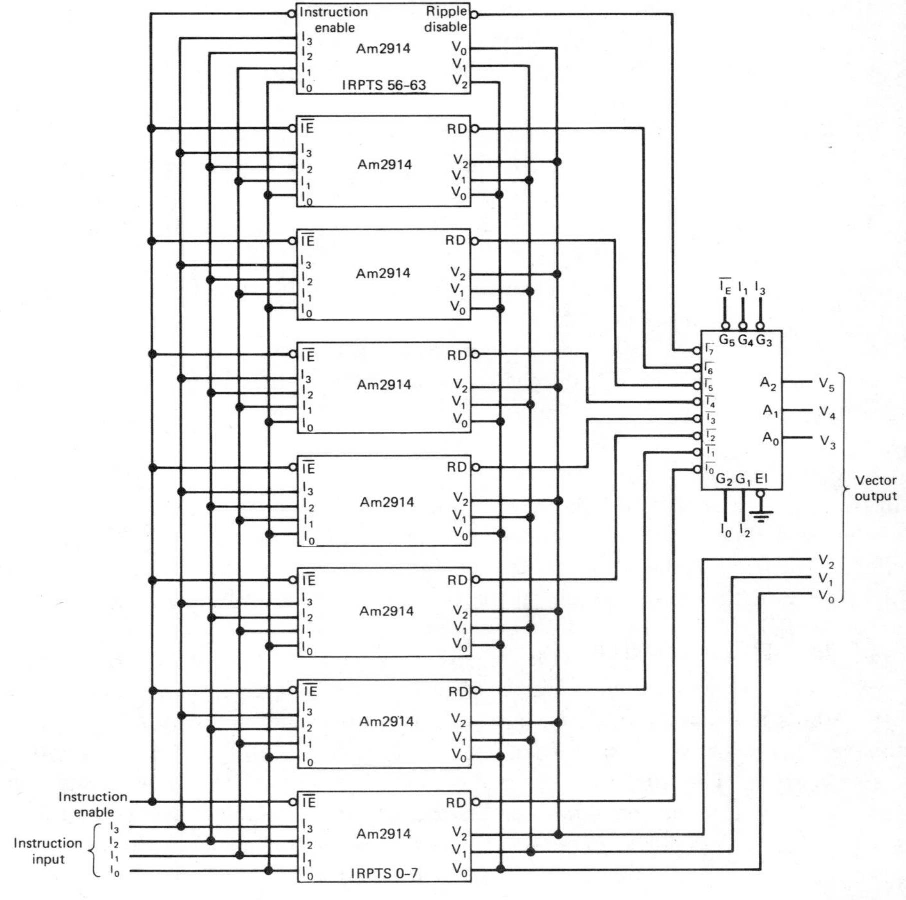

Figure 6.12. V-bus connections with the 2913 for cascaded 2914 interrupt controllers (Copyright 1978 Advanced Micro Devices, Inc. Reproduced with permission of copyright owner.).

In addition to the connections of Figures 6.12 and 6.13, the ripple or parallel cascade connections must be made.

2914 Timing

The longest paths through the 2914 (guaranteed times) are

I inputs to V outputs	—55 ns
Clock input to IR output	—82 ns

THE 3214 INTERRUPT CONTROL UNIT

Another example of an interrupt controller is Intel's 3214 [5]. Although somewhat simpler that the 2914, the 3214 is a useful component in designs that do not require the power of the 2914.

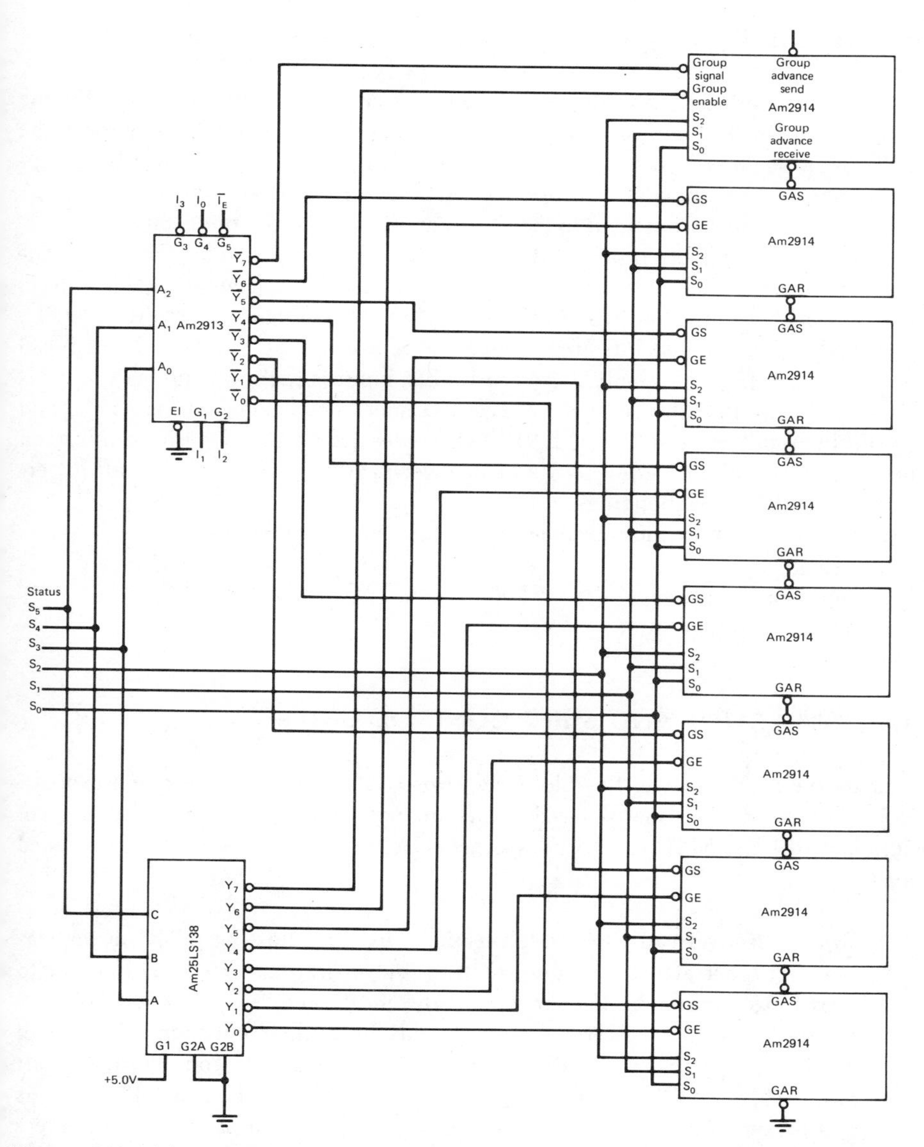

Figure 6.13. S-bus connections with the 2913 for cascaded 2914 interrupt controllers (Copyright 1978 Advanced Micro Devices, Inc. Reproduced with permission of copyright owner.).

Figure 6.14 depicts the organization of the 3214. Like the 2914, the 3214 contains an interrupt latch, a status (priority-level) register, a comparator, and logic to cascade multiple units. Unlike the 2914, the 3214 contains no mask register to selectively disable interrupts independent of their priority, and no logic to automatically update the status register when an interrupt occurs.

3214 External Connections

The external connections of the 24-pin 3214 are shown in Figure 6.15. Unlike the 2914, the 3214 does not, on its own, signal a pending interrupt. One strobes the 3214 for a pending interrupt by raising input ISE; if an interrupt has been captured, the 3214 responds by dropping IA. Input ELR is used to enable the A output bus. Input ECS is used to load the status register from the B bus.

The SGS and ETLG inputs and the ENLG output are provided for the cascading of multiple 3214s. SGS is used to inform a 3214 that the current priority level does not belong to it; that is, it disables the 3214. In a design with multiple 3214s, the upper n-3 status-bus bits are decoded and sent to the SGS inputs. The ENLG output of a 3214 is high only if (1) the ETLG input is high, (2) the SGS input is low (indicating that the current priority level does not belong to this 3214), and (3) no interrupt request exists. ENLG is rippled from a 3214 with an interrupt pending to the ETLG input of the next-lower-priority 3214 to inhibit all lower-priority units from responding.

Representative maximum propagation delays for Intel's 3214 are

R inputs to A outputs	—100 ns
Clock input to IA output	—25 ns

THE 2904 STATUS AND SHIFT CONTROL UNIT

Despite the availability of flexible ALU/register slices for the design of a processing section, one is usually faced with the need for a substantial amount of auxiliary SSI and MSI logic. Among other things, this surrounding logic is needed for

1. *Shift control.* Because of multiplication, division, floating-point normalization, and other algorithms, and rotating shift operations, one usually needs to interconnect the ends of shifter(s) in the most- and least-significant ALU/register slices in a variety of ways. In addition to interconnecting the ends of the shifters, there are usually requirements to gate, onto one or more shift lines, a 0, 1, ALU result sign, ALU carry output, and so on. This entails multiplexers and decoding logic (to decode the processing-section microorders to control the multiplexers). (Note that a few ALU/register slices, for example the 74S481 and SBP0400, contain some of the necessary shift multiplexer logic).

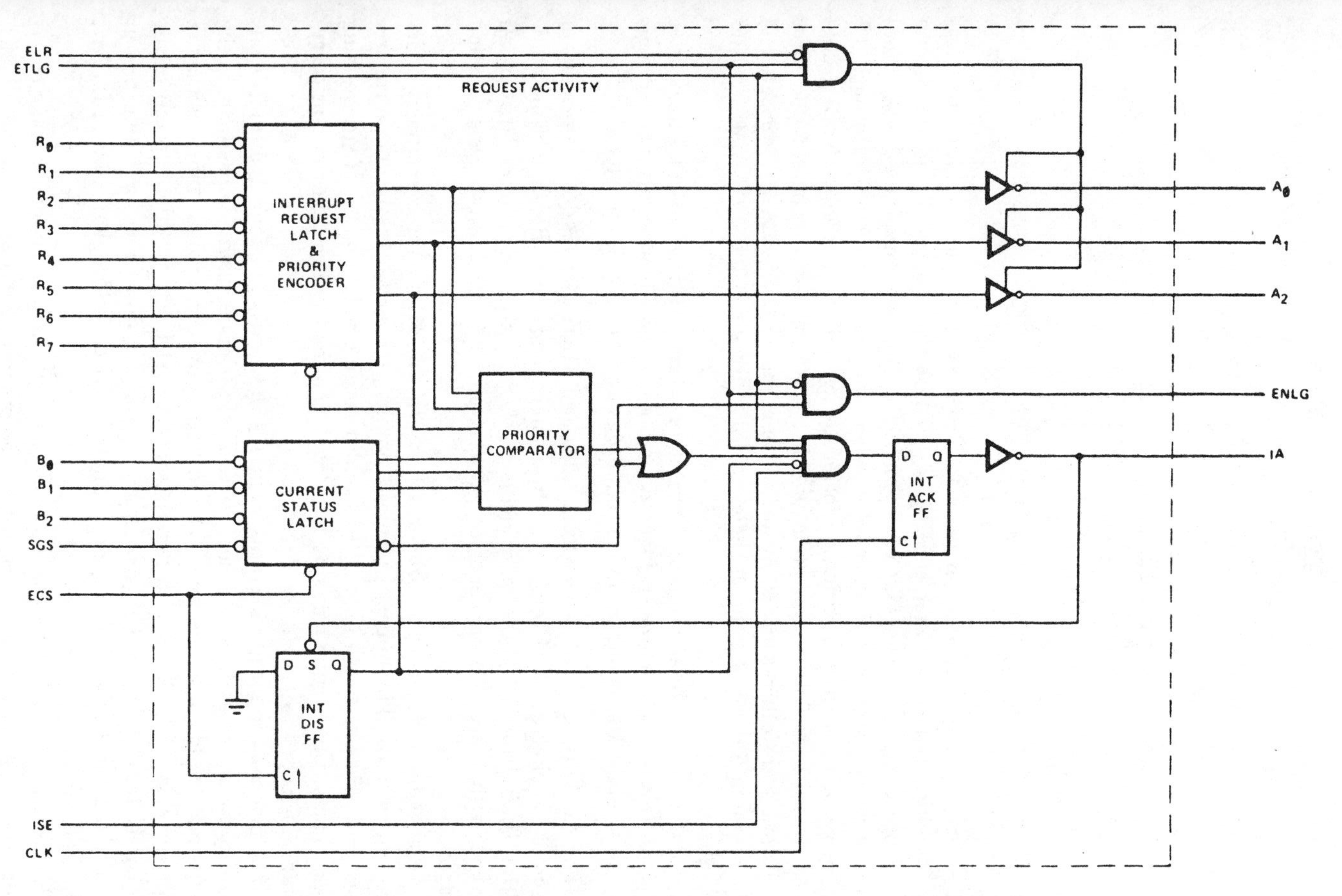

Figure 6.14. Organization of the 3214 (Reproduced with permission of Intel Corp.).

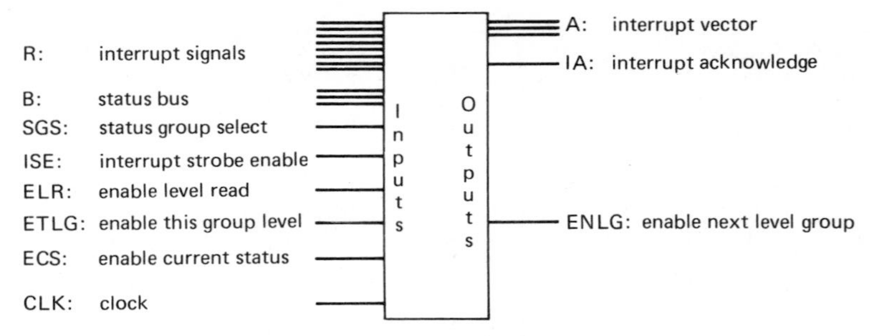

Figure 6.15. 3214 external connections.

2. *Carry control.* Although earlier examples showed the carry input of the least-significant ALU/register slice being fed directly from a 1-bit microorder, a more complex scheme is usually needed. One may need logic that automatically generates a 0 carry input when an addition is being performed, a 1 when a subtraction is being performed, uses a flipflop as input when a double-length arithmetic operation is being performed, and generates the carry input from other sources for special operations. (For instance, the latter is required for use of some of the extended functions in the 2903 ALU/register slice.)
3. *Status control.* In addition to the need for a status register and multiplexers to test status, one often needs logic to manipulate individual status bits and logic to combine status bits into such conditions as A>B, A≥B, and so on.

The Am2904, a 40-pin device from Advanced Micro Devices [6], represents an attempt to collect most of this logic into a single component. Because of the motivation behind the 2904, the manufacturer refers to it as "LSI glue," that is, an LSI chip that provides much of the glue required to hold a processing section together.

As shown in Figure 6.16, the 2904 is organized into three relatively independent sections. The upper, and most complex, section is the status-control section. It consists of two 4-bit registers and a considerable amount of surrounding logic. The second section contains shift logic, and the third section contains carry-control logic.

The principal input to the status section is the I bus. The four lines in this bus will be referred to as Iz, Ic, In, and Iovr. These I inputs normally originate at the status outputs of the processing section. The motivation of these labels is that Iz represents an ALU zero result, Ic represents the ALU carry output, In represents the ALU result sign, and Iovr represents ALU overflow. The individual bits in both status registers will be correspondingly labelled Uz, Uc, Un, Uovr, Mz, Mc, Mn, and Movr.

The reason that the 2904 contains two status registers is that one of them (U) is intended to hold the status generated by the processing section at the end of each machine cycle, and the other (M) is intended to hold status or "condition-code" information that is defined at the machine-language level. As shown, the U register can be loaded from the I bus or the M register, and the M register can be loaded from the I bus, the U register, or the bidirectional Y bus. Control lines are available to selectively load the four bits of the M register.

The test logic gates any bit or its complement from the U or M register, or on the I bus, to the CT output. Also, a variety of exclusive-OR, exclusive-NOR, AND, and OR operations can be performed with the test logic.

The shift logic connects the four bidirectional shift lines in a variety of ways. As shown in Figure 6.16, some of the bits in the M register and on the I bus can be gated to the shift pins.

The carry logic is a 7-input multiplexer, allowing one to gate the C_x input, Uc, Mc, their complements, 0, or 1 to the C_o output.

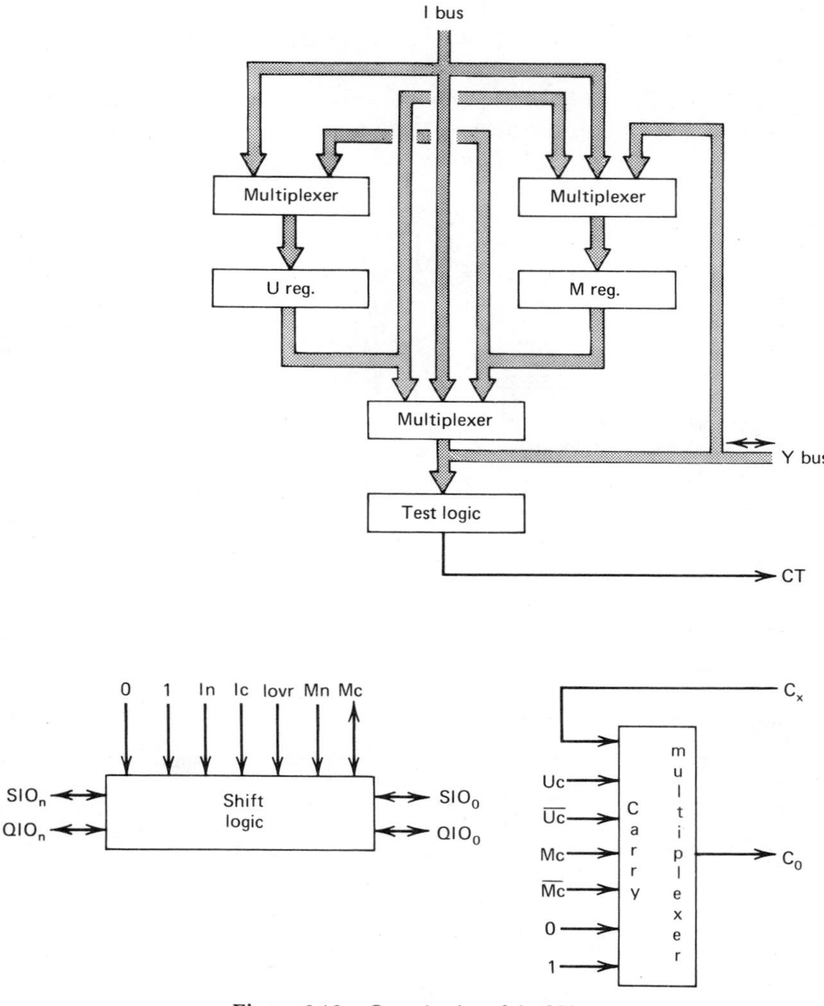

Figure 6.16. Organization of the 2904.

2904 External Connections

Figure 6.17 shows the external connections of the 2904. With a few exceptions, each connection is associated with one of the three logic sections of the 2904. In the status section, the I inputs are normally connected to the status outputs of the processing section. If the processing section is composed of cascaded 2901 or 2903 slices, Ic, Iovr, and In are normally connected to the C_{n+4}, OVR, and N (F_3 for the 2901) outputs of the most-significant ALU/register slice, and Iz is connected to the wire-ORed Z (F=0 for the 2901) outputs of the ALU/register slices.

The CT output is usually connected to the test-condition input of the control section (e.g., the CC input of a 2910 sequencer). CT is a 3-state output, allowing one to maintain additional status conditions outside of the 2904.

The bidirectional Y bus can be used to gate the U or M register to another location, or as another source for loading the M register. Using the Y bus as an output allows one to save the machine status when an interrupt is to be processed.

The CE_u and CE_m control lines allow one to disable the U and/or M registers regardless of the operation specified on the function-control lines. The Ez, Ec, En,

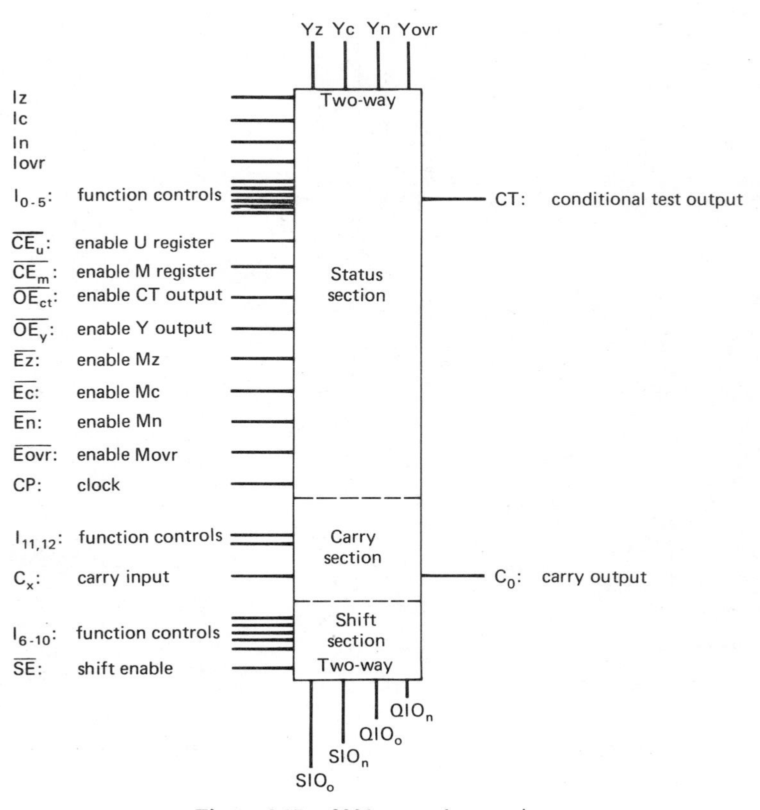

Figure 6.17. 2904 external connections.

and Eovr lines allow one to selectively load the M register. If one of these lines is high, it causes the corresponding bit in the M register to remain unchanged, regardless of the operation specified on the function-control lines.

The carry-control section contains only four external connections. The C_o output is normally connected to the carry-input pin of the least-significant ALU/register slice. If the system needs to perform double-length arithmetic (e.g., the processing section has a width of 16 bits, but 32-bit arithmetic must be performed), the C_x input might be driven by an external carry flipflop. In the case of the 2903 ALU/register slice, the C_x input is usually connected to the Z outputs of the 2903s to facilitate carry control for some of the 2903's extended functions. The carry section is controlled by its two function-control lines and, if these lines are both high, by controls lines 1, 2, 3, and 5 of the status section.

In the shift section, the four bidirectional shift lines are usually connected to the two shifters in the processing section. Control input SE can be used to place all four shift lines in the high-impedence state.

2904 Status Functions

The functions performed in the status section are specified by six function-control lines, although the functions may be constrained by the register- and bit-enable control lines mentioned above. The function-control lines control three operations: (1) the loading of values into the U and M registers, (2) the generation of an output on CT via the test logic, and (3) the gating of the I bus or the U or M register to the Y bus. Table 6.1 defines the source for each U- and M-register bit for each value of the function-control lines. The value of the lines is represented in octal. A hyphen designates no change; apostrophes indicate that this entry is identical to the one above it.

Operation 02, if both registers are enabled, allows one to swap the values in U and M. Operations 06 and 07 OR Iovr into Uovr. This is useful when one wants to perform a series of arithmetic operations and be able to determine, at the end, whether an overflow occurred in any operation.

The six function-control lines also control the status test logic; that is, they determine the value gated to the CT output line. The CT outputs produced are defined in Figure 6.18 (note that there is a slight difference in typography). The logical operations in some of the situations are provided to facilitate relational tests after subtraction (A–B) operations in the processing section. For instance, if the microinstruction allows one to branch on the condition $A \leq B$, where A and B are two's complement numbers and 2903s are used in the processing section, one must test the value of

(N exclusive-OR OVR) OR Z

after a subtraction.

The function-control lines of the status section also control the use of the Y bus as an output bus. For the situation where I_5 is 0, U is gated to the Y bus. If I_5 is 1

I_{3-0} HEX	I_3 I_2 I_1 I_0	$I_5 = I_4 = 0$	$I_5 = 0, I_4 = 1$	$I_5 = 1, I_4 = 0$	$I_5 = I_4 = 1$
0	0 0 0 0	$(\mu_N \oplus \mu_{OVR}) + \mu_Z$	$(\mu_N \oplus \mu_{OVR}) + \mu_Z$	$(M_N \oplus M_{OVR}) + M_Z$	$(I_N \oplus I_{OVR}) + I_Z$
1	0 0 0 1	$(\mu_N \odot \mu_{OVR}) \cdot \bar{\mu}_Z$	$(\mu_N \odot \mu_{OVR}) \cdot \bar{\mu}_Z$	$(M_N \odot M_{OVR}) \cdot \bar{M}_Z$	$(I_N \odot I_{OVR}) \cdot \bar{I}_Z$
2	0 0 1 0	$\mu_N \oplus \mu_{OVR}$	$\mu_N \oplus \mu_{OVR}$	$M_N \oplus M_{OVR}$	$I_N \oplus I_{OVR}$
3	0 0 1 1	$\mu_N \odot \mu_{OVR}$	$\mu_N \odot \dot{\mu}_{OVR}$	$M_N \odot M_{OVR}$	$I_N \odot I_{OVR}$
4	0 1 0 0	μ_Z	μ_Z	M_Z	I_Z
5	0 1 0 1	$\bar{\mu}_Z$	$\bar{\mu}_Z$	$\bar{M}_Z$	$\bar{I}_Z$
6	0 1 1 0	μ_{OVR}	μ_{OVR}	M_{OVR}	I_{OVR}
7	0 1 1 1	$\bar{\mu}_{OVR}$	$\bar{\mu}_{OVR}$	$\bar{M}_{OVR}$	$\bar{I}_{OVR}$
8	1 0 0 0	$\mu_C + \mu_Z$	$\mu_C + \mu_Z$	$M_C + M_Z$	$\bar{I}_C + I_Z$
9	1 0 0 1	$\bar{\mu}_C \cdot \bar{\mu}_Z$	$\bar{\mu}_C \cdot \bar{\mu}_Z$	$\bar{M}_C \cdot \bar{M}_Z$	$I_C \cdot \bar{I}_Z$
A	1 0 1 0	μ_C	μ_C	M_C	I_C
B	1 0 1 1	$\bar{\mu}_C$	$\bar{\mu}_C$	$\bar{M}_C$	$\bar{I}_C$
C	1 1 0 0	$\bar{\mu}_C + \mu_Z$	$\bar{\mu}_C + \mu_Z$	$\bar{M}_C + M_Z$	$\bar{I}_C + I_Z$
D	1 1 0 1	$\mu_C \cdot \bar{\mu}_Z$	$\mu_C \cdot \bar{\mu}_Z$	$M_C \cdot \bar{M}_Z$	$I_C \cdot \bar{I}_Z$
E	1 1 1 0	$I_N \oplus M_N$	μ_N	M_N	I_N
F	1 1 1 1	$I_N \odot M_N$	$\bar{\mu}_N$	$\bar{M}_N$	$\bar{I}_N$

Notes: 1. ⊕ Represents EXCLUSIVE-OR ⊙ Represents EXCLUSIVE-NOR or coincidence.

Figure 6.18. Determination of the 2904 CT output value. (Copyright 1978 Advanced Micro Devices, Inc. Reproduced with permission of copyright owner).

and I_4 is 0, M is gated to Y. If I_5 and I_4 are 1, the I bus is gated to the Y bus. In all other cases, Y is in the high-impedence state.

2904 Carry Functions

Function-control lines I_{11} and I_{12} control the carry multiplexer. For the values 00, 01, and 10, the values 0, 1, and C_x are gated to the C_o output. For the value 11, a variety of values on lines I_1, I_2, I_3, and I_5 cause Uc, Mc, or their complements to be gated to the C_o output.

2904 Shift Functions

The 2904 provides 32 combinations of uses of the four bidirectional shift connections. Figure 6.19 defines the 32 functions. Note that some of the functions gate a value from a shift line into the Mc register, and others gate various U-register, M-register, and I-bus bits onto the shift lines.

Using the 2904

Despite its flexibility, the 2904 presents the system designer with a number of problems, many of which stem from its flexibility. The 2904's major drawback is that it attempts to be all things to all people and, in doing so, becomes a device that may be too complex for most applications.

If the full capabilities of the 2904 are to be exploited, it requires 25 control inputs. It is extremely unlikely that one would devote 25 bits in the microinstruction to the 2904. One reason, of course, is the expense of doing so. A second reason is that the 2904 should not be controlled independent from the processing section. The 2904 is an auxiliary device; thus one would like the 2904 to "see" the microorders being sent to the processing section and perform the appropriate auxiliary operations, rather than require separate microorders.

The latter is feasible, but unfortunately the 2904's control signals do not directly correspond to those of the 2901 or 2903, the manufacturer's ALU/register slices. For instance, the use of the 2903's extended functions requires one to switch the connections of the carry input of the least-significant slice and the four shift lines on the most-significant and least-significant slices. The 2904 was intended to do this, but the encoding of its control signals does not match those of the 2903. Table 6.2 illustrates examples of four 2903 extended functions, showing, for each,

TABLE 6.1. 2904 Status-register Control

	Value loaded into the U and M registers							
I_{543210}	Uz	Uc	Un	Uovr	Mz	Mc	Mn	Movr
00	Mz	Mc	Mn	Movr	Yz	Yc	Yn	Yovr
01	1	1	1	1	1	1	1	1
02	Mz	Mc	Mn	Movr	Uz	Uc	Un	Uovr
03	0	0	0	0	0	0	0	0
04	Iz	Ic	In	Iovr	Iz	Movr	In	Mc
05	"	"	"	"	$\overline{Mz}$	$\overline{Mc}$	$\overline{Mn}$	$\overline{Movr}$
06, 07	"	"	"	Iovr OR Uovr	Iz	Ic	In	Iovr
10	0	—	—	—	"	$\overline{Ic}$	"	"
11	1	—	—	—	"	"	"	"
12	—	0	—	—	"	Ic	"	"
13	—	1	—	—	"	"	"	"
14	—	—	0	—	"	"	"	"
15	—	—	1	—	"	"	"	"
16	—	—	—	0	"	"	"	"
17	—	—	—	1	"	"	"	"
20–27	Iz	Ic	In	Iovr	"	"	"	"
30, 31	"	$\overline{Ic}$	"	"	"	$\overline{Ic}$	"	"
32–47	"	Ic	"	"	"	Ic	"	"
50, 51	"	$\overline{Ic}$	"	"	"	$\overline{Ic}$	"	"
52–67	"	Ic	"	"	"	Ic	"	"
70–71	"	$\overline{Ic}$	"	"	"	$\overline{Ic}$	"	"
72–77	"	Ic	"	"	"	Ic	"	"

I_{10}	I_9	I_8	I_7	I_6	M_C RAM Q	SIO_o	SIO_n	QIO_o	QIO_n	Loaded into M_C
0	0	0	0	0	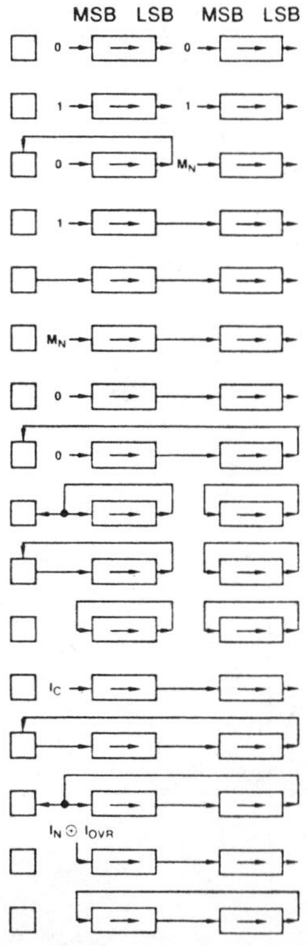	Z	0	Z	0	
0	0	0	0	1		Z	1	Z	1	
0	0	0	1	0		Z	0	Z	M_N	SIO_o
0	0	0	1	1		Z	1	Z	SIO_o	
0	0	1	0	0		Z	M_C	Z	SIO_o	
0	0	1	0	1		Z	M_N	Z	SIO_o	
0	0	1	1	0		Z	0	Z	SIO_o	
0	0	1	1	1		Z	0	Z	SIO_o	QIO_o
0	1	0	0	0		Z	SIO_o	Z	QIO_o	SIO_o
0	1	0	0	1		Z	M_C	Z	QIO_o	SIO_o
0	1	0	1	0		Z	SIO_o	Z	QIO_o	
0	1	0	1	1		Z	I_C	Z	SIO_o	
0	1	1	0	0		Z	M_C	Z	SIO_o	QIO_o
0	1	1	0	1		Z	QIO_o	Z	SIO_o	QIO_o
0	1	1	1	0		Z	$I_N \oplus I_{OVR}$	Z	SIO_o	
0	1	1	1	1		Z	QIO_o	Z	SIO_o	

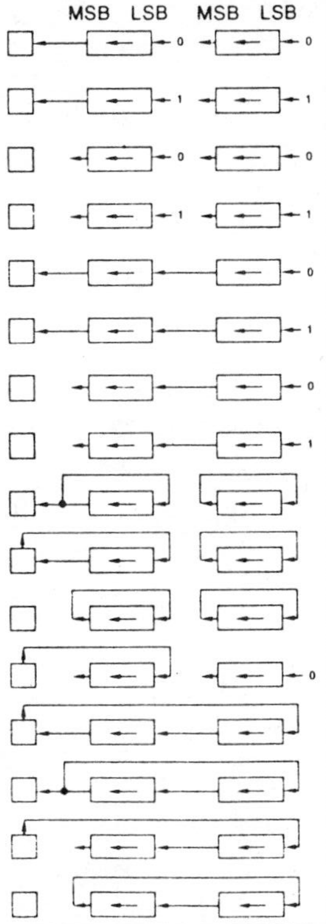

					MSB LSB MSB LSB					
1	0	0	0	0		0	Z	0	Z	SIO_n
1	0	0	0	1		1	Z	1	Z	SIO_n
1	0	0	1	0		0	Z	0	Z	
1	0	0	1	1		1	Z	1	Z	
1	0	1	0	0		QIO_n	Z	0	Z	SIO_n
1	0	1	0	1		QIO_n	Z	1	Z	SIO_n
1	0	1	1	0		QIO_n	Z	0	Z	
1	0	1	1	1		QIO_n	Z	1	Z	
1	1	0	0	0		SIO_n	Z	QIO_n	Z	SIO_n
1	1	0	0	1		M_C	Z	QIO_n	Z	SIO_n
1	1	0	1	0		SIO_n	Z	QIO_n	Z	
1	1	0	1	1		M_C	Z	0	Z	
1	1	1	0	0		QIO_n	Z	M_C	Z	SIO_n
1	1	1	0	1		QIO_n	Z	SIO_n	Z	SIO_n
1	1	1	1	0		QIO_n	Z	M_C	Z	
1	1	1	1	1		QIO_n	Z	SIO_n	Z	

Notes: 1. Z = High impedance (outputs off) state.
2. Outputs enabled and M_C loaded only if $\overline{SE}$ is LOW.
3. Loading of M_C from I_{10-6} overrides control from I_{5-0}, $\overline{CE}_M$, $\overline{E}_C$.

Figure 6.19. 2904 shift-control operations (Copyright 1978 Advanced Micro Devices, Inc. Reproduced with permission of copyright owner.).

TABLE 6.2. Correspondence Between 2903 and 2904 Function Controls

	2903 inputs	2904 inputs	
2903 function	$I_{8\text{-}5}$	$I_{12\text{-}11}$	$I_{10\text{-}6}$
Multiply	0010	00	00110
Multiply correction	0101	10	00110
Divide	1100	10	11101
Double-length normalize	1010	00	10110

the value that must be gated to control lines $I_{5\text{-}8}$ of the 2903 and the values that must be gated to the 2904 to switch the carry and shift lines properly.

Unfortunately, there is no pattern between the 2903 function controls and those for the 2904. Hence one needs a substantial amount of decoding logic (or several PROMs) to direct the 2904 from the microorders controlling the processing section. In some cases one could find that the number of auxiliary components needed to control the 2904 is not much less that the number of components that one was trying to save by using the 2904.

A few representative propagation times (typical) for the 2904 are listed below.

I-bus input to CT output	—30 ns
$I_{0\text{-}5}$ input to CT output	—30 ns
C_x input to C_o output	—12 ns
any shift input to any shift output	—16 ns
$I_{6\text{-}10}$ input to shift output	—19 ns

THE 2925 CLOCK GENERATOR AND DRIVER

Another area of a digital system that could be subject to some LSI assistance is the system clocking. Rarely is a system so simple that only a single, constant clock signal is sufficient; one may desire clocking logic that is able to produce variable machine-cycle times, or that produces timing signals within the machine cycle. Furthermore, one might also need the ability to suspend the timing signals, for instance when the system is waiting for an external action, such as the completion of a memory operation. Also, it is desirable that the timing logic be capable of generating outputs for only one machine cycle, so that the system can be manually cycled during debugging processes.

The Am2925 has the above objectives. Its organization is shown in Figure 6.20. At the time of writing this book, Advanced Micro Devices was in the process of designing the 2925; therefore only an overview of the device's operations are presented here.

The 2925 produces four different output waveforms for controlling operations within a machine cycle. Also, the L lines allow one to select one of eight different cycle lengths. If a variable-length cycle is needed, as discussed in Chapter 5, the L lines can be driven by a microorder or by decoding other microorders.

The $\overline{\text{HALT-REQ}}$ line, if dropped to 0, halts the clocks. $\overline{\text{SINGLE-STEP}}$ allows one to cause the 2925 to generate the clock signals for only a single machine cycle, providing that the 2925 had been previously stopped by the $\overline{\text{HALT-REQ}}$ input. A pulse on the WAIT-REQ input suspends the clock signals and produces an indication of this on the WAIT-ACK output. The 2925 is started again by a pulse on the READY input.

Advanced Micro Devices has developed an equivalent circuit for the 2925 which shows that the 2925 can replace 39 SSI and MSI components.

Motorola also provides a similar device, the MC10802, for the ECL M10800 family.

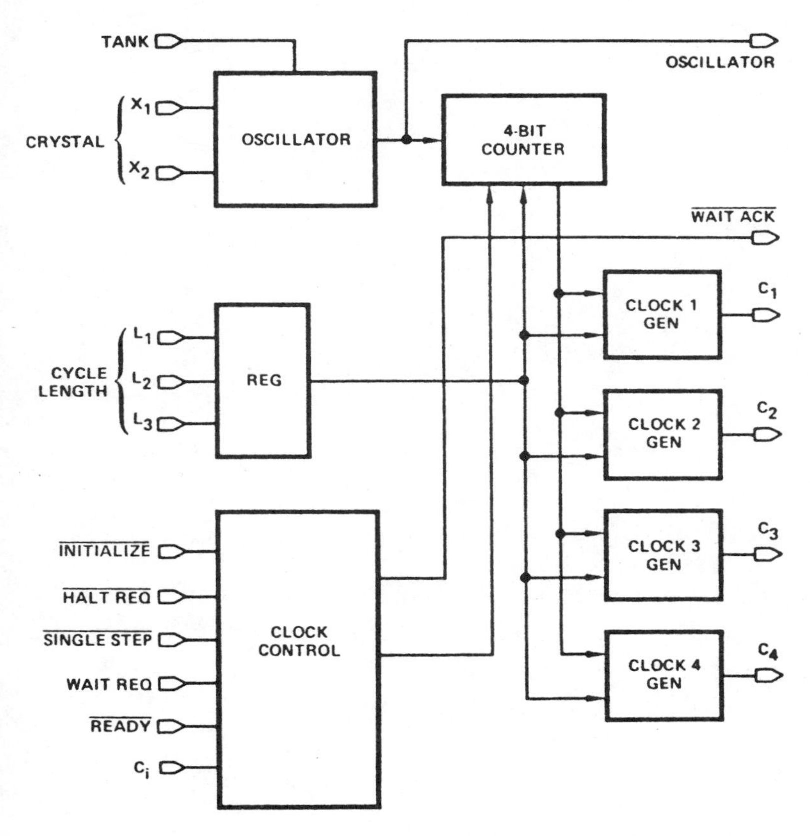

Figure 6.20. Organization of the 2925 (Copyright 1978 Advanced Micro Devices, Inc. Reproduced with permission of copyright owner.).

REFERENCES

1. *The Am2900 Family Data Book.* Sunnyvale, Cal.: Advanced Micro Devices, 1978.
2. *Macrologic Bipolar Microprocessor Databook.* Mountain View, Cal.: Fairchild Camera and Instrument Corp., 1978.
3. *MC10803 MECL LSI Memory Interface Function.* Phoenix: Motorola, 1977.
4. *Am2914 Vectored Priority Interrupt Controller.* Sunnyvale, Cal.: Advanced Micro Devices, 1976.
5. *Series 3000 Reference Manual.* Santa Clara, Cal.: Intel, 1976.
6. *Am2904 Status and Shift Control Unit.* Sunnyvale, Cal.: Advanced Micro Devices, 1978.

7

Programmable Logic

Another solution to the problems impeding the development of LSI building blocks is programmable or customizable logic. The primary example is the programmable logic array.

The programmable logic array, or PLA, is a universal array of logic gates that can be customized, normally by blowing fuses, to particular applications. Although not a bit-slice device, it is discussed here because, like bit-slice devices, it is a newer and powerful digital building block.

STRUCTURE OF A PLA

Although various PLAs differ slightly in organization, most of them have a structure similar to that of Figure 7.1. The PLA consists of a set of input lines (typically 8–16), a set of output lines (typically 6–8), a set of AND gates (typically 48–96), a set of OR gates (one per output line), a set of inverters on each output line, and two arrays of alterable connections. If these connections are made via a customized mask in the manufacturing process, the PLA is called a factory-programmed PLA. If these connections are made by selectively blowing fused links, the device is known as a field-programmable logic array (FPLA). Throughout this chapter, the term PLA will be used to denote both.

Since an actual PLA might contain over 100 gates and over 1000 fuses, an illustration of such a device is infeasible. However, to explain the concept, the hypothetical PLA in Figure 7.2 will be used. Remember throughout the discussion that actual PLAs are much larger.

The hypothetical PLA contains two input lines and two output lines. The input lines are passed through inverters, making each input signal and its complement available. These signals are then made available to four AND gates. Each input and its complement is connected, through a fuse (in the case of an FPLA) to each AND gate. Each fuse is depicted as an X in the diagram. By retaining or blowing these links, one can obtain different product terms on the output of each AND.

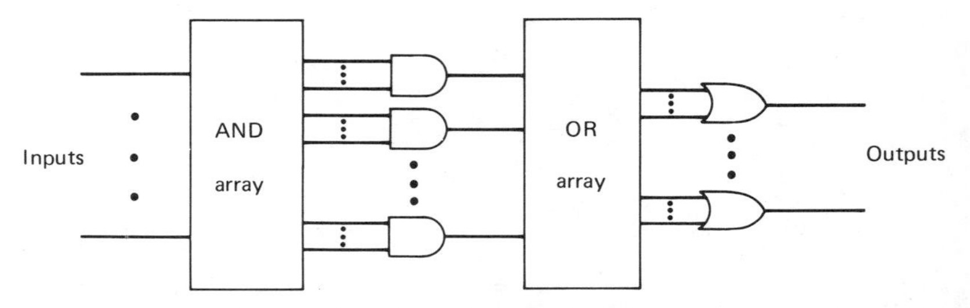

Figure 7.1. General structure of a PLA.

These inverters, connections, fuses, and AND gates correspond to the AND array in Figure 7.1.

The product terms, or AND outputs, are then made available to two OR gates. Again, each of the OR inputs passes through a fused link. The output of each OR gate passes through an exclusive-OR with a fused connection to ground, allowing one to pass the output of each OR, or its complement, to each output line. The circuit designs are such that unconnected (blown) AND and exclusive-OR inputs assume the logical high state; unconnected OR inputs assume the low state.

Although Figure 7.2 is immensely oversimplified, one can see the flexibility that even this circuit provides. Excluding useless and ambiguous product terms, each of the four ANDs can be configured to produce one of the following partial products

$$\begin{array}{ll} A & AB \\ \overline{A} & A\overline{B} \\ B & \overline{A}B \\ \overline{B} & \overline{A}\overline{B} \end{array}$$

Likewise, each OR can be configured to sum one, two, three, or four of the partial products. Given the 26 fuses, there are 2^{26} possible configurations of this PLA, although most of them are not unique, since there are 256 unique truth tables for two inputs and two outputs.

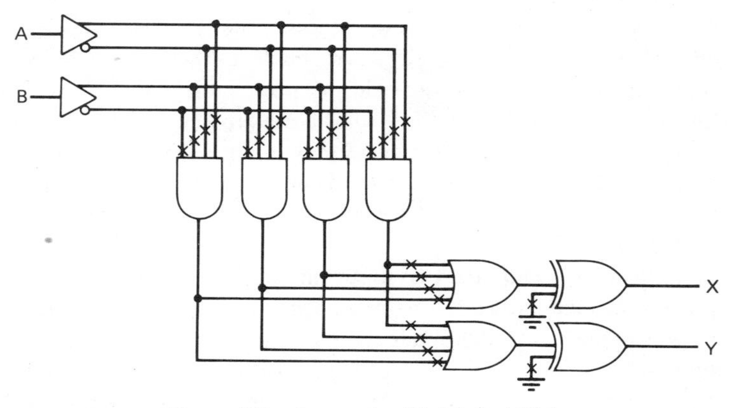

Figure 7.2. An oversimplified 2×2×4 PLA.

Actual PLAs have the structure of Figure 7.2, but the number of circuits is much larger. Each AND is usually connected (through fuses) to each input and its complement, but the number of input lines may be eight or more, meaning that each AND has 16 or more inputs. The number of AND gates is considerably higher, typically 48 or more. Hence the fuses can be blown to form 48 or more product terms. Each OR gate is connected to each AND output, meaning that each OR has 48 or more inputs, and the number of OR gates (and output lines) is typically six to eight. In a typical PLA, the number of fuses is given by

$$2PI + O(1+P)$$

where P is the number of product terms (AND gates) and I and O represent the number of input and output lines.

Depending on the model, PLAs may also include 3-state outputs (programmable via a fuse), open-collector outputs, registers, and feedback connections.

Because of the difficulty in depicting a PLA, a different logic notation is sometimes used. Figure 7.3 is the equivalent of Figure 7.2 (although the output-line inverters have not been shown). Rather than drawing a large number of lines into each gate, only one line is used. The X is used to indicate the presence of an unblown fuse. For instance, the presence of four X's on the common line to an AND indicates that the four physical inputs to the AND are connected. Hence, Figure 7.3 represents an unprogrammed device, a device with all fuses intact.

Factory-programmed PLAs are programmed by designing a mask to be used during the final step in the fabrication process. The mask includes or excludes a transistor or diode at each connection point. FPLAs are programmed by melting metallic fuses at the connection points. Typically, one selects a fuse by specifying its "address" on the input and/or output lines and then melts it by raising the supply voltage to a higher-than-usual value for approximately 100 microseconds. The programming of FPLAs is a tedious process, usually requiring the services of the manufacturer, although some simple FPLAs can be programmed with PROM programmers.

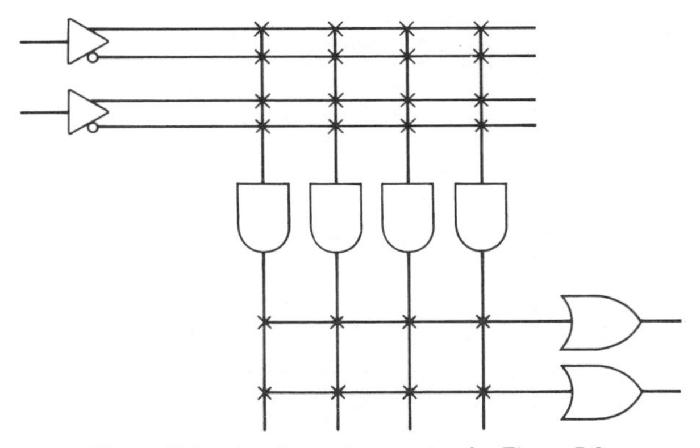

Figure 7.3. An alternative notation for Figure 7.2.

USE OF THE PLA

The obvious way to view a PLA is as a universal 2-level logic element that can be programmed as a custom logic element to replace discrete logic. However, one gains more insight to its use by viewing the PLA as a conditionally addressable memory device [1]. When viewed this way, the PLA is similar to a ROM or PROM, but the PLA's unique properties make it a superior device in many situations.

As illustrated in earlier chapters, the PROM can be used as a combinatorial logic element, that is, a replacement for discrete logic. For instance, a PROM can be used to implement truth-table logic, where the argument-value, or input, is placed on the address lines and the function-values, or outputs, are stored in its words. A PROM is seen to have two components: an address decoder (which is fixed) and a data matrix (which is alterable). Similarly, a PLA can be said to have an address decoder (the AND array) and a data matrix (the OR array). However, both the address decoder and data matrix in a PLA are alterable.

It follows, then, that one can view a PROM as a special case of a PLA, a PLA with a fixed AND array. This is illustrated in Figure 7.4, which is a PLA representation of an 8-word by 4-bit PROM. The fuses in the AND array have been blown to cause each AND to decode a unique input value from 000–111. The programming of the PROM is analogous to the selective blowing of the fuses in the OR array.

Given this view of a PROM and the fact that the address decoder (the AND array) cannot be altered, one can make the following observations about the PROM

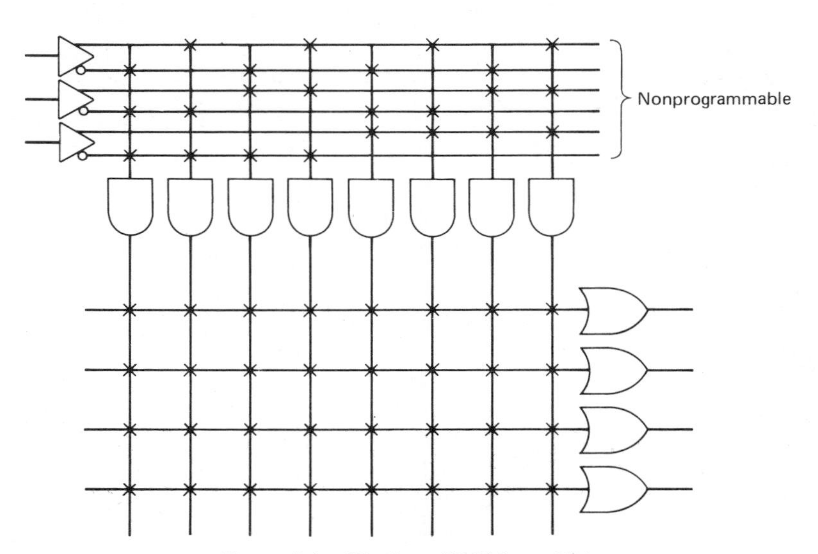

Figure 7.4. Viewing a PROM as a PLA.

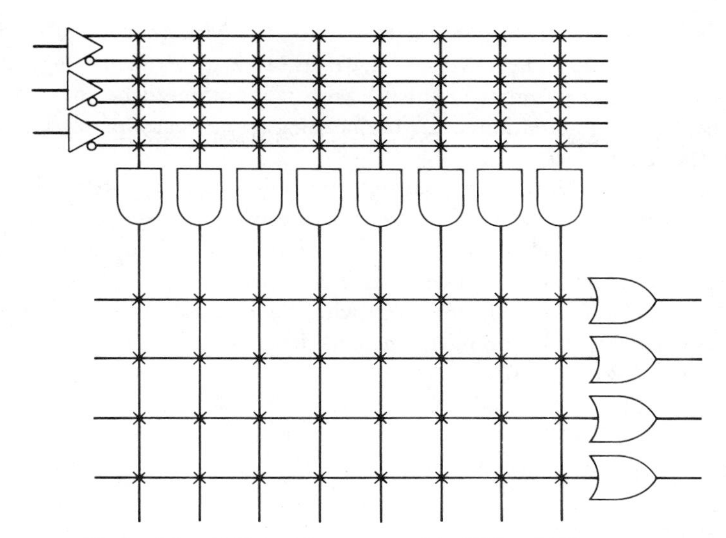

Figure 7.5. A PLA with programmable AND and OR arrays.

1. Each AND is always connected to three of the six lines in the AND array.
2. Each AND computes a minterm, or complete product, of the input. This is equivalent to saying that a PROM cannot have multiple addresses for the same word.
3. At any time, one and only one of the AND gates has an output value of 1. This is equivalent to saying that a PROM can address only one word at a time.
4. At any time, at most one input to each OR gate is a 1.
5. Because of the properties above, the PROM is a single circuit. That is, it cannot act as two or more smaller independent PROMs.

The PLA, because it does not have these restrictions, is a more general device. In contrast to the restrictions above, the PLA, as shown in Figure 7.5, has the following properties.

1. Each AND may be connected to any combination of the six lines in the AND array.
2. Each AND computes a partial or complete product. As a result, the AND can have don't-care inputs, and multiple inputs or addresses can select the same word.
3. None, one, or multiple AND outputs can be 1 simultaneously. Hence multiple words in the OR array can be selected simultaneously. All such words are ORed together. Also, if no AND output is a 1, no word in the OR array is selected.
4. At any time, each OR gate can have multiple 1-valued inputs.
5. Because of the properties above, the PLA can be partitioned into multiple, smaller, independent PLAs.

The principal difference between a PLA and PROM is that a PROM involves exhaustive addressing. A PROM has 2^n words, and, to perform a logic function, all words must be programmed. A PLA with *n* inputs need not have 2^n words, since many input combinations may be meaningless and groups of combinations may have the same effect.

As a simple illustration of this, assume one is designing a system with a 16-bit memory-address bus. Assume that memory-mapped I/O is being employed, meaning that predefined memory addresses correspond to I/O devices rather than memory words. The following addresses and address ranges will be assumed to correspond to I/O device addresses, and when such an address is encountered on the bus, logic is needed to inhibit the memory from responding and send the listed 3-bit code to an I/O controller.

000A	000
000B	001
000C	010
000D-000F	011
4FFE	100
4FFF	101
5000	110
FFF0-FFFF	111

Three alternatives come to mind. One is discrete logic, which would result in a considerable number of components. The second is a PROM of 64k 4-bit words. One of the bits in the word is a memory-inhibit signal and the other three bits are the device code (which is only meaningful when the memory-inhibit output is active). The PROM solution requires 256k memory gates, most of which are wasted because of the enormous redundancy among the PROM words. The third alternative is a PLA.

As an example of a programmed PLA, Figure 7.6 represents a small hypothetical PLA (or a portion of a real PLA) programmed to produce a 1 output when the 4-bit input value represents a number that is not an integral multiple of 2 or 3 (i.e., the 4-bit values 1, 5, 7, 11, and 13). The low-order input bit corresponds to the upper input line in the diagram. Only four ANDs are used, since the logic function was minimized to a sum of four partial products. Note that some of the ANDs have don't-care inputs. Since only four of the AND gates and one OR gate were used, the remainder of the gates and fuses are available for implementing other functions.

PROGRAMMABLE ARRAY LOGIC

Reverting to Figure 7.1, we have seen the PLAs have programmable AND and OR arrays (and usually programmable output inverters), and PROMs have fixed AND arrays and programmable OR arrays. The remaining possibility is a device

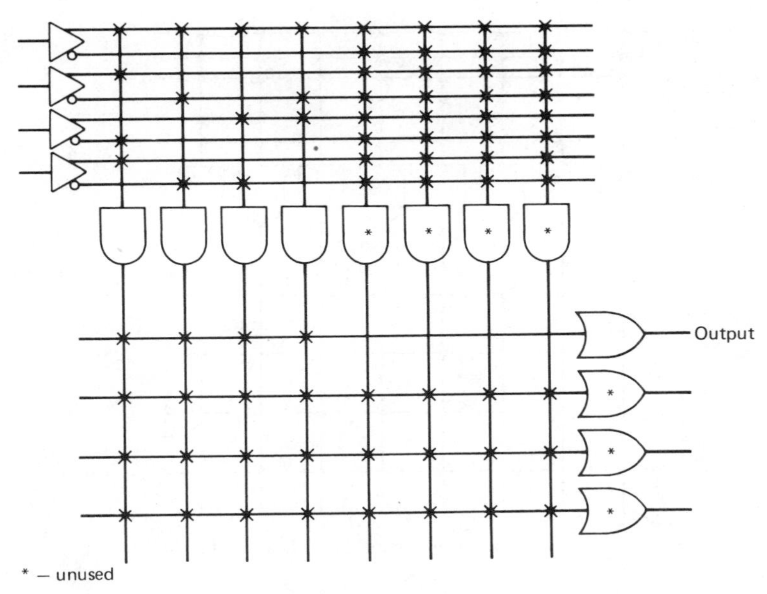

Figure 7.6. A PLA programmed to find nonmultiples of 2 and 3.

with a programmable AND array and fixed OR array, which is analogous to a memory device with preestablished word values but an alterable address-decode matrix. This alternative is represented by the family of programmable array logic devices from Monolithic Memories [2], or PAL, a trademark of Monolithic Memories.

The family of devices available from Monolithic Memories include several of the form of Figure 7.7 (with variations in the number of input and output lines), devices with inverted outputs, and devices with flipflops and outputs that are internally gated back as inputs. The available variations are shown in Figure 7.8.

The first nine devices are similar to one another, all with 16 product terms (ANDs) but with different fixed OR arrays and output polarities. The PAL16L8 has 64 product terms, eight per OR gate, with its 3-state outputs gated back into the the AND array. The PAL16R8 is similar, except that its outputs are stored in eight flipflops. The logic diagram of the PAL16R8 is shown in Figure 7.9. As described earlier, one would place an X on a line intersection in the AND array to indicate the presence of a connection.

The PAL is considerably easier to program than most PLAs; they can be programmed on a standard PROM programmer equipped with an appropriate personality card and socket. To the PROM programmer, the PAL appears as a 512 × 4 PROM, although most of the devices have less than 2048 fuses. The PAL contains verification logic to be used after programming, although the verification logic can be disabled by blowing one or two special fuses, thus making copying of the PAL difficult (a security feature).

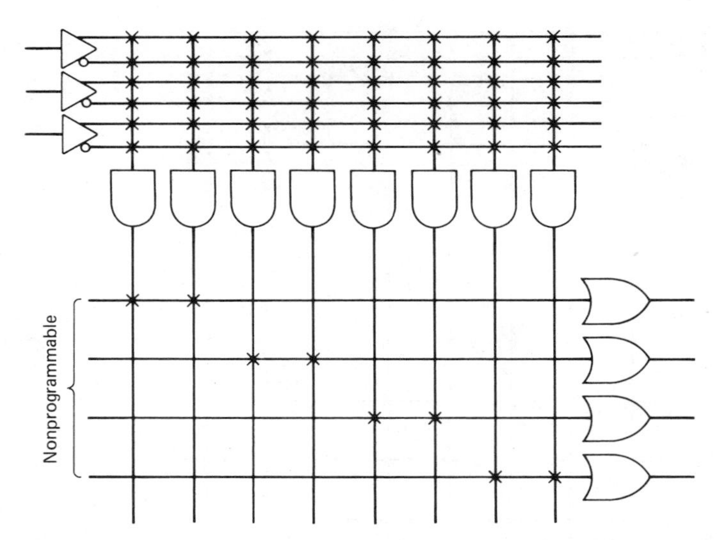

Figure 7.7. Structure of a programmable AND array, fixed OR array.

The PAL devices have a propagation delay from input to output of 25 ns (typical) and 40 ns (maximum). The maximum supply current for most of the devices is 90 mA (225 mA for the PAL16L8 and the PALS with registers).

THE 74S330 FPLA

The 74S330 is a 20-pin FPLA produced by Texas Instruments [3]. The 74S330 has 3-state outputs; the 74S331 has an identical organization with open-collector outputs.

The structure of the 74S330 is outlined in Figure 7.10. It consists of 12 input lines, 6 output lines, and 50 product terms. Each AND has 24 fused inputs (the 12 input lines and their complements). Each OR has 50 fused inputs, the outputs of the ANDs. One may connect each OR output, or its complement, to its output line via a fused exclusive-OR gate. By blowing one of two additional fuses, one can cause the outputs to be automatically enabled by a true product term or by an output-enable input line. The total number of fuses is 1508.

One must program the 74S330 in a series of steps. The first sequence involves programming the AND array, the second the programming of the OR array, and the third the programming of the output levels (the exclusive-ORs).

The AND array is programmed, fuse by fuse, by (1) applying 10.5 V to the 11 input lines not associated with the fuse, (2) addressing the AND gate by placing an address value on the output lines, (3) applying the logic level to be programmed at the input pin corresponding to the fuse, and (4) raising the supply voltage to 10.5 V for 100–1000 microseconds.

The OR array is programmed, fuse by fuse, by (1) applying to the input pins a unique product term previously programmed into the AND array, (2) applying 0.25 V to an output pin corresponding to the OR gate containing the fuse, and (3) raising the supply voltage to 10.5 V.

The output inverters and output-enable option are programmed in a similar way.

Because of their programming awkwardness, the manufacturers of FPLAs normally supply programming sheets that can be returned to the manufacturer to have the FPLAs programmed to the user's specifications before they are shipped.

The propagation delays of the 74S330 and 74S331 are 35 ns (typical) and 60 ns (maximum). The maximum supply current is 165 mA.

THE 82S100 FPLA

Another example of an FPLA is the 82S100 (3-state outputs) and 82S101 (open-collector outputs) from Signetics [4]. The FPLA has 16 inputs, eight outputs, and 48 product terms and is packaged as a 28-pin DIP.

The organization of the 82S100 is almost identical to that of the 74S330, the difference being that its 3-state output control is not programmable. Programming of the 82S110 is similar to that of the 74S330.

OTHER PROGRAMMABLE LOGIC

Two other common forms of programmable logic are programmable gate arrays (PGA) and programmable logic sequencers (PLS). A PGA is simply a PLA minus the OR array, that is, an array of programmable AND gates. A PLS is a PLA with some internal flipflops and an external clock signal. Rather than all OR outputs being connected to output pins, some of them are connected to the flipflops. Rather than all inputs to the AND array being connected to input pins, some of them are connected to the flipflop outputs. This facilitates the use of the device as a sequential controller.

Another form of programmable logic is the programmable multiplexer, as illustrated by the example in Figure 7.11. Initially, the fuse arrays connect each multiplexer input to the OR of the 10 input signals and their complements. The fuses can be blown to gate a particular input (or possibly the OR of several values) to particular multiplexer inputs.

Another approach, exploited by Motorola, IBM, Exar, and others, is the "macrocell" or "master slice." As with PLAs, the macrocell is an array that can be customized. However, rather than being an array of gates, it is an array of unconnected transistors and resistors. By creating a custom metallization pattern, it can be configured into a large number of unique devices.

Motorola's Macrocell Array is a 68-pin ECL device that can be configured into

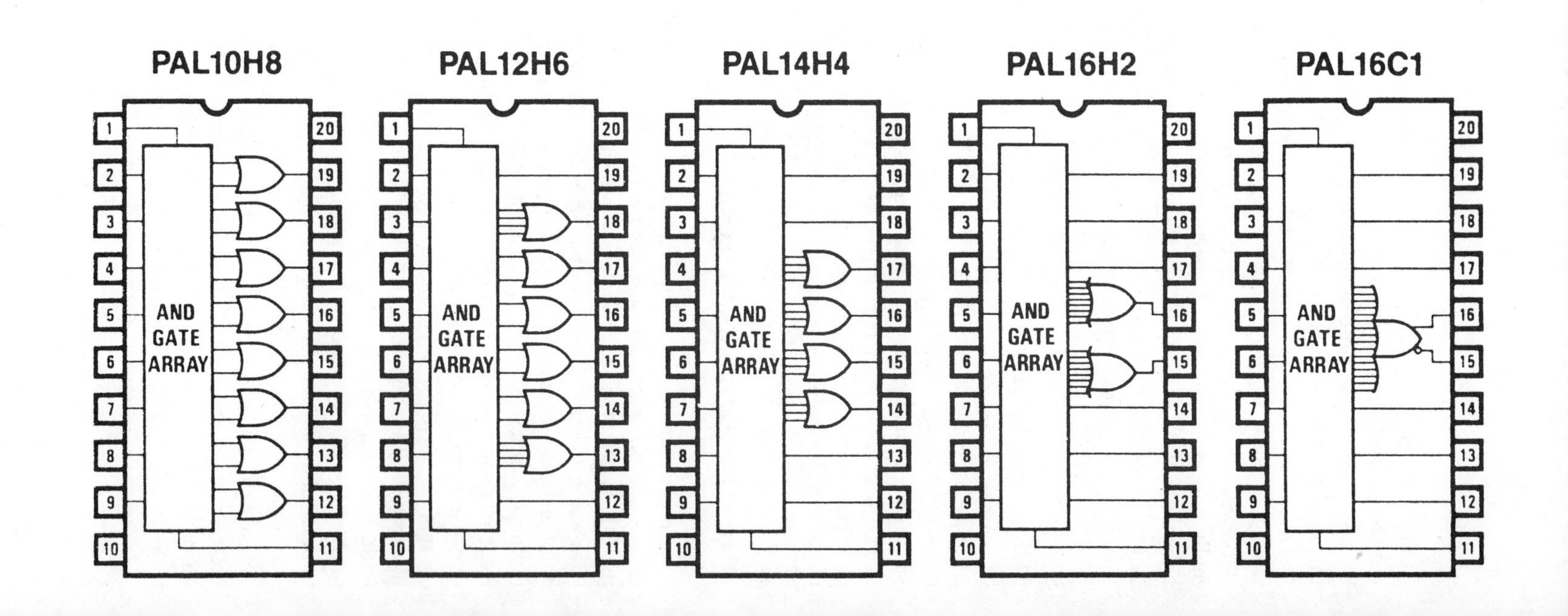

PAL10H8
AND GATE ARRAY
PAL12H6
AND GATE ARRAY
PAL14H4
AND GATE ARRAY
PAL16H2
AND GATE ARRAY
PAL16C1
AND GATE ARRAY

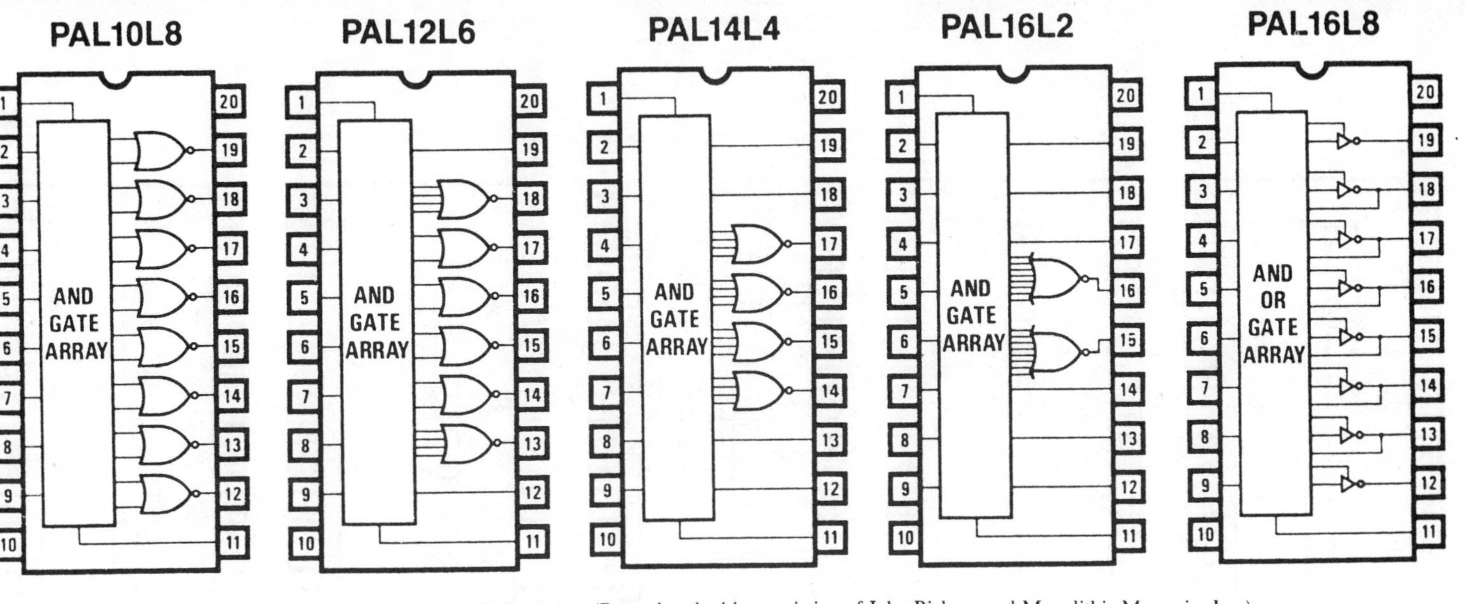

Figure 7.8. External connections of the PAL devices (Reproduced with permission of John Birkner and Monolithic Memories Inc.).

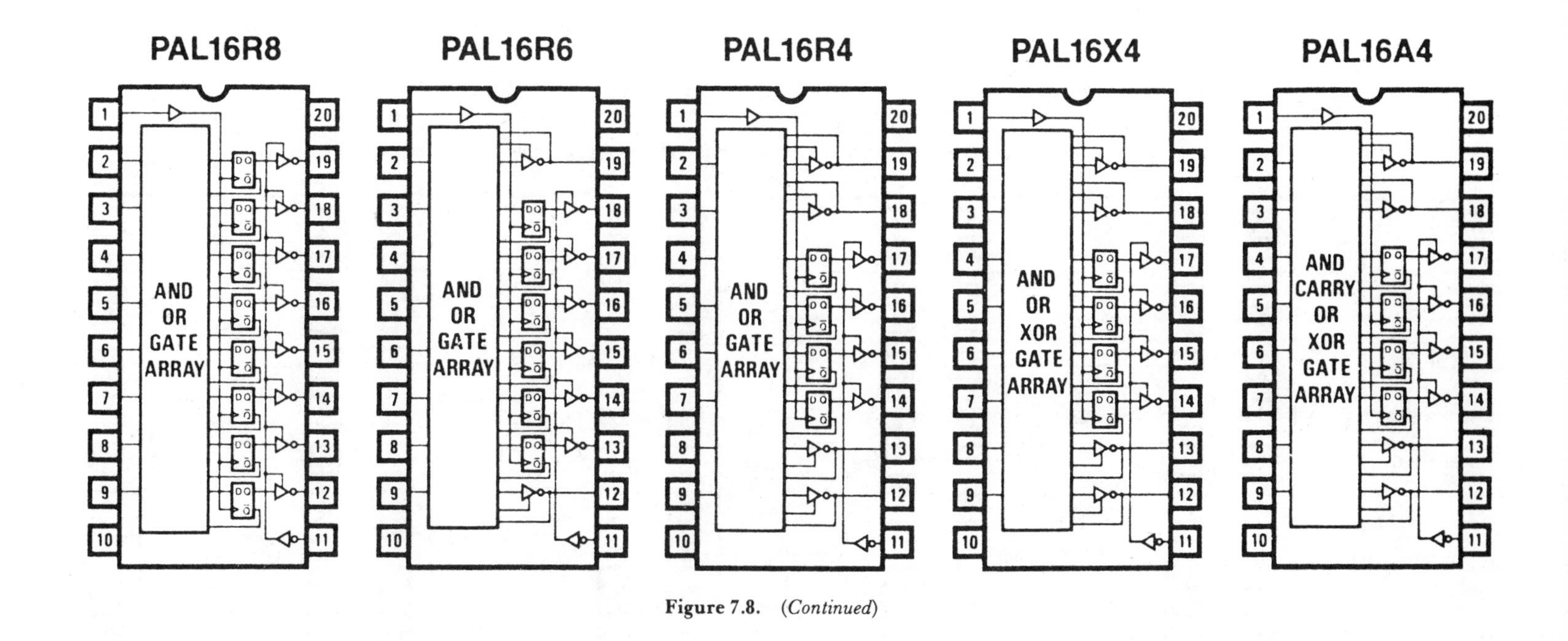

Figure 7.8. *(Continued)*

as many as 1192 gates [5]. To ease the design process, the manufacturer has developed a library of 85 "macros," or predesigned logic cells. Examples of these cells are flipflops, adders, ANDs, and ORs. The user of a Macrocell can select cells from this library, placing as many as 106 cells, with a customized interconnection pattern, on a Macrocell chip. Typical examples of custom Macrocell designs are an 8-bit binary and BCD ALU, a 16 × 4 dual-port register array, and an 8 × 8 multiplier.

EXPANDING THE PLA

Returning to the PLA, one occasionally encounters the situation where a PLA is useful, but the available PLAs have an insufficient number of inputs, outputs, or product terms to fit the application.

The cascading of PLAs to increase their dimensions is straightforward. To expand the number of product terms, one takes several open-collector PLAs and connects all of their inputs and outputs in parallel. To expand the number of outputs, one connects the inputs of several PLAs but leaves their outputs separate. Expansion of the number of input lines is not as easy; it usually involves a decoder and the control of the output-enable inputs of several PLAs.

APPLICATIONS OF THE PLA

The PLA and, in many situations, the PAL device, has an enormous number of applications, limited only by the designer's ingenuity. Listed below are a set of common applications of PLAs.

ROMs and PROMs

As mentioned earlier, the PROM is a special case of a PLA, implying that the PLA can serve as a PROM. An example of where a PLA might be used as a PROM is an application requiring a fast PROM of unusual dimensions, say a PROM of 40 5-bit words. Also, for a PROM holding a large number of non-unique values, the use of a PLA, with its ability to be addressed with don't-care input values, may be a more economical alternative.

If secrecy is a concern, PLAs are more difficult to copy than PROMs. A PROM can be easily copied by a PROM programmer. With a PLA, one must try all input combinations and examine the outputs for each, and even then one is faced with the difficulty of attempting to discern the programmed logic equations.

Code Conversion

The PLA is also suited to many code-conversion applications, particularly those situations where not all input combinations are meaningful. Typical applications are ASCII-EBCDIC, BCD-binary, and keyboard encoding. Related applications

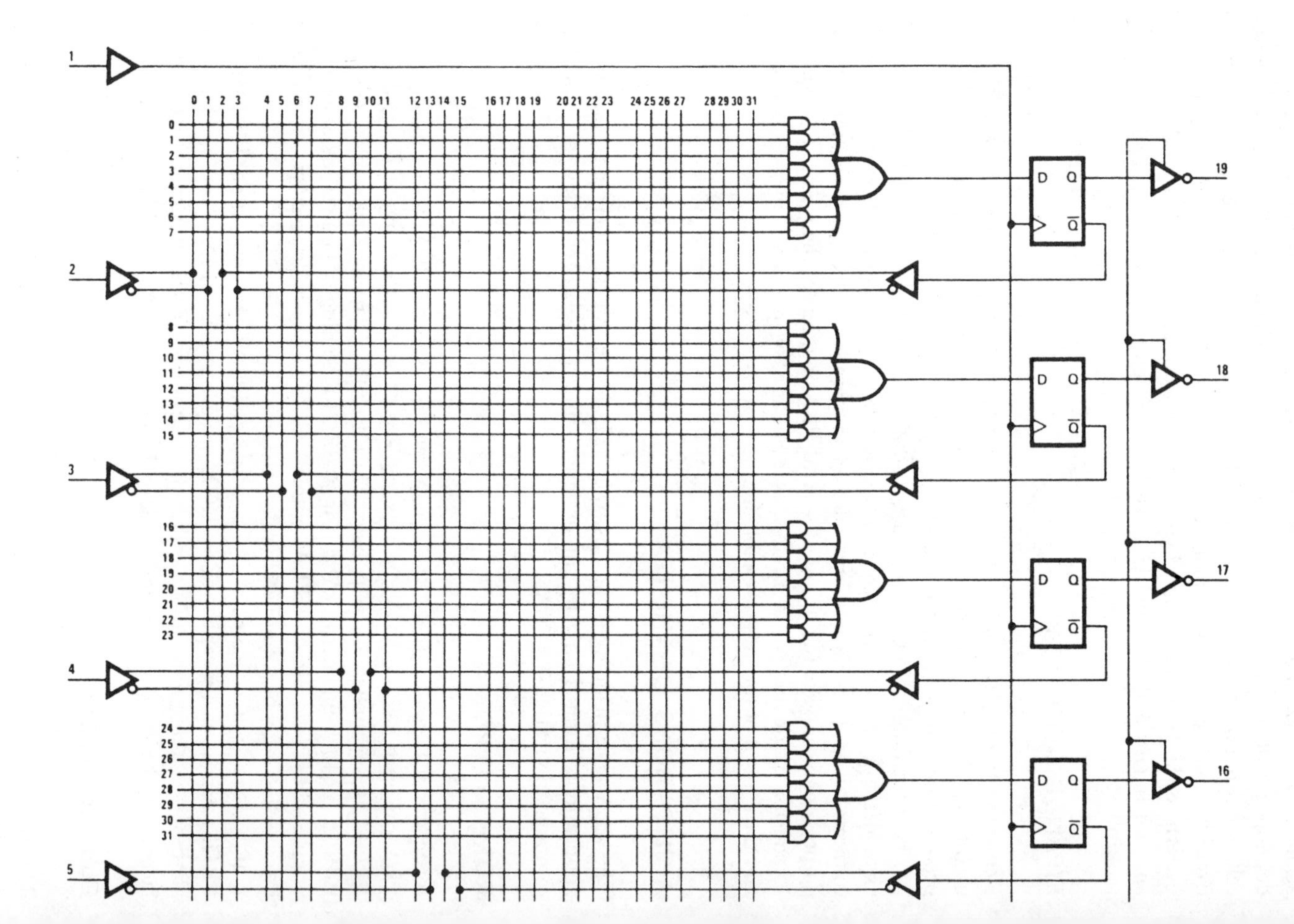
1
2
3
4
5
0 1 2 3 4 5 6 7 8 9 10 11 12 13 14 15 16 17 18 19 20 21 22 23 24 25 26 27 28 29 30 31
0
1
2
3
4
5
6
7
8
9
10
11
12
13
14
15
16
17
18
19
20
21
22
23
24
25
26
27
28
29
30
31
D Q
$\overline{Q}$
19
18
17
16

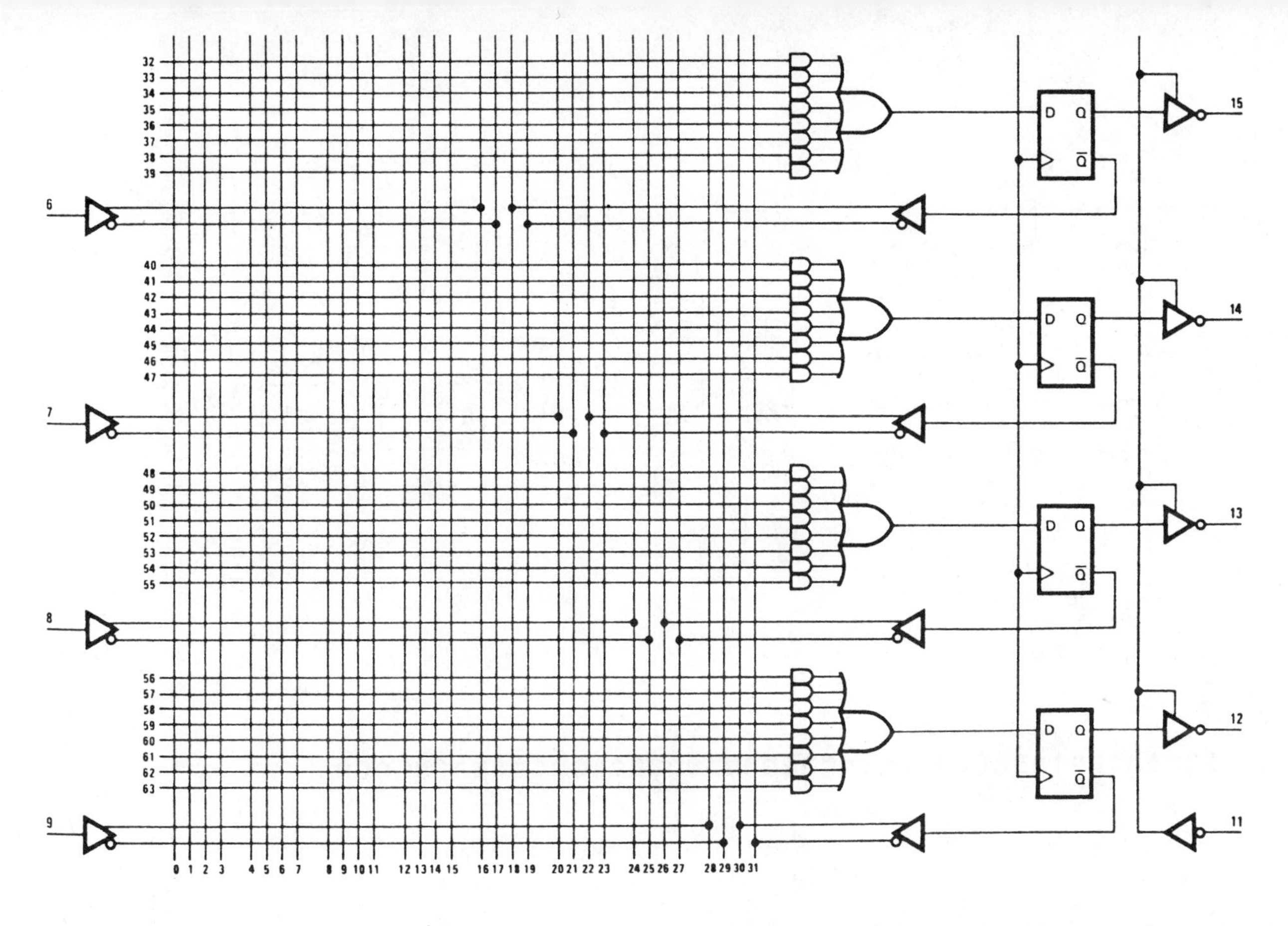

Figure 7.9. Logic diagram of the PAL16R8 (Reproduced with permission of John Birkner and Monolithic Memories Inc.).

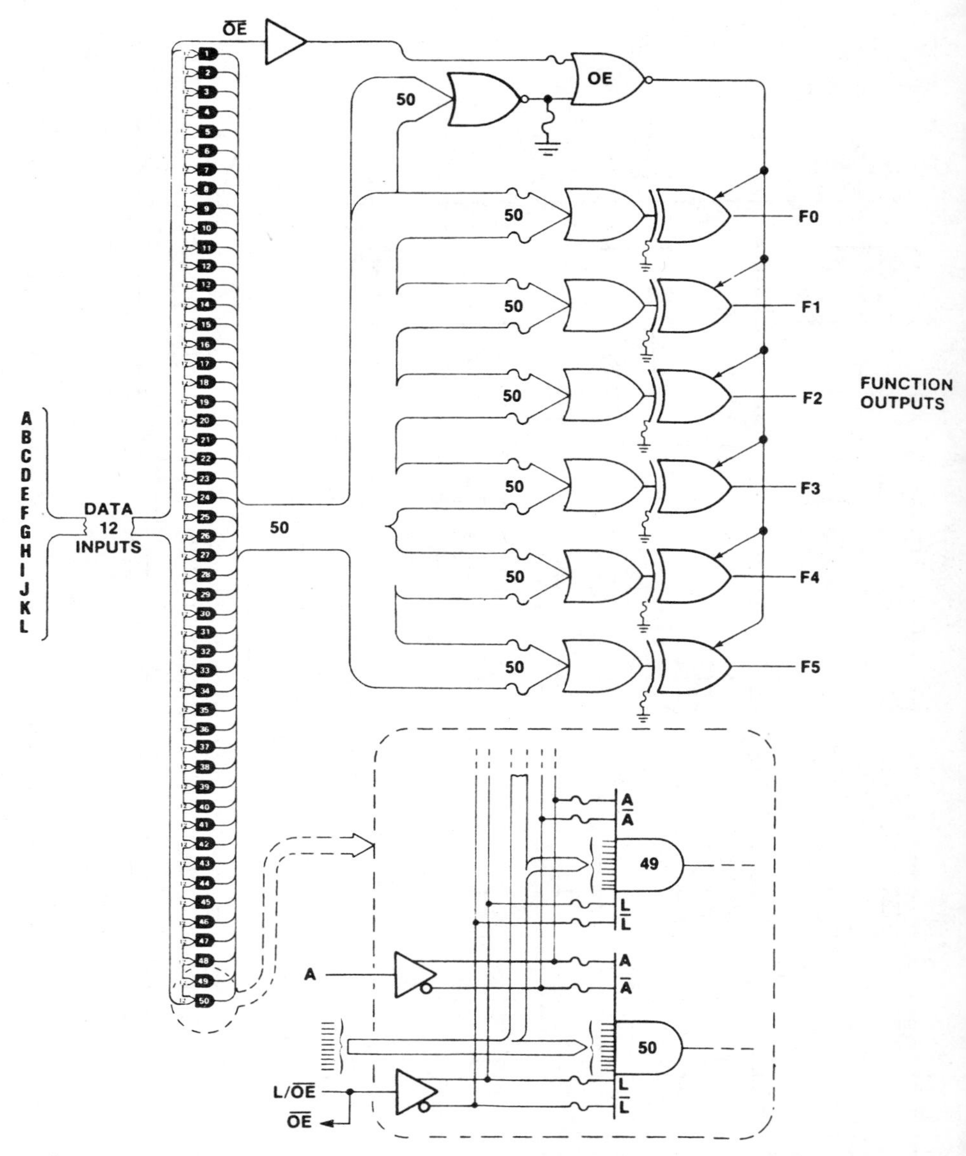

Figure 7.10. Organization of the SN74S330 FPLA (Copyright 1977 Texas Instruments Inc. Reproduced with permission.).

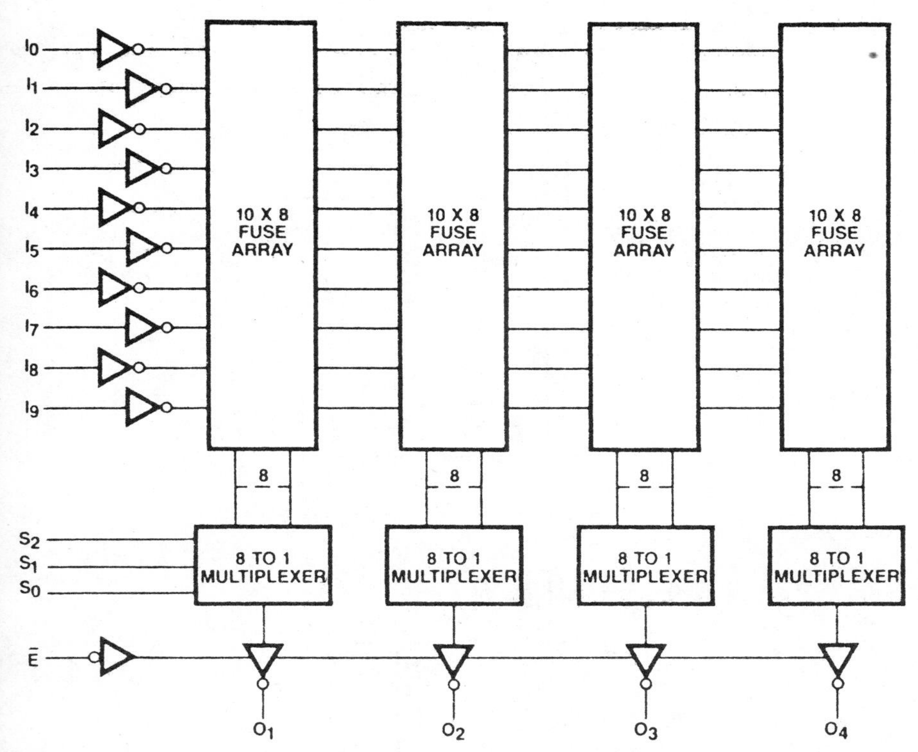

Figure 7.11. 29693 Programmable Multiplexer (Reproduced with permission of John Birkner and Monolithic Memories Inc.).

include character generation for high-resolution CRTs, where the production of large dot matrices would require large ROMs.

Controllers

A PLA in which some of the outputs are fed back to its inputs can be used as a state machine for control applications, such as an I/O device controller, machine controller (e.g., vending machine, traffic light), bus arbiter [6], and microprogram sequencer.

Figure 7.12 illustrates a microprogram sequencer built with three Monolithic Memories PAL devices. It is a 10-bit sequencer with three operations: unconditional branch, sequence, and conditional skip over the next microinstruction. The upper device serves as a status multiplexer, feeding a test condition to the middle device via the COND line. Input TF serves as a polarity control for the status condition. The use of a PLA as a status multiplexer gives one more flexibility in terms of defining test conditions as ANDs and ORs of individual status conditions. For instance, since the multiplexer has 12 status inputs and four control inputs, one could define one of the 16 selection possibilities as the branch condition "overflow AND zero ALU output."

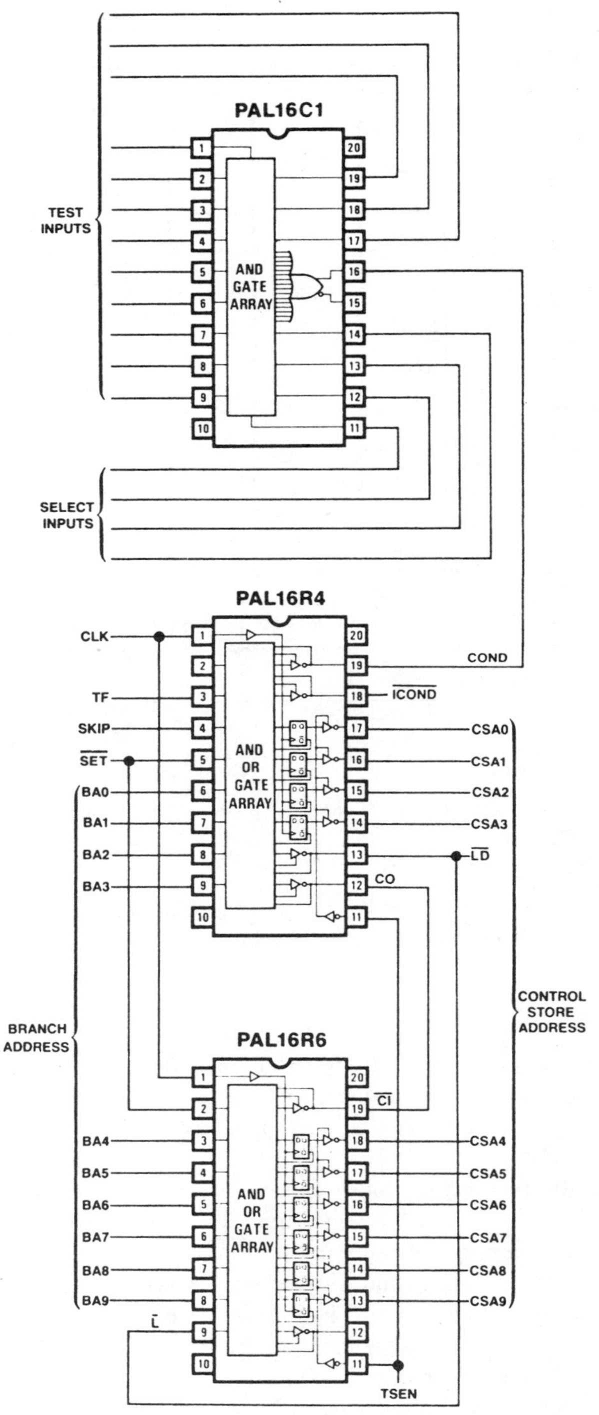

Figure 7.12. A microprogram sequencer using PLAs (Reproduced with permission of John Birkner and Monolithic Memories Inc.).

The bottom two PAL devices have their flipflops programmed as a 10-bit counter for sequence operations.

Discrete Logic Replacement

One of the most obvious applications of a PLA is simply as a replacement for a large amount of discrete combinatorial logic within a design. Also, one can program a PLA to create a tailored TTL device in situations where the available TTL devices do not exactly match the requirements of a particular design.

Monolithic Memories illustrates many such applications of their devices [2], such as memory interfaces, memory-mapped I/O-port decoders, microprocessor interfaces, limit and range comparators, priority encoders, multiplexers, shift registers, counters, and carry-lookahead generators.

Operation-Code Decoders

As mentioned earlier in the book, a PROM can be used to decode the operation code of a machine instruction by gating the opcode to the address inputs and storing in the PROM the addresses of the first microinstructions corresponding to each machine instruction. In the situation where the opcodes are variable in size (e.g., some are four bits in length, others are eight bits), this technique becomes unwieldy. However a PLA, because of its ability to handle don't-care situations, can be programmed to decode instructions with variable-length opcodes.

ROM Patch

Consider a system with a ROM and the situation where one wishes to alter the contents of several words in the ROM. As an alternative to remanufacturing a new ROM, one can use a PLA to decode the addresses of the words which need new values and overrride the ROM when any of these words are selected. For instance, if words 7 and 26 are incorrect, the PLA inputs would be connected to the ROM address lines and its AND array would be programmed to recognize the values 7 and 26. The OR array would be programmed to emit the new values for words 7 and 26, as well as to generate a signal to inhibit the ROM's output when words 7 or 26 are selected.

REFERENCES

1. E. R. Hnatek, *A User's Handbook of Semiconductor Memories*. New York: Wiley, 1977.
2. *PAL Programmable Array Logic Handbook*. Sunnyvale, Cal.: Monolithic Memories, 1978.
3. *The Bipolar Microcomputer Data Book for Design Engineers*. Dallas: Texas Instruments, 1977.
4. *Introducing the Series 3000 Bipolar Microprocessors*. Sunnyvale, Cal.: Signetics, 1977.
4. *Motorola MECL 10,000 Macrocell Array, Design Manual*. Phoenix, Motorola, 1979.
5. S. J. Durham, "Fast LSI Arbiters Supervise Priorities for Bus Access in Multiprocessor Systems," *Electronic Design*, **27**(11), 128–132 (1979).

8

Microprogram Support Tools

Given that the use of bit-slice devices usually involves the use of microprogrammed control, one must take an interest in the tools that are needed to facilitate the development and debugging of a microprogrammed system. The digital system designer is likely to be familiar with tools that apply to hardware development, such as logic-analysis programs and logic analyzers, but may be unfamiliar with the tools needed to support the concept of microprogramming. This chapter discusses the tools most likely to be needed, and, where applicable, illustrates available tools.

The tools discussed fall into four categories

1. *Microassemblers,* software programs that allow one to encode a microprogram in a symbolic language and translate such a representation of the microprogram into an absolute representation for loading into control storage.
2. *PROM Formatters,* programs that facilitate the programming of PROMs that are used for control storage by taking the microprogram to be placed in control storage and slicing it up among multiple PROMs.
3. *Development and Instrumentation Systems,* a general category consisting of both hardware and software tools that allow one to store the representation of a microprogram under development on disk files and edit the microprogram from a terminal, as well as providing instrumentation and emulation devices for the system under test.
4. *Hardware Simulators,* programs that simulate the details of the data flow of a hardware design, allowing one to test the microprogram logic in parallel with the development of the hardware system.

MICROASSEMBLERS

As mentioned in Chapter 2, one does not write a microprogram by specifying 0 and 1 values for every bit in every control-storage word. Although this may be feasible for designs containing an extremely small number of short microinstructions, it is usually much more economical to have the ability to write the microprogram in a symbolic language, and have a program, called a microassembler, translate the symbolic representation into the bit patterns for control storage.

In justifying the cost of a microassembler (i.e., the cost of purchasing or developing one), there is a tendency to look at only the short-term benefits provided by a symbolic microprogramming language. The short-term benefits are often ample justification, but the longer-term benefits usually justify the cost of a microassembler many times over. A symbolic language allows one to write symbolic representations of microorder values, as well as symbolic representations of branch addresses. The latter means that rather than writing a microinstruction containing a microorder that specifies location 104 as a branch target, one would write the microinstruction specifying a label (e.g., "INTREQ4") as the branch target. The microassembler has the job of converting symbolic microorder values to specific bit values, as well as the job of assigning microinstructions to control-storage locations and converting symbolic labels to absolute branch addresses.

The primary short-term benefit, a benefit occurring even under the unrealistic assumption that one will never make an error, is increased productivity in writing the microprogram, since the microassembler divorces one from the myriad of details concerning the specific bit representations of microorder values, and divorces one from the need to lay out the microprogram in control storage and assign absolute branch addresses. The other benefits, many of which are longer-term benefits, are

1. A degree of independence from the hardware structure. If the microprogram is encoded symbolically, one may alter such things as the encoding of microorder values (e.g., swap the definitions of two values of the microorder in order to simplify the decoding logic) without having to change the entire microprogram. One would simply alter the microassembler and reassemble the symbolic microprogram.
2. Increased understandability. A symbolic representation of a microprogram (depending somewhat on the flexibility of the microassembler language) should be considerably easier to read and understand. This increases productivity during the debugging of the microprogram and while making changes to the microprogram in the future.
3. Ease of change. If one has to encode binary branch-address values in microinstructions, the insertion of a microinstruction (e.g., to fix a design error) might upset the entire microprogram, causing one to have to alter all the branch addresses. In a symbolic language, branch addresses are represented as symbolic labels associated with microinstructions; thus one would simply insert the new microinstruction and reassemble the microprogram, causing the microassembler to reassign all the addresses.
4. Error checking. The microassembler might be capable of detecting certain types of errors in the microprogram, such as branches to nonexistent microinstructions and the use of mutually exclusive or contradictory micro-

order values. If certain microorders have values that have been left undefined (e.g., there is a 4-bit ALU-function microorder but only 12 ALU functions are defined), the use of a symbolic language would prevent one from encoding one of the undefined values.

5. Cross-reference information. Most microassemblers produce a listing showing all microinstructions that reference particular symbolic values. For instance, the cross-reference listing might list the microinstructions that branch to a particular labelled microinstruction, as well as the microinstructions that reference specific registers. This information is invaluable during the debugging and alteration of the microprogram.

Microassemblers fall into two categories. The first consists of general or definition-driven microassemblers, usually available from software companies and manufacturers of bit-slice devices, which may be tailored for a particular microinstruction design. These microassemblers are depicted in Figure 8.1. The microassembler normally has two inputs: a definition of the microinstruction format and the interpretation of the symbolic values, and the symbolic microprogram itself. In other words, the microassembler is a table-driven type of program. One first constructs a set of definitions that specify the length of the microinstruction, the position of all the fields in the microinstruction, and the bit values that are to correspond to each symbolic microorder value. Depending on the microassembler, the microinstruction definitions are either placed on a disk file and processed once, or are placed in front of the symbolic representation of the microprogram and are processed each time the microprogram is assembled.

Most assemblers make two passes through the symbolic microprogram. The first pass converts the symbolic values to the binary encoding of the microinstruction and assigns sequential control-storage addresses to each microinstruction, and the second pass fills in the branch-address fields in the microinstructions. One output is a listing of the microprogram showing, for each microinstruction, each symbolic representation, its translated binary representation, and its assigned control-storage address. Another output is the cross-reference listing. The third output is the absolute microprogram.

The second category of microassemblers describes programs developed for a specific microassembly language and a specific microinstruction design. Such microassemblers are developed, either from scratch or by using software tools available for the development of compilers or assemblers, for the specific design at hand.

In general, the second type of microassembler tends to be superior to the general, definition-driven, microassembler. The definition-driven microassemblers, because of their generality, tend to have rather cryptic language representations. The specialized microassembler tends to have a more flexible and readable language, as well as the advantage of additional error checking (because one knows in advance the specific hardware target of the assembler). The cost difference is not necessarily large. The definition-driven microassembler must be purchased or licensed from a manufacturer, but the specialized microassembler must be

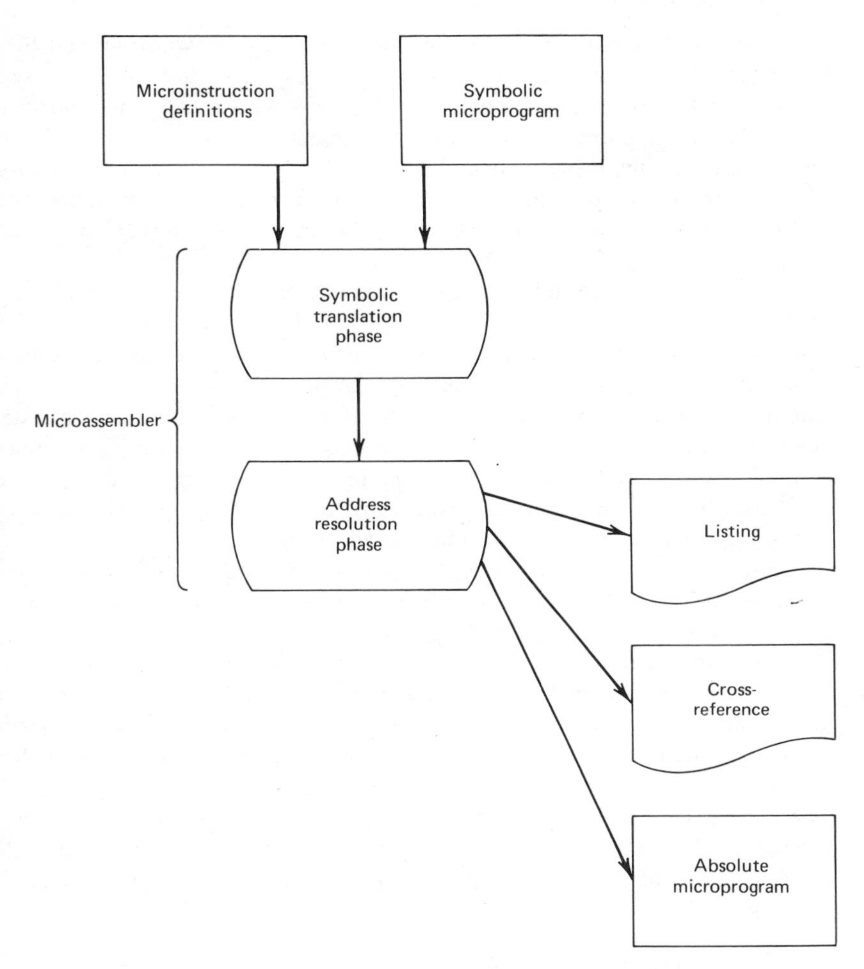

Figure 8.1. Structure of a definition-driven microassembler.

developed. However, the existence of powerful compiler-writing tools can allow one to develop a specialized microassembler in a month or two. (Of course, the disadvantage is the requirement of the programming talent to develop the specialized microassembler.)

DEFINITION-DRIVEN MICROASSEMBLERS

A definition-driven microassembler has two symbolic languages. The first language, the definition language, is used to describe the syntax and semantics of the second language: the customized microprogramming assembly language. A set of desirable attributes of both languages is listed below.

1. The definition language must be flexible, yet straightforward enough to be learned in a limited amount of time.
2. With the definition language, one should be able to describe a microinstruction of any size and any format (within reasonable limits).
3. One should be able to specify default microorders with the definition language (e.g., the sequence operation as the default branch-operation microorder) so that the microprogrammer need not code, in each symbolic microinstruction, a value for each field.
4. The symbolic microassembly language should have a flexible, simple, and readable syntax.
5. The symbolic microassembly language should allow the interspersing of explanatory comments within the microprogram.
6. The microassembly language should contain predefined statements that allow one to align a microinstruction on a specific boundary (e.g., position the next microinstruction on a 16-word boundary in control storage to allow it to be the target of a 16-way multiway branch), and to position sections of the microprogram at specific locations in control storage.
7. The microassembly language should contain predefined statements that allow one to control the formatting of the output listing, such as statements for inserting blank lines or skipping to a new page.
8. The microassembler should detect such error conditions as the use of an undefined symbol, the use of the same microinstruction label (symbolic address) on more than one microinstruction, and the absence of a symbolic microorder value in a microinstruction where no default has been specified in the definition language.
9. The microassembler should contain no dependencies on specific devices (e.g., specific ALU/register slices), but it is convenient if the microassembler contains optional "canned" symbolic definitions for the control fields of popular bit-slice devices, such as those in the 2900 family.

A large number of definition-driven microassemblers are available from software companies and semiconductor manufacturers. Several representative microassemblers are surveyed in this section. To illustrate each language, the microinstruction design in Figure 8.2 will be used. This microinstruction design is intended only for illustrative purposes. It is an unrealistic, simplified microinstruction format for a hypothetical design.

The first field in the microinstruction can specify a branch address or an 8-bit emit or literal value that may be gated into the ALU. Most of the other fields should be self-explanatory. The last field specifies both a second input to the ALU and the destination of the ALU output.

The following microprogram of three microinstructions will be illustrated in the symbolic language of some of the microassemblers

1. Add R0 plus R1 and store the result in R1. Conditionally branch to the microinstruction after the next one if overflow occurs.

2. Add the emit value of 2 to R1. Sequence on to the next microinstruction.
3. Perform a subroutine call and OR R2 into R3.

The AMDASM Micro Assembler

AMDASM is a microassembler produced by Advanced Micro Devices [1–3]. Versions of it are available on Intel's Intellec Microprocessor Development System, Advanced Micro Devices' System 29 Microprogram Development System (discussed later in this chapter), and on a commercial timesharing service [4]. Also, Microtec Corp. produces an assembler that is compatible with AMDASM but is written in the Fortran language, allowing the assembler to be executed on most computer systems [5].

AMDASM does not process the definition language with the assembly language. The definition language is normally processed once (unless it is changed) and the output is stored on a disk file, which is used each time the assembly phase is executed.

A defintion of a symbolic language for the microinstruction design of Figure 8.2 is given in Figure 8.3. The first statement defines the microinstruction size. The subsequent statements define the symbolic microorder values. Note that the registers (R0,...) have been defined as 3-bit values, although, in one of the

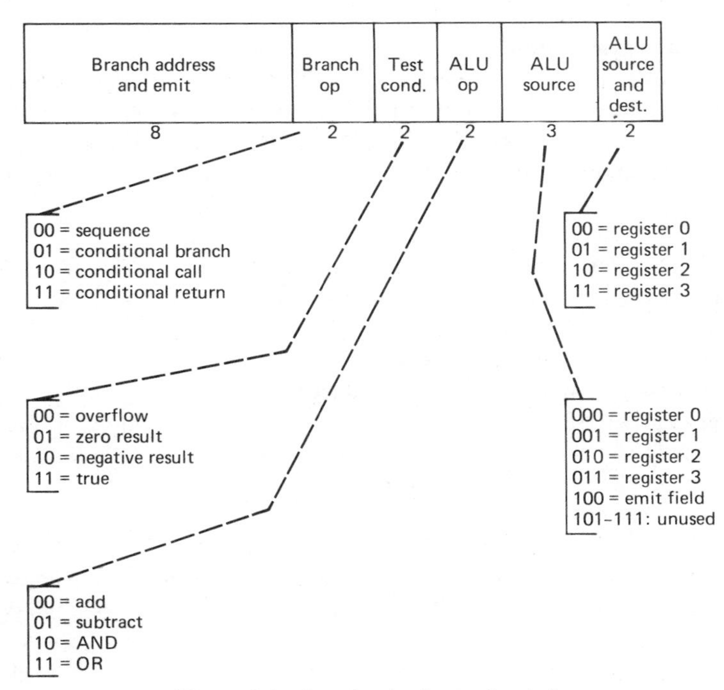

Figure 8.2. Sample microinstruction design.

```
WORD 19
; REGISTER DEFINITIONS FOLLOW
R0:   EQU B#000
R1:   EQU B#001
R2:   EQU B#010
R3:   EQU B#011
EMIT: EQU B#100
; BRANCHING OPERATIONS FOLLOW
GO:     EQU B#00
CBR:    EQU B#01
CCALL:  EQU B#10
CRET:   EQU B#11
; CONDITIONS FOLLOW
OFLOW:   EQU B#00
ZERO:    EQU B#01
NEG:     EQU B#10
TRUE:    EQU B#11
; ALU OPERATIONS FOLLOW
ADD: EQU B#00
SUB: EQU B#01
AND: EQU B#10
OR:  EQU B#11
; MICROINSTRUCTION DEFINITION FOLLOWS
M: DEF 8VH#00,2VB#00,2VB#11,2VB#11,
/       3VB#000,2VB#00:
END
```

Figure 8.3. AMDASM definition of the sample microinstruction.

microorders, they are specified by 2-bit values. This is handled in the DEF statement at the end of the definition file.

The DEF, or definition, statement defines the microinstruction format. Since AMDASM allows one to have multiple formats or definitions of microinstructions, each DEF statement must be labelled. The single DEF statement here is labelled M. Reading across the DEF statement, the microinstruction fields are defined as

1. An 8-bit field with a default value of zero.
2. A 2-bit field with a default value of 00 (sequence).
3. A 2-bit field with a default value of 11 (the condition TRUE).
4. A 2-bit field with a default value of 11 (OR).
5. A 3-bit field with a default value of 000 (register 0).
6. A 2-bit field with a default value of 00 (register 0). The colon (:) is an added attribute of the field indicating that values assigned to this field having lengths greater than two bits will be truncated on the left.

The defaults indicate that if no explicit branching operation is coded in a symbolic microinstruction, the default value will be the sequence operation, if no condition is coded, the default is "true," and if the ALU operation and register designators are not coded, the default is ORing R0 with itself.

The symbolic microprogram corresponding to the example mentioned earlier is given in Figure 8.4. A microinstruction is coded with an optional address label, a format (definition) name, and a positional sequence of microorder values, separated by commas. If a microorder value is omitted (i.e., the default value is desired), the comma must still be present to indicate the absence of the value.

This example omits many aspects of AMDASM; it is intended only to give one a flavor of the language. The definition language also contains a mechanism to define subfields within fields. The assembly language contains statements for printer control (e.g., TITLE, SPACE, EJECT) and for the control of address assignment (e.g., "ORG 24" to set the assembler's program counter or address of the next microinstruction to 24, and "ALIGN 4" to set the address of the next microinstruction to the next value that is an integral multiple of 4). One can also insert comment statements in both the definition language and the symbolic assembly language.

In addition to producing a listing of the assembly program with the corresponding assembled absolute microinstructions and their addresses, AMDASM produces a symbol table listing the value of each label. It does not produce a cross-reference listing.

The biggest criticisms of the AMDASM language are that it is cryptic and difficult to read, and that it does not associate symbolic micoorder values with particular fields, making it prone to errors. For instance, if one transposed two fields in the first microinstruction in Figure 8.4 by coding

M XYZ,OFLOW,CBR,ADD,R0,R1

the error is not detected and the microinstruction becomes one that performs a sequence operation (OFLOW = 00 = sequence).

A program associated with AMDASM is AMPROM. If PROMs are being used as control storage, it is likely that control storage will consist of many PROMs. For instance, a possible configuration is that of Figure 8.5, where the 28-bit by 512-word control storage consists of six 256×8 PROMS and two 256×4 PROMs. AMPROM takes the output of AMDASM and, given a description of the PROM configuration across control storage, punches the proper paper tape (for input to a PROM programmer) for each PROM.

The Signetics Micro Assembler

Signetics produces a definition-driven microassembler [1,6,7], which is available on commercial timesharing services and as a Fortran program. It consists of both a definition language and the symbolic microprogramming language defined by the definition language. The definition statements must precede the microprogram statements through the assembler.

```
      M   XYZ,CBR,OFLOW,ADD,R0,R1
      M   02,,,ADD,EMIT,R1
XYZ:  M   ZZZ,CCALL,,OR,R2,R3
      END
```

Figure 8.4. Sample AMDASM program.

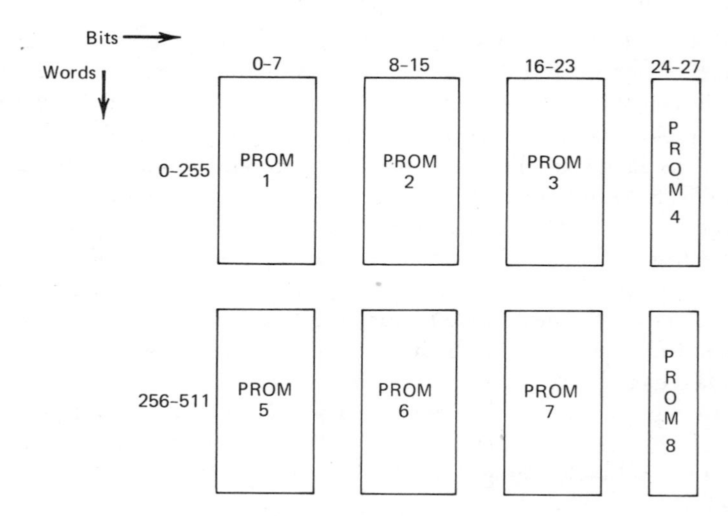

Figure 8.5. Hypothetical control-storage configuration.

Figure 8.6 shows the definition statements corresponding to the example used earlier. The INSTRUCTION statement defines the microinstruction width. The FIELD statements define the size of each field in the microinstruction, assign the field a symbolic name, and assign the field a default value.

The MICROP statements, whose use is optional, allow one to assign values to fields in the symbolic microprogram by encoding keywords. For instance, if a symbolic microinstruction specifies the keyword "SEQ," field BOP is assigned the value 00. MICROP statements also allow the definition of arguments. For instance, the MICROP defining ADD will cause the ALUOP field to be 00, the RSRC field to be 010, and the RSCRD field to be 11 if the following expression is coded in a microinstruction

ADD (R2,R3)

Figure 8.7 is the symbolic microprogram corresponding to the example. The language tends to be more readable and less error-prone than that defined by AMDASM, principally because of its keyword orientation, its not being positional (the requirement in AMDASM to order the microorders and the need for extra commas to mark omitted microorders), and the ability to combine related microorders as arguments. Microinstructions are delimited by semicolons, and comments can be placed anywhere by enclosing them in quotation marks.

Like AMDASM, the Signetics microassembler assigns each microinstruction consecutive addresses starting with zero, except where the assignment of addresses is explicitly changed via ORG statements. The Signetics product contains predefined definitions (MICROP statements) for the 3002 and 2901 ALU/register slices and the 8X02 sequencer. For instance, by including the statement

INTRINSIC '2901';

in the set of definition statements, one obtains a large set of predefined MICROP statements for control fields for the 2901.

```
INSTRUCTION WIDTH 19;
FIELD BA     WIDTH 8 DEFAULT 0;
FIELD BOP    WIDTH 2 DEFAULT 0;
FIELD COND   WIDTH 2 DEFAULT 3;
FIELD ALUOP  WIDTH 2 DEFAULT 3;
FIELD RSRC   WIDTH 3 DEFAULT 0;
FIELD RSRCD  WIDTH 2 DEFAULT 0;
MICROP GO     ASSIGN BOP=0;
MICROP CBR    ASSIGN BOP=1;
MICROP CCALL  ASSIGN BOP=2;
MICROP CRET   ASSIGN BOP=3;
MICROP OFLOW  ASSIGN COND=0;
MICROP ZERO   ASSIGN COND=1;
MICROP NEG    ASSIGN COND=2;
MICROP TRUE   ASSIGN COND=3;
MICROP ADD(RA,RB) ASSIGN ALUOP=0
       RSRC=RA RSRCD=RB;
MICROP SUB(RA,RB) ASSIGN ALUOP=1
       RSRC=RA RSRCD=RB;
MICROP AND(RA,RB) ASSIGN ALUOP=2
       RSRC=RA RSRCD=RB;
MICROP OR(RA,RB) ASSIGN ALUOP=3
       RSRC=RA RSRCD=RB;
R0: EQU 0;
R1: EQU 1;
R2: EQU 2;
R3: EQU 3;
EMIT: EQU 4;
END INSTRUCTION;
```

Figure 8.6. Signetics Microassembler definition of the sample microinstruction.

Another useful feature of the assembler is the DCL (declare) statement, which allows one to specify the contents of mapping PROMs (e.g., op-code decoding PROM, interrupt-vector PROM) in the assembly language. For instance, if ZADD, ZBRANCH, ZMOVE, and ZMULT are statement labels within the symbolic microprogram, the statements in Figure 8.8 define addresses for the first four locations in a 256×12 op-code mapping PROM.

In addition to producing a listing, the assembler produces a cross-reference listing, showing for each symbol, its value, type, number of the line in which it is defined (line in the listing), and numbers of the lines in which it is referenced. Signetics also provides an additional program, the Micro Format program, which is similar in concept to AMPROM (i.e., it allows one to specify the PROM configuration of control storage and punches the appropriate PROM programming tapes).

```
PROGRAM QQQ WIDTH 19;
      BA=XYZ CBR OFLOW ADD(R0,R1);
      BA=2 ADD(EMIT,R1);
 XYZ: BA=ZZZ CCALL OR(R2,R3);
      END;
```

Figure 8.7. Sample Signetics Microassembler program.

```
PROGRAM OPDECODE WIDTH 12 LENGTH 256;
DCL ZADD;
DCL ZBRANCH;
DCL ZMOVE;
DCL ZSUB;
```

Figure 8.8. Signetics Microassembler definitions for a mapping PROM.

The XMAS Micro Assembler

XMAS is a microassembler from Intel Corp. [1,8]. XMAS is a program in the CROMIS package, which also contains a PROM formatting program (XMAP). CROMIS is available on a commercial timesharing service.

XMAS is used by placing a set of definition statements ahead of the symbolic microprogram, but the definition statements have a different purpose than those in other microassemblers; they extend the predefined symbolic assembly language, rather than completely define it. XMAS is intended for use with designs incorporating the 3001 sequencer and the 3002 ALU/register slice. As such, it assumes that the first four fields of the microinstruction, occupying 18 bits, contain microorders controlling the 3001 and 3002. The definitions of these four fields are fixed, including their symbolic representations, and cannot be changed by the user of XMAS. The XMAS definition statements allow one to define the remaining fields of the microinstruction.

The first field, a 7-bit field, is assumed to be gated to the control inputs of the 3002 slices. Symbolic microorders are predefined in XMAS; for instance "ILR(R3)" sets the field to 0000011, indicating that R3 is added to the 3002 carry input and stored in R3 and AC. The second and third fields, two bits each, are assumed to control the 3001's flag-control inputs. The fourth field, a 7-bit field, is assumed to control the addressing function of the 3001. For instance, the symbolic value "JCR(L)" generates the "jump in current row" control value, where L is a symbolic label of a microinstruction in the current row. The address of L is analyzed to complete the absolute value of this field. A typical symbolic microinstruction in XMAS might be

08H: ZADD: LMI(R7) HCZ FFO JZR(X) other-microorders

Recall that when using the 3001 sequencer, one must carefully lay the microprogram out in control storage. The 08H specifies that this microinstruction shall be stored in control-storage location 8; location 8 is symbolically labelled as ZADD. The microinstruction gates R7 into the 3002 MAR register and increments R7 by the carry input, causes the C and Z flipflops in the 3001 to be held, sets the 3001 FO output to zero, and performs a jump-to-row-zero operation to the location labelled X. (If X is not in row zero, the assembler generates an error message.)

XMAS contains a definition statement, FIELD, to allow one to define the remainder of the microinstruction. For instance to define field four of Figure 8.2, one might specify

ALUOP FIELD LENGTH=2 MICROPS(ADD=0,SUB=1,AND=2,OR=3)

A useful feature of XMAS is the ability to define macros that replace frequently used combinations of symbolic microorders. For instance, one can specify

EXIT STRING 'JZR(IFETCH) HCZ LMI(R7)'

defining EXIT as the specified string. Then, a microinstruction which is coded as

83H: FFO STORE EXIT;

is equivalent to

83H: FFO STORE JZR(IFETCH) HCZ LMI(R7);

In addition to producing an output listing and a file containing the absolute microprogram, XMAS produces (1) a cross-reference listing showing each statement label and the statements in which the label is referenced, and (2) a graphic representation of the contents of control storage, illustrating each microinstruction as a box and printing, in each box, the branch-operation microorder, the target address, the sequential line number of the symbolic microinstruction in the output listing, and the number of microinstructions that branch to this microinstruction.

SPECIALIZED MICROASSEMBLERS

Another alternative solution to the need for a microassembler is the development of a specialized microassembler, that is, the specification of the desired symbolic language and then the development of a program to translate this language into the absolute control-storage representation of the microprogram. Such a microassembler has the following advantages over definition-driven microassemblers

1. The language can be made more natural and readable. One is not constrained to the cryptic syntactic characteristics often seen in definition-driven microassemblers.
2. One is not constrained by restrictions that definition-driven microassemblers might have in such areas as microinstruction size, microorder sizes, microorder placement, assembler control statements, and so on.
3. Given that the microassembler is being developed for a specific microinstruction design, one can build additional error checking into the microassembler, such as detecting inconsistent microorder values (e.g., a microinstruction initiates a memory read into the memory data register and also gates the ALU output into the same register) and some cases of inconsistent microinstruction sequences (e.g., a microinstruction is testing the condition "ALU A input is less than ALU B input" but the ALU operation in the previous cycle was not a subtraction).

The obvious disadvantage of specialized microassemblers are the expense of developing them and the requirement for the programming talent to do so. The development expense for a specialized microassembler need not be excessive; in fact, they can often be developed at a cost that is not much higher than the cost of purchasing a definition-driven microassembler. For instance, the microassemblers discussed in this section were developed in three to eight manweeks.

One example of a specialized microassembler is AMAS, a free-format assembler that is in use by the author. AMAS was developed for the microinstruction design described at the end of Chapter 5 and illustrated in Figure 5.23, a design incorporating the 2903 and 2910 devices. AMAS was developed and tested by an expert programmer, using a tool developed for the writing of high-level-language compilers, in three weeks.

A sample of the symbolic language processed by AMAS is shown in Figure 8.9. The system has four microinstruction formats; the first two characters (e.g., F0) specify the microinstruction format. The remainder of the microinstruction is keyword oriented, allowing related sets of microorders to be associated with a single keyword for ease of readability (e.g., the ALU keyword contains three microorder values within the surrounding parentheses). Comments may be freely interspersed within and between microinstructions. The language also includes printer-control statements and address-control statements (e.g., ORG and ALIGN).

Another specialized microassembler is Microbe [9], produced for a design having a 48-bit microinstruction with 12 control fields and using 2900-family logic. Microbe is unusual in that the language is terse and positional, but the assembler generates, on the output listing, comments for each microinstruction in a form similar to a high-level language.

Figure 8.10 represents a sample from an output listing produced by Microbe. The symbolic representation of the microinstruction is the symbols written under the column headings; the expressions under the COMMENTARY heading are generated by Microbe and represent the higher-level description of each microinstruction.

Microbe was written in the PL/I language, requiring an investment of six weeks for the design, development, and testing of the assembler.

A third example is the Mirager language [10]. An example of the language is shown in Figure 8.11. Symbolic microinstructions are delimited by ellipses. The microorders show the movement of data over specific busses. For instance, the first

```
SUM:
  F0 BACE=4 ALU(EMIT PASSL * TO RC)        /SET LOOP COUNTER TO 5-1/.
SUMLOOP:
  F0 M(READ6 MAR1 MDR1) ALU(MAR1 + 6 TO MAR1)
     /READ NEXT ELEMENT INTO MDR1 AND POINT MAR1 TO NEXT ELEMENT/.
  F0 BACE=SUMLOOP BR(ITERATE) WAIT
     ALU(MDR1 + B TO B) B(R5).

  F1 DALU(P11 & P12 TO P12) I(MAR2 BY+1).
  F2 BACE=MASKIT SETB(P12 ¬PRINTER) ALU(R5 - B TO B,LSHIFTB) B(R6).
  F2 BACE=JUMPTABLE
     MWB(P11 TO123)              /TAKE 16-WAY BRANCH/
     ALU(MDR1 - Q TO MAR2)   /MAR2 = DEVICE ADDRESS/
     M(READ1 MAR1 MDR2)          /READ CONSOLE CODE/.
```

Figure 8.9. Sample of the AMAS language.

```
LAB OUT XB- DDRY -A- -B- RS FCN C SEQF JCND  SEQT ADDR      COMMENTARY
*** *** *** **** *** *** ** *** * **** ***** **** **** ****************

    XRG LIT BQ*F DE    E  AB XOR   POPC -10203            XRG=XB=(-10203);
                                                          Q=ROTL(Q);
                                                          E=ROTL(DE#E);
                                                          POPC;
    XRG     B..A  X  CNT OB S-R 0 JMP  =0     RTS  $      XRG=XB=X;
                                                          CNT=FCN=(-0)-
                                                                 1+CNT;
                                                          IF FCN=0
                                                            THEN RTS;
                                                            ELSE JMP $;
```

Figure 8.10. Sample of a Microbe output listing with generated commentary.

line in Figure 8.11 means "apply operation BIPASS3 to register RB, move the result along bus B1 into SPA19." Mirager was implemented in two man-months.

Other examples of specialized microassembler languages include one for a 3000-series design [11] and another for the AN/UYK-17 military computer [12].

HIGH-LEVEL MICROPROGRAMMING LANGUAGES

Given the low-level nature of microassembly languages, one is inclined to consider the possibility of encoding microprograms in higher-level languages (e.g., languages similar to Fortran, PL/I, and Pascal) and the development of compilers for these languages.

Vertical microinstruction designs, being similar to traditional machine instructions, allow one to readily develop a higher-level microprogramming language; for instance, one such language is EMPL [13]. However, such is not the case for horizontal designs. Conventional higher-level languages do not allow one to specify parallel operations (e.g., the IF, GO TO, and DO statements specify branching operations, the assignment statement specifies a series of zero or more computations followed by a store operation). Hence, a direct compilation of a program written in a high-level language would result in a highly inefficient use of the microinstruction, resulting in excessively large and slow microprograms.

The compilation of a high-level microprogramming language to a horizontal microinstruction requires a compiler than can detect operations that can

```
       <BIPASS3>RB-B1->SPA19
       <SET15>SPB14-B2->SPB14
       IF B1=0 GO TO #243
       ...
#230:  <SET13>SPB14-B2->SBP14
       RA-B4->BMAR
       WRITE
       WAIT
       ...
       <UNPASS5>RB-B1->SPA19
       INCREMENT SPAR
       ...
```

Figure 8.11. Sample of the Mirager language.

potentially be performed in parallel, as well as recognize the timing dependencies among them. In other words, the compiler would attempt to repackage the algorithmic representation of the microprogram in the high-level language into as few horizontal microinstructions as possible. To the author's knowledge, no such compiler is in use today, although papers abound on techniques such compilers would use (e.g., [14–17]) and on design principles for microprogramming languages [18–19].

A compiler for a high-level microprogramming language might be a moot point for two reasons. First, contrasted with computer systems where thousands or more application and system programs are written, thus justifying the cost of compilers, one usually writes only one microprogram for a machine, and the microprogram tends to be small (in relation to programs). The cost of developing a compiler for a high-level microprogramming language is likely to exceed significantly the cost of developing the microprogram, thus making it an uneconomical venture. Second, microprogrammers are usually interested in the design of time- and space-optimal microprograms, and the microprogram produced by a compiler is likely to be less efficient than a microprogram produced in a low-level microassembly language.

Conceivably, these problems could be overcome with a generic compiler having (1) the capability of being driven with definitions of a specific target microinstruction design, thus allowing the compiler to be used for more than one specific machine, and (2) highly sophisticated optimization techniques. At least one encouraging result in this area has been reported [20].

DEVELOPMENT AND INSTRUMENTATION SYSTEMS

Another set of tools available for the support of microprogrammed systems falls into the general category of development and instrumentation systems. These tools, usually consisting of both hardware and software, provide some or all of the following functions

1. Editing and storing, on disk files, microprograms and test data.
2. Simulation of control storage in the system being developed.
3. Debugging support for the system under development, including clock control, address breakpoints, and state analysis.
4. Instrumentation of the system under development, such as traces of specified signals.

Several commercially available tools are surveyed below.

System 29

System 29 is a flexible self-contained development system produced by Advanced Micro Devices [2,21]. As shown in Figure 8.12, System 29 contains two sections: (1) a support system, containing software and I/O devices on a microprocessor

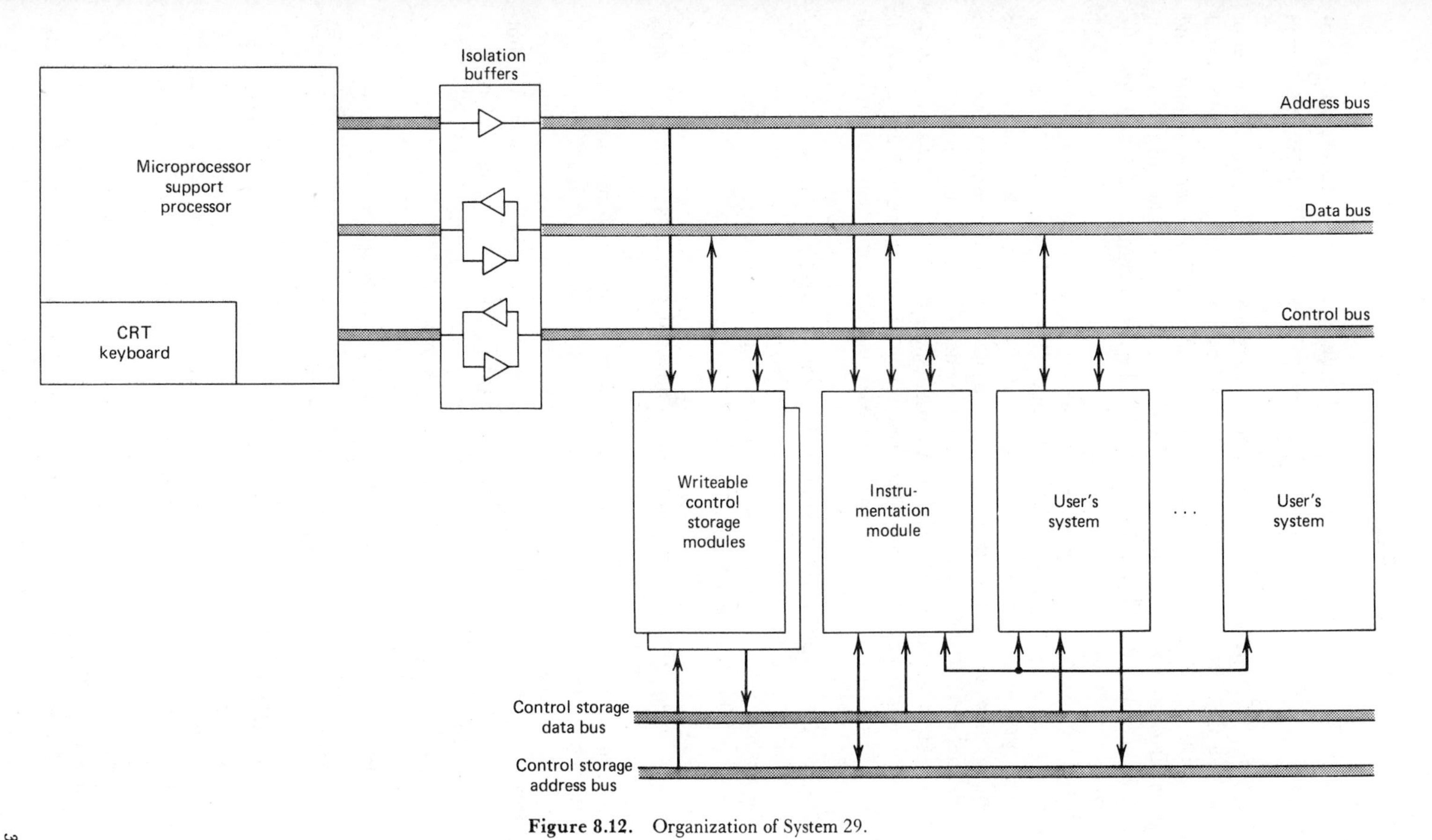

Figure 8.12. Organization of System 29.

base, and (2) interfaces to the system under development, including an instrumentation module and one or more control-storage modules provided with System 29.

The support system, built around an 8080 microprocessor, contains 32k bytes of memory, a CRT console, dual flexible-disk drives, and several serial and parallel I/O ports, allowing the attachment of a printer, PROM programmer, and modem. It contains a disk operating system with user commands to manage and edit files on the diskettes and interface with the instrumentation module and the system under development.

The second section consists of a backplane holding several busses, as well as cards that can be plugged into the backplane. One card provided is a universal wire-wrapped prototype card, allowing the designer to build a prototype of his system on one or more cards within the System 29 mainframe. Other cards include memory (RAM) cards, which are intended for use as substitutes for control storage in the user's design. The card supplied is a 2k by 64-bit 250-ns memory. The amount of memory (control storage) is expandable to 2048 by 128 bits, or 4096 by 64 bits. An optional card is available to represent a 50 ns control storage of 1k by 64 bits.

Another card supplied with System 29 is the instrumentation card, which is discussed later.

The designer has three options for interfacing the system under develoment with System 29

1. The system can be constructed on up to nine universal cards, using System 29 RAM cards as the control storage.
2. The manufacturer provides a card containing a general 12-bit control section. It contains two 2911 and one 2909 sequencer slices, 29803 multiway branch controller, 29811 next-address controller, status multiplexer, and pipeline register. One can build a prototype within the System 29 mainframe, but using this card as the control section and the RAM cards as control storage.
3. The system can be constructed outside of System 29, using cables to allow the System 29 RAM cards to substitute for control storage.

The instrumentation card, along with controlling software and user commands in the support system, provides most of the debugging function. The functions provided are

1. Clock control, allowing the user to stop or single cycle the system under development.
2. Addressing trapping. The user can specify a control-storage address, which is stored in the instrumentation card. If this control-storage word is selected during system operation, the system is stopped.
3. Address control, allowing the user to force a branch to any control-storage address.
4. Signal tracing. The instrumentation card can monitor up to 100 test points in the system under development. These points are periodically sampled and

stored in a high-speed memory. At any time, the user can display, on the CRT, the last 256 states. This function of the instrumentation card is similar to the basic function of a digital logic analyzer.

The terminal user, and programs in the support system, address the remainder of the system in the following way. The microprocessor has an addressing range of 64k bytes, but only 32k bytes of memory exists in the support system. By manipulating a page register, different sections of the system can be mapped into the upper 32k bytes of the address space. For instance, one value of the page register causes the simulated control storage to be mapped into this area. Another value causes the facilities of the instrumentation card to be memory mapped into this area. User-defined facilities (e.g., registers in the prototype system) can also be memory mapped.

MACE 29/800

The MACE 29/800 Microcode Analyzer and Control Storage Emulator [22], a product of Motorola Corp., is similar to System 29. However, rather than being a self-contained unit, it is designed to interface with Motorola's EXORcisor microprocessor development system, taking advantage of its user terminal and software.

MACE 29/800 consists of a card cage, power supply, RAM modules for emulation of control storage, and instrumentation logic. Its organization is shown in Figure 8.13. The user has the option of building his prototype within the MACE card cage (the interfaces support both ECL and TTL logic) or connecting cables from MACE to his system.

A variety of writeable control-storage configurations are available, ranging in

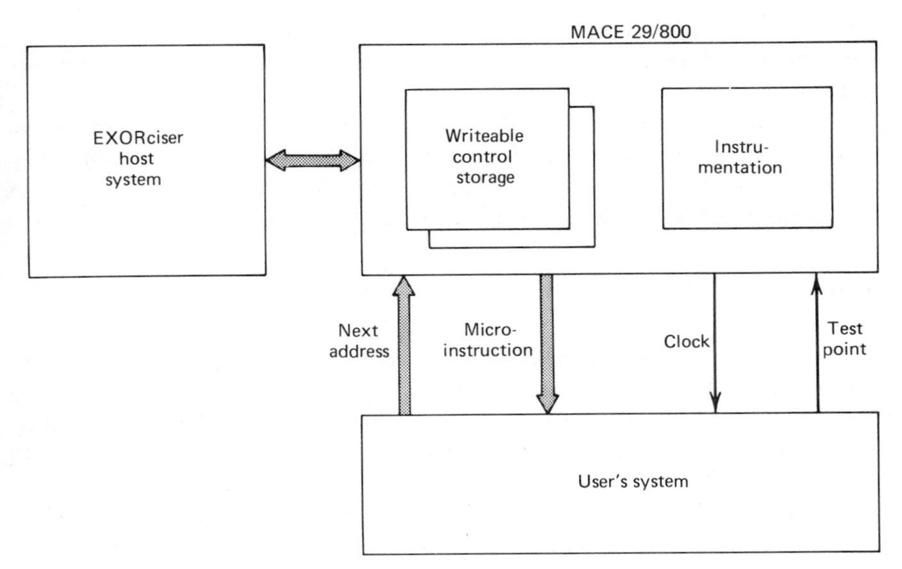

Figure 8.13. Organization of MACE 29/800.

depth, width, and speed. The maximum configurations are 2k by 112 bits, 4k by 48 bits, 6k by 32 bits, and 8k by 16 bits. Access times range from 50 to 80 ns.

The instrumentation logic provides clock control to the user's system, allowing one to stop the clock and run in single-cycle mode. It also provides the user with breakpoint functions when a designated control-storage word is addressed. The instrumentation logic contains a 256-word RAM, allowing one to trace control-storage address references and test points in the system under development.

STEP-2

Another similar tool is the STEP-2 Firmware Integration and Test Station [23]. The STEP-2 chassis contains a small built-in CRT and keyboard, microprocessor, instrumentation logic, and slots for RAM modules. The functions available through the keyboard, however, are not as extensive as those in System 29; the user can display and modify memory contents and control the instrumentation logic.

As shown in Figure 8.14, STEP-2 emulates control storage and provides instrumentation for the system under development. STEP-2 interfaces to the user's system via a set of optional ROM/PROM simulators that plug into the control-storage DIP sockets in the system under development. These simulators are useful in situations where control storage in the system resides on the same board as other logic, making use of a separate control-storage simulation board awkward. Seventeen different ROM/PROM simulators are available, simulating over 200 types of ROMs and PROMs.

Up to 96k bits of control storage can be obtained, configured in widths of 8 to 96 bits. This storage can be segmented into several independent addressable storage arrays, allowing one, for instance, to simulate both control storage and a mapping PROM.

The instrumentation logic provides breakpoint (address-comparison) functions. As an option, storage can be added to trace the last 250 control-storage addresses referenced.

STEP-2 contains no support for the editing, storage, and assembly of microprograms. Instead, it contains a serial interface that can be connected to a development system; to this interface, STEP-2 appears as a terminal. The microprograms would be prepared on the development system and transmitted to STEP-2 for testing.

ICE-30

The Intel ICE-30 in-circuit emulator [24] represents another approach to the problem. Rather than simulating control storage, ICE-30 simulates the operation of the 3001 control-storage sequencer. Rather than being a self-contained facility, ICE-30 consists of a logic board and software that are inserted into one of Intel's development systems.

As shown in Figure 8.15, the principal interface from ICE-30 is a cable and plug that is inserted into a 40-pin socket for a 3001 sequencer in the system under

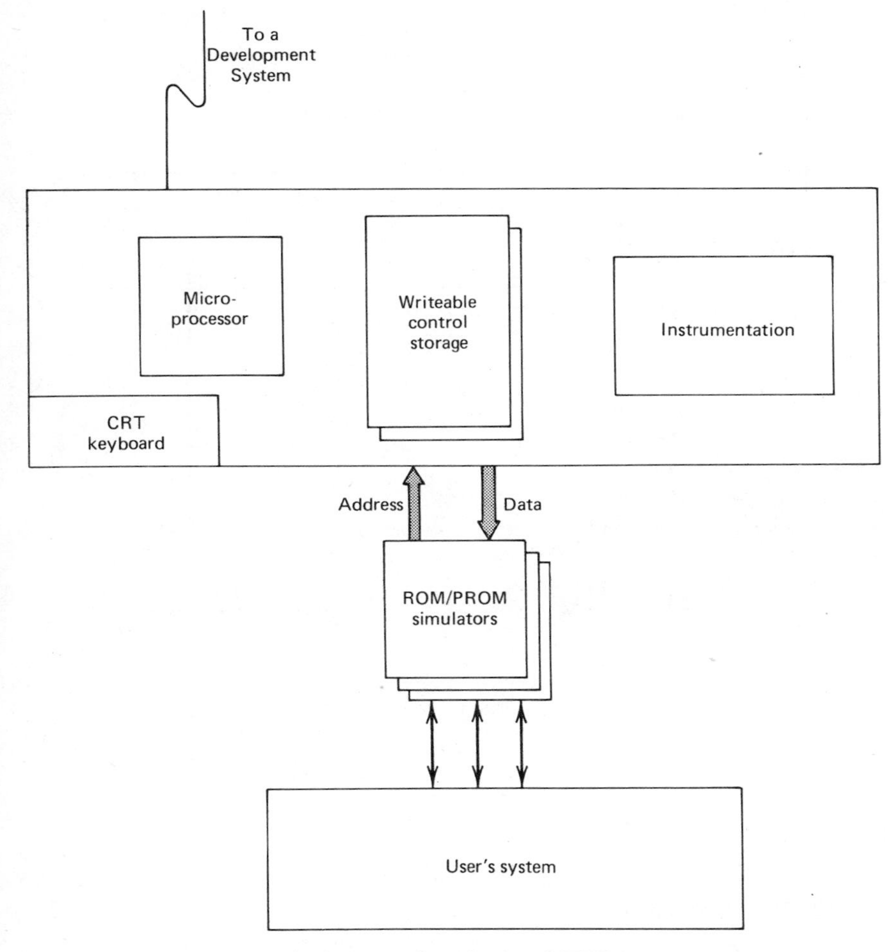

Figure 8.14. Organization of STEP-2.

development. This cable is terminated in a buffer unit, which is connected to the ICE-30 board in the development system. The ICE-30 board contains an actual 3001 and instrumentation logic. The 3001 on the ICE-30 board takes the place of the 3001 in the system under development.

In addition to the 3001 cable, one may also connect a clock-control line and three test probes to the system under development.

The user has the ability to stop and single cycle the system. One can display the values on the address, status, and control lines of the 3001. One can set two address breakpoints, as well as breakpoints on specific values on the three logic probes.

ICE-30 can also operate in conjunction with an optional ROM/PROM simulator.

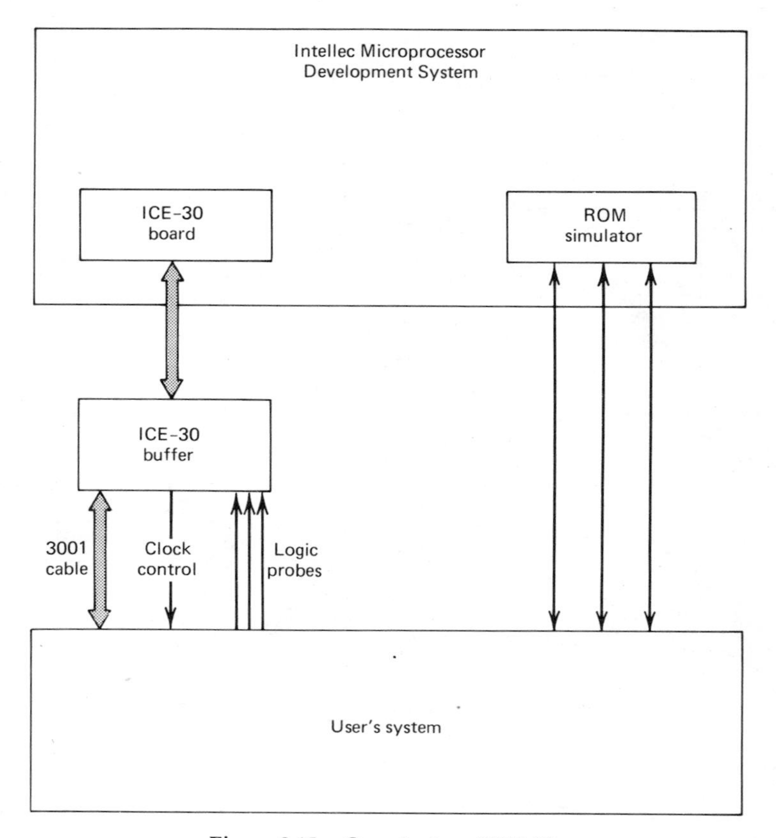

Figure 8.15. Organization of ICE-30.

SOFTWARE SIMULATORS

Another useful tool for the development of microprogrammed systems is the software simulator, a program written to simulate the precise behavior of the data flow of a design from the point of view of the microprogram. That is, the principal input to the simulator is the microprogram; the program simulates the underlying hardware design by "executing" the microprogram. Hence, the simulator allows one to test and debug the microprogram before the hardware system is available.

An example is a simulator that has been used by the author for the testing and debugging of a large amount of microcode for a 250-chip experimental processor employing 2910 and 2903 devices. The simulator makes full use of the screen of a CRT terminal. When one invokes the simulator, the CRT displays a "power-on" frame. One can select functions from this frame, such as loading control storage or main storage from a file and altering timing parameters in the simulator. Other frames that may be displayed on the terminal are

1. Traps and traces, which allows one to set breakpoints on up to 10 ranges of addresses in control storage or main storage, and/or request that the simulator maintain a trace of control-storage or main-storage references.
2. Recording functions, which allows one to print the current machine state, print dumps of main storage or control storage, or record the current machine state on a file for a future resumption of the simulation.
3. Performance measurements, which displays the simulated elapsed time and certain statistics regarding memory operations.
4. Alter/display, which allows one to examine or alter main storage, control storage, and the internal processor registers and flipflops, as well as alter the current microprogram address, examine the last 64 microprogram addresses, and "execute" the microprogram in single-cycle mode, for a specified number of machine cycles, or until a breakpoint is encountered.

The advantages of such simulators are

1. They allow one to test, debug, and optimize the microprogram before the hardware is available.
2. They give one a more flexible and convenient vehicle for microprogram testing.
3. They can contain debugging and instrumentation functions that are more powerful than those in development and instrumentation systems (e.g., System 29).
4. They can contain checks for error situations, such as the detection of timing errors and simultaneous uses of busses.
5. By containing parameters defining the cycle time and memory-access and cycle times, they can be used to predict system performance, such as the elapsed times of machine instructions.
6. By allowing the testing of the microprogram to begin earlier, the use of the simulator can provide valuable feedback on the hardware design, both in terms of errors and possible optimizations.

The disadvantage of the software simulator is cost, since it must be developed from scratch for each hardware design. For instance, the simulator program mentioned above contains approximately 10,000 PL/I-language statements and was developed at a cost of approximately nine man-months. However, where a substantial amount of microcode must be developed, the investment in a software simulator is usually a good one.

REFERENCES

1. V. M. Powers and J. H. Hernandez, "Microprogram Assemblers for Bit-Slice Microprocessors," *Computer,* **11**(7), 108–120

2. *The Am2900 Family Data Book.* Sunnyvale, Cal.: Advanced Micro Devices, 1978.
3. J. H. Hernandez, "AMDASM, A Microprogram Assembler for Bit-Slice Microprocessors," *MIDCON/77 Conference Record.* El Segundo, Cal.: Electrical and Electronics Exhibitors, 1977.
4. *CSC AMDASM Reference Manual.* Los Angeles: Computer Sciences Corp., 1976.
5. *Meta Assembler Program.* Sunnyvale, Cal.: Microtec, 1978.
6. L. Fesperman, "Micro Assemblers—Advancing the State of the Microprogramming Art," *MIDCON/77 Conference Record.* El Segundo, Cal.: Electrical and Electronics Exhibitors, 1977.
7. *Signetics Micro Assembler Reference Manual.* Sunnyvale, Cal.: Signetics, 1977.
8. *Intel Series 3000 Microprogramming Manual.* Santa Clara, Cal.: Intel, 1976.
9. B. A. Laws, Jr., "Microbe: A Self-Commenting Microassembler," *MICRO10 Proceedings.* New York: ACM, 1977, pp. 61–65.
10. R. C. Clark, "Mirager, the 'Best-Yet' Approach for Horizontal Micropramming," *Proc. 1972 ACM Annual Conference.* New York: ACM, 1972, pp. 554–571.
11. D. Willen, "An Intel 3000 Cross Assembler," *SIGMICRO Newsletter,* **7**(4), 87–94 (1976).
12. T. G. Rauscher, "Towards a Specification and Semantics for Languages for Horizontally Microprogrammed Machines," *Proc. ACM SIGPLAN-SIGMICRO Interface Meeting.* New York: ACM, 1974, pp. 98–111.
13. D. J. DeWitt, "Extensibility—A New Approach for Designing Machine Independent Microprogramming Languages," *MICRO9 Proceedings.* New York: ACM, 1976, pp. 33–41.
14. C. V. Ramamoorthy and M. Tsuchiya, "A High-Level Language for Horizontal Microprogramming," *IEEE Trans. on Computers,* **C-23**(8), 791–801 (1974).
15. G. Wood, "On the Packing of Micro-Operations into Micro-instruction Words," *MICRO11 Proceedings.* New York: ACM, 1978, pp. 51–55.
16. P. W. Mallett and T. G. Lewis, "Considerations for Implementing a High Level Microprogramming Language Translation System," *Computer,* **8**(8), 40–52 (1975).
17. M. Tokoro, T. Takizuka, E. Tamura, and I. Yamaura, "A Technique of Global Optimization of Microprograms," *MICRO11 Proceedings.* New York: ACM, 1978, pp. 41–50.
18. G. R. Lloyd and A. Van Dam, "Design Considerations for Microprogramming Languages," *Proc. 1974 National Computer Conference.* Montvale, N. J.: AFIPS, 1974, pp. 537–543.
19. K. Malik and T. Lewis, "Design Objectives for High Level Microprogramming Languages," *MICRO11 Proceedings.* New York: ACM, 1978, pp. 154–160.
20. T. Baba, "A Microprogram Generating System - MPG," *Proc. 1977 IFIP Congress.* Amsterdam: North-Holland, 1977, pp. 739–744.
21. J. R. Mick and R. Schopmeyer, "MOS Support Microprocessor Teams with Bit-Slice Prototypes for Easier Microprogram Debugging," *Electronics,* **50**(19), 127–130 (1977).
22. *Motorola MACE 29/800.* Phoenix: Motorola, 1978.
23. *Step-2 Firmware Integration and Test Station.* Sunnyvale, Cal.: STEP Engineering, 1978.
24. *MDS-ICE-30 3001 MCU In-Circuit Emulator.* Santa Clara, Cal: Intel, 1976.

9

Firmware Engineering

In the 1970s, widespread recognition of the problems associated with the production of software led to the initiation of a field of study known as software engineering. Software engineering, although still in its infancy, is a body of theories, methodologies, and tools for the design, development, documentation, testing, debugging, maintenance, and management of software systems.

Given the recent growth of the use of microprograms or firmware and the recognition that microprogram development is similar in many ways to software development, the term "firmware engineering" was coined [1]. Onlookers have warned that the use of microprogramming is evolving in much the same way that software development began in the 1950s, and that techniques and tools similar to those of software engineering must be invented and used [2].

The motivation for firmware-engineering principles is much the same as for software engineering, namely, to enable microprograms to be produced cheaply, quickly, and reliably. Errors in microprograms can have serious consequences, normally more so than the typical error in a typical software program. Also, microprogram errors are more pervasive. Consider the consequences of an error in the floating-point arithmetic microcode on a processor controlling a nuclear reactor, laboratory equipment, or an air-traffic-control system. Consider the consequences of an error in the microprogram in an I/O controller within a banking system. Even an error in the microcoded logic of a desk calculator can prove to be enormously expensive to its manufacturer.

In addition to the consequences of errors, the cost of correcting such errors can be enormous. Consider the discovery of an error in the microprogram of a processor after 1000 models have been installed in businesses throughout the world, or an error in a microprogrammed microprocessor after an initial manufacturing run of 100,000 chips has been completed.

Although the emphasis above has been on reliability, one must also be concerned with production costs and timeliness. Given the tendency to migrate functions that have traditionally been performed by software into microprograms [3], microprograms are becoming larger and therefore more expensive (nonlinearly) to develop.

Unfortunately, firmware engineering, as a separate topic, has received little attention. However, as microprogramming has much in common with programming, the microprogrammer can learn much from the programmer. The aim of this chapter is to survey aspects of the software-engineering field that are relevant to microprogramming, particularly because the typical microprogrammer is an electrical engineer and probably unfamiliar with the field of software engineering. Since literally tens of thousands of pages of literature have been written in the last decade on software engineering, only a brief survey is presented here.

THE DEVELOPMENT CYCLE

A model of the development cycle for a microprogram is illustrated in Figure 9.1. The first two steps are not normally part of the development of the actual microprogram, but they are shown here to illustrate where the development of the microprogram begins. The first step is the specification of the architecture of the system. For a processor, this is the specification of the computer architecture. For a disk controller, this might include the specification of the commands processed by the controller. The second step is the design of the data-flow or register-transfer level of the system, including the design of the microinstruction.

At this point, the microprogram design and hardware design (e.g., logic design) often proceed in parallel, although only the microprogram development processes have been shown in Figure 9.1. The next step is the design of the structure of the microprogram. This normally includes the partitioning of the microprogram into subroutines and establishing system-wide conventions in such areas as register usage. The following step is the precise definition of the interfaces among the sections or subroutines of the microprogram. The last design step is the detailed

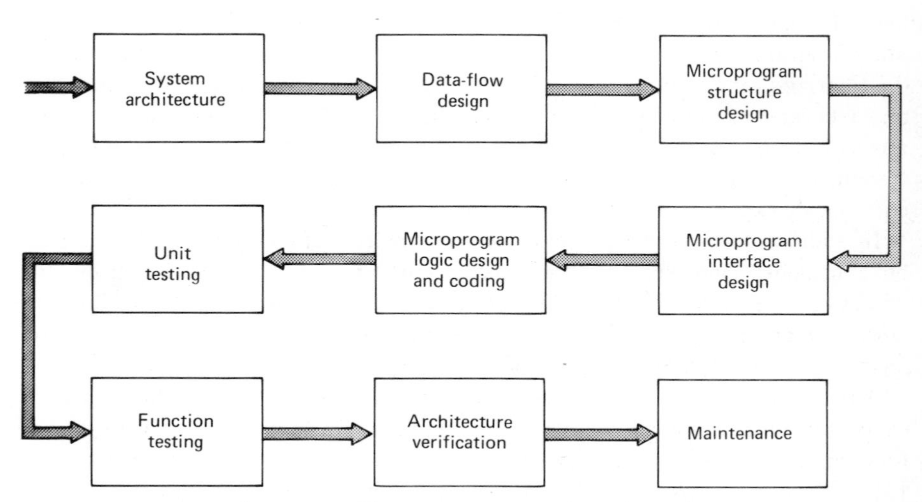

Figure 9.1. The microprogram development cycle.

design of the microprogram algorithms and logic and the encoding of the microprogram in a symbolic language.

After this, the testing (looking for errors) and debugging (correcting errors) begins. This testing can be done on a prototype hardware system, if available, or a simulator. If the microprogram has been partitioned into subroutines, it is usually more efficient to test each subroutine first as an independent entity. This form of testing is known as unit testing. The next step is the testing of the external interfaces and function of the microprogram as a whole. In the case of a processor, this would include the testing of the processor with respect to the architecture definition (e.g., the instruction set).

Often, the system being developed is a processor in a family of processors (e.g., a member of the S/370 or PDP-11 series). In these situations, the manufacturer usually develops an extensive set of test programs to verify whether the system adheres to its architecture. These programs concentrate on known idiosyncratic aspects of the architecture, explore relationships (e.g., side effects) among sequences of instructions, and so on. The last step, often overlooked, is the maintenance of the microprogram after the system is placed into production (e.g., the correction of newly discovered errors).

There are at least two other processes that must be considered, although they cannot be clearly depicted in Figure 9.1 because they pervade many of the steps. They are performance measurement and optimization, usually a matter of significant importance in microprogram development, and debugging (the correction of discovered errors throughout the process).

This development cycle is quite similar to that for programs, and the large microprogram project might benefit from advice about the management of the software development cycle [4–6].

MICROPROGRAM DESIGN

Unfortunately, many of the principles of software design are not applicable to microprogram design. The reasons are (1) efficiency or performance is normally of utmost concern to the microprogrammer, (2) the microprogrammer (particularly in a horizontal design) must cope with high degrees of parallelism, and (3) microprograms do not have the advantages most programs have, namely the ability to make use of the underlying facilities of operating systems and compilers. Rather than surveying the software design literature, a few basic principles and ideas for microprogram design are presented below.

As mentioned earlier, the first design step is one of structure, that is, partitioning the microprogram into small sections, such as subroutines. Although not all the principles are applicable, the designer of a large microprogram might wish to examine the program-design methodologies of composite or structured design [7–9].

Simple partitioning of a microprogram can have several positive effects. If a microprogram consists of one large body of code, its understandibility is severely reduced (understandibility during its design, debugging, and maintenance). One objective of partitioning is to reduce the number of factors that one must keep track of to understand the microprogram. Another advantage is that partitions create a number of well-defined and documented boundaries or interfaces within the microprogram, which are invaluable in the comprehension of the microprogram. A third advantage is that subroutines, if defined carefully, can be used in multiple places or contexts within the microprogram, thus reducing the number of microinstructions that must be written and reducing the control-storage requirements. However, the latter should not be the sole criterion for the establishment of subroutines; subroutines should be employed to add clarity to the microprogram.

Composite and structured design recognize that *any* partitioning of a program is not likely to be the best. They are based on the concept of *high independence,* the concept that a program should be partitioned such that the interactions among partitions are at a minimum ("low module coupling") and the interactions within any partition are maximized ("high module strength"). This concept has many parallels to the criteria one would use to partition a large system design across customized LSI components. Composite and structured design include recommendations about desirable and undesirable partition definitions and interface techniques, as well as iterative thought processes for developing the partitions.

Once the microprogram has been structured, the next step is the development of a precise specification of each interface. The motivations for doing this are (1) it allows the sections of the microprogram to be developed by many people, (2) the act of defining the interfaces can point out errors in the design, and (3) the specifications serve as an excellent form of documentation about the design. An example of an interface specification for a microprogram subroutine is shown in Figure 9.2. It is also a good idea to place these specifications directly within the symbolic microprogram as commentary statements.

One significant area in which program and microprogram development differ is the amount of time spent on detailed algorithmic design and coding. On a software project, this phase consumes a relatively small percentage of the total effort, but in the development of a microprogram, this phase consumes a larger percentage of the total effort. The reasoms are (1) the usual requirement to produce microprograms that are highly time and space optimized, (2) the need to deal with a high

```
/COPYLR
Function: Copies one area of main storage into another
Inputs:   MAR1 -> start of source area
          MAR2 -> start of target area
          R5    =  length of area in bits divided by 4
          R0    =  zero
Outputs:  MAR1 -> end of source area + 1
          MAR2 -> end of target area + 1
          R5    =  zero
Further levels of subroutine calls: None
Other registers modified: MDR2
Side effects: none
Notes: This logic is highly optimized
Author: J. Jones
```

Figure 9.2. Sample subroutine interface specification.

```
IF(requested size ≤ quickcell size)
  THEN dequeue a quickcell block.
       IF(found one) THEN return its address.
                     ELSE signal no-storage fault.
  ELSE search free storage pool for a free block ≥
       requested size.
       IF(found one)
         THEN IF(equal to requested size)
                THEN dequeue the block.
                     return its address.
                ELSE dequeue the block.
                     take upper unneeded bytes and
                     enqueue them as a free block.
                     return block's address.
         ELSE signal no-storage fault.
```

Figure 9.3. Partial refinement of a microcoded algorithm.

degree of parallelism in microprogramming a horizontal design, and (3) the lack of higher-level microprogramming languages.

The detailed microprogram design can be done with the traditional flowchart, but this approach has at least three drawbacks. First, one usually finds that as much effort went into the drafting of a "nice" flowchart as went into the design of the logic expressed by the flowchart. Second, the flowchart does not allow one to readily perform the design in a number of iterative steps, each step refining the information in the previous step. Third, modifications to flowcharts are difficult, meaning that they usually become obsolete as the design changes.

A superior approach is the use of stepwise refinement [6,10]. Here one writes a high-level description of the design in a hypothetical language resembling traditional high-level programming languages, but at a higher level. One then takes this representation of the design and refines it into a description of the design in another hypothetical, but somewhat lower-level, language. A high-level refinement for a storage-allocation microprogram subroutine is shown in Figure 9.3. After several refinements, the design is usually detailed enough to translate the last refinement into microprogram logic.

This refinement process allows one to break the algorithmic design problem iteratively into smaller and smaller detailed problems at each level, and thus is usually found to be a less-error-prone thought process than the "all at once" approach of flowcharting.

Of course, a major task in detailed microprogram design and coding is achieving optimal parallelism, both in terms of trying to overlap operations (e.g., starting memory operations as far in advance as possible) and trying to squeeze as much function into every microinstruction as possible. This task is unique to microprogramming.

MICROPROGRAM TESTING

In both microprogram and software development, a considerable amount of time and energy is spent in the testing phase. Surprisingly, very little literature exists on the subject; for instance, there is only one text that is devoted to the subject [11]. Most of the information in this text is applicable to microprogram testing and is summarized below.

One can place the major considerations of testing into four categories: economics, psychology, technology (test-case design), and management (e.g., answering such questions as "When do I stop testing?"). The important economic lesson to learn is that one can never afford to test a program to the extent that one can guarantee that no errors exist. For instance, exhaustive testing (trying every possible circumstance) of even a simple machine instruction usually requires an astronomical number of test cases.

The psychological issues have been shown to have a strong bearing on the effectiveness of a testing effort. Some of the psychological issues discussed in reference 11, slightly paraphrased for the microprogrammer, are

1. Testing must be viewed as the process of exercising a microprogram *with the intent of finding errors,* not with the intent of showing the absence of errors.
2. A vital part of each test case is a definition of the correct result (to avoid the "eye seeing what it wants to see" problem).
3. A microprogrammer should avoid attempting to test his or her own microprogram. A substantial percentage of errors are due to misunderstandings, such as misunderstandings of the architecture specification. If one attempts to test one's own program, it is likely that the same misunderstandings will be reflected in the test cases, thus causing the errors to be missed. Furthermore, successful testing requires a destructive frame of mind (i.e., the desire to expose the flaws in the program), but it is extremely difficult to be destructive against one's own work.
4. Test cases must be developed to cover invalid and unexpected situations. A large percentage of the latent errors in microprograms are associated with out-of-the-normal circumstances; typical examples in processor microprograms are overflow conditions, I/O errors, invalid data representations, and unorthodox uses of machine instructions.
5. In addition to testing to see if the microprogram does not do what it is supposed to do, one must also test to see if it does what it is not supposed to do. In other words, one must examine the results for unwanted side effects (e.g., a machine instruction that is not supposed to alter the condition codes does).

The third area of concern is the effective design of test cases. Test-case design methodologies can be grouped into two categories: black-box and white-box methodologies [11]. White-box methodologies involve the derivation of test cases to meet criteria with respect to the logic of the microprogram. Such criteria are

1. Sufficient test cases to cause each microinstruction to be active at least once (a necessary, but woefully insufficient, objective).
2. Sufficient test cases to cause each branching operation in the microprogram to be exercised in all possibile directions at least once.

3. Sufficient test cases to cause every unique path within the microprogram to be traversed at least once (a criterion that is usually unachievable in programs, but sometimes achievable in microprograms).

Black-box methods, which should be used in concert with the white-box methods, involve the derivation of test cases based on an analysis of the external interfaces of a system, such as the instruction-set architecture in the case of a microprogrammed processor. Such analysis methods include:

1. *Equivalence partitioning,* methodology based on the measures of an effective test case being (1) it reduces, by a high value, the number of other test cases that must be developed to achieve some predefined goal of "reasonable" testing, and (2) it tells one something aboout the presence or absence of errors beyond the specific set of input values in this test case. The methodology involves partitioning the input state of the microprogram into "equivalence classes" such that there is a high probability that one test case in an equivalence class is representative of all other test cases in the equivalence class.
2. *Boundary-value analysis,* a methodology that recognizes that the most fruitful test cases are those at, and just beyond, the boundaries or edges of equivalence classes in both the input and output states of the program.
3. *Cause-effect graphing,* a methodology that involves the translation of a specification into a combinatorial-logic graph and then the application of path-sensitizing techniques to reduce the graph to a set of test cases.

Finally, many hints about designing software test cases also apply to those for microprograms. For instance, given the tedious nature of verifying results of microprogram tests, it is important to strive toward self-checking test cases. For instance, if one is developing a set of test cases for the ADD machine instruction in a microprogrammed processor, the test cases should be written to verify the expected result (e.g., each ADD instruction is followed by a comparison instruction with a preestablished expected result).

MICROPROGRAM WALKTHROUGHS AND INSPECTIONS

Another error-detection technique employed successfully in software development is the code inspection or walkthrough [11–13]. This technique is usually employed after the program has been written but before computer-based testing begins. The program, or a segment of it, is manually inspected by a group of people in a meeting, the objective being to find errors. The program is inspected either with respect to a checklist of historically common errors, or by mentally tracing the execution of test data through the logic of the program.

Although, on the surface, the concept seems too simple to be effective, it has proven to be an effective method for finding errors. Although it has not been widely applied to microprograms, the idea would seem to be of equal value in microprogram development.

MICROPROGRAM CORRECTNESS PROOFS

A lively area of research in the software-engineering field is the concept of attempting to prove mathematically the correctness of a program. Most of the techniques in this area are based on the notion that one can write assertions about the desired behavior of the program and develop theorems, from the actual semantics of the program logic, stating that the program logic matches the assertions [14]. If the theorems can be proved, the program is "correct" (within certain limitations); if not, the program is incorrect or the proof techniques are faulty. The assertions can take the form of (1) logic statements about conditions that should be true at the beginning, end, and intermediary points in the program, or (2) a high-level representation of the desired logic of the program, to which one wants to prove the program logic equivalent. A related approach is *symbolic execution,* where the execution of the program is simulated using symbolic, rather than actual or specific, data values [15].

Such techniques are not used in practice in software development today because of several obstacles, but it appears that these obstacles are less severe in microprograms. One obstacle is the large size of most programs, but microprograms tend to be considerably smaller. (Although there are more small programs than large ones, the large programs tend to be more critical and in need of proofs.) Also, a microinstruction tends to have a more easily defined effect on the system state, and also a more limited effect on the system state, another advantage.

One significant research effort in this area is the Language for Symbolic Simulation (LSS) and the Microprogram Certification System (MCS) [16–18]. LSS is a language that is used to define a high-level abstract machine. LSS is used to (1) describe a precise, but high-level, model of the system to be evaluated, and (2) describe the hardware machine on which the microprogram to be evaluated resides. At this point, one has two representations: the actual microprogram and the LSS model. The MCS tool is then used to attempt to prove equivalence between the two representations. MCS does this by symbolically executing both and attempting to prove their equivalence at various points during execution.

As an example of its utility, MCS has been used to evaluate the microprogram in a computer produced by IBM's Federal Systems Division. The evaluation, performed after conventional testing had been completed, found the following types of errors

1. Several subtle timing problems (I/O related) were present in the system-reset microcode.
2. If an I/O interrupt occurs during a machine instruction generating an arithmetic overflow, the overflow condition is ignored.
3. The system fails when a halfword instruction is fetched from the last halfword in memory.
4. The branch-and-link-register instruction did not function correctly.
5. The specification stated that a particular bit in one machine instruction was unused and ignored, but the microprogram required it to be zero.

Other approaches (e.g., [19]) have employed languages in which one states assertions about the microprogram, and programs that attempt to prove (with human assistance) the theorems developed from the microprogram logic. Another approach [20] involves the statement of input/output functions of sections of the microprogram and an analysis of these functions using the algebra of relations. Another involves a language into which microprograms are translated, allowing standard program-proof techniques to then be applied to the microprogram [21,22]. A further approach is a technique for verifying the equivalence of two microprograms, providing that both contain no loops [23].

REFERENCES

1. S. Dividson and B. D. Shriver, "An Overview of Firmware Engineering," *Computer,* **11**(5), 21–33 (1978).
2. M. M. Lehman, "Microprogramming Trend Considered Dangerous," *SIGMICRO Newsletter,* **6**(3), 37–39 (1975).
3. J. Stockenberg and A. van Dam, "Vertical Migration for Performance Enchancement in Layered Hardware/Firmware/Software Systems," *Computer,* **11**(5), 35–49 (1978)
4. E. B. Daly, "Management of Software Development," *IEEE Trans. on Software Engineering,* **SE-3**(3), 229–242 (1977).
5. W. C. Cave and A. B. Salisbury, "Controlling the Software Life Cycle—The Project Management Task," *IEEE Trans. on Software Engineering,* **SE-4**(4), 326–334 (1978).
6. G. J. Myers, *Software Reliability: Principles and Practices.* New York: Wiley-Interscience, 1976.
7. G. J. Myers, *Composite/Structured Design.* New York: Van Nostrand Reinhold, 1978.
8. G. J. Myers, *Reliable Software Through Composite Design.* New York: Van Nostrand Reinhold, 1975.
9. E. Yourdon and L. L. Constantine, *Structured Design.* Englewood Cliffs, N. J.: Prentice-Hall, 1979.
10. N. Wirth, "Program Development by Stepwise Refinement," *Comm. of the ACM,* **14**(4), 221–227 (1971).
11. G. J. Myers, *The Art of Software Testing.* New York: Wiley-Interscience, 1979.
12. M. E. Fagan, "Design and Code Inspections to Reduce Errors in Program Development," *IBM Systems Journal,* **15**(3), 182–211 (1971).
13. G. J. Myers, "A Controlled Experiment in Program Testing and Code Walkthroughs/Inspections," *Comm. of the ACM,* **21**(9), 760–768 (1978).
14. R. B. Anderson, *Proving Programs Correct.* New York: Wiley, 1979.
15. J. C. King, "A New Approach to Program Testing," *Proc, of the 1975 Int. Conf. on Reliable Software.* New York: IEEE, 1975, pp. 228–233.
16. G. B. Leeman, Jr., "Some Problems in Certifying Microprograms,"*IEEE Trans. on Computers,* **C-24**(5), 545–553 (1975).
17. W. H. Joyner, W. C. Carter, and G. B. Leeman, "Artomated Proofs of Microprogram Correctness," *MICRO9 Proceedings.* New Youk: ACM, 1976 pp. 51–55.
18. W. C. Carter, W. H. Joynes, Jr., and D. Brand, "Microprogram Verification Considered Necessary," *Proc. 1978 National Computer Conf.* Montvale, N.J.: AFIPS Press, 1978, pp. 657–664.

19. D. A. Patterson, "Strum: Structured Microprogram Development System for Correct Firmware," *IEEE Trans. on Computers* **C-25**(10), 974–985 (1976).

20. A. Blikle and S. Budkowski, "Certification of Microprograms by an Algebraic Method," *MICRO9 Proceedings*. New York: ACM, 1976, pp. 9–14.

21. S. Budkowski and P. Dembinski, "Firmware versus Software Verifications," *MICRO10 Proceedings*. New York: IEEE, 1978, pp. 119–127.

22. P. Dembinski and S. Budkowski, "An Introduction to the Verification Oriented Microprogramming Language 'MIDDLE'," *MICRO10 Proceedings*. New York: IEEE, 1978, pp. 139–143.

23. C. V. Ramamoorthy and K. S. Shankar, "Automatic Testing for the Correctness And Equivalence of Loopfree Microprograms," *IEEE Trans. on Computers,* **C-23**(8), 768–782 (1974).

Index